METHODS IN MOLECULAR BIOLOGY™

Series Editor
John M. Walker
School of Life Sciences
University of Hertfordshire
Hatfield, Hertfordshire, AL10 9AB, UK

For further volumes:
http://www.springer.com/series/7651

Fungal Secondary Metabolism

Methods and Protocols

Edited by

Nancy P. Keller

Department of Medical Microbiology and Immunology
University of Wisconsin-Madison, Madison, WI, USA

Geoffrey Turner

Department of Molecular Biology and Biotechnology
University of Sheffield, Sheffield, UK

Humana Press

Editors
Nancy P. Keller
Department of Medical Microbiology
and Immunology
University of Wisconsin-Madison
Madison, WI, USA

Geoffrey Turner
Department of Molecular Biology
and Biotechnology
University of Sheffield
Sheffield, UK

ISSN 1064-3745 ISSN 1940-6029 (electronic)
ISBN 978-1-4939-5960-0 ISBN 978-1-62703-122-6 (eBook)
DOI 10.1007/978-1-62703-122-6
Springer New York Heidelberg Dordrecht London

Printed on acid-free paper

Humana Press is a brand of Springer
Springer is part of Springer Science+Business Media (www.springer.com)

Preface

Filamentous fungi have long been known for their ability to produce an enormous range of unusual chemical compounds known as secondary metabolites, many of which have potentially useful antibiotic or pharmacological properties. Recent focus on fungal genomics coupled with advances in detection and molecular manipulation techniques has galvanized a revitalization of this field. *Fungal Secondary Metabolism: Methods and Protocols* is aimed at providing the key methodologies currently in use and necessary for accessing and exploiting the natural product information provided by the genomes of this large and varied kingdom. Written by active researchers in the field, the chapters deal with all the steps necessary, from optimization of fungal culture conditions for metabolite production through rapid genome sequencing and bioinformatics and genetic manipulations for functional analysis, for the detection and testing of metabolites. In addition, chapters on basic science address approaches to the genetic regulation, protein biochemistry, and cellular localization of the biosynthetic pathways. Presented in the usual practical and detailed "hands-on" approach of the series, *Methods in Molecular Biology, Fungal Secondary Metabolism: Methods and Protocols* encourages new investigators to enter the field and expands upon the expertise and range of skills of those already researching fungal natural products.

Madison, WI, USA — *Nancy P. Keller*
Sheffield, UK — *Geoffrey Turner*

Contents

Preface . . . . *v*
Contributors . . . . *ix*

1 Library Preparation and Data Analysis Packages for Rapid Genome Sequencing 1
Kyle R. Pomraning, Kristina M. Smith, Erin L. Bredeweg, Lanelle R. Connolly, Pallavi A. Phatale, and Michael Freitag

2 Bioinformatics Approaches and Software for Detection of Secondary Metabolic Gene Clusters 23
Natalie D. Fedorova, Venkatesh Moktali, and Marnix H. Medema

3 Media and Growth Conditions for Induction of Secondary Metabolite Production 47
Jens C. Frisvad

4 Manipulating Filamentous Fungus Chemical Phenotypes by Growth on Nutritional Arrays 59
José R. Tormo, Francisco J. Asensio, and Gerald F. Bills

5 The "FERMEX" Method for Metabolite-Enriched Fungal Extracts 79
Gerald F. Bills, Anne W. Dombrowski, and Michael A. Goetz

6 The Chemical Identification and Analysis of *Aspergillus nidulans* Secondary Metabolites 97
James F. Sanchez and Clay C.C. Wang

7 Detection and Purification of Non-ribosomal Peptide Synthetase Products in *Neosartorya fischeri* 111
Beate Wollinsky and Shu-Ming Li

8 Production, Detection, and Purification of Clavine-Type Ergot Alkaloids 121
Christiane Wallwey and Shu-Ming Li

9 Analysis of Volatile Compounds Emitted by Filamentous Fungi Using Solid-Phase Microextraction-Gas Chromatography/Mass Spectrometry 133
Ludmila V. Roze, Randolph M. Beaudry, and John E. Linz

10 Tools for Manipulation of Secondary Metabolism Pathways: Rapid Promoter Replacements and Gene Deletions in *Aspergillus nidulans* 143
C. Elizabeth Oakley, Heather Edgerton-Morgan, and Berl R. Oakley

11 Fast and Easy Method for Construction of Plasmid Vectors Using Modified Quick-Change Mutagenesis 163
Jin Woo Bok and Nancy P. Keller

12 Reconstitution of a Secondary Metabolite Biosynthetic Pathway in a Heterologous Fungal Host 175
Takayuki Itoh, Tetsuo Kushiro, and Isao Fujii

13 Multi-Copy Genetic Screen in *Aspergillus nidulans* 183
Hee Soo Park and Jae-Hyuk Yu

14 Identification of Protein Complexes from Filamentous Fungi with Tandem Affinity Purification 191
Özgür Bayram, Özlem Sarikaya Bayram, Oliver Valerius, Bastian Jöhnk, and Gerhard H. Braus

15 Correlating Secondary Metabolite Production with Genetic Changes Using Differential Analysis of 2D NMR Spectra 207
Ry R. Forseth and Frank C. Schroeder

16 Chromatin Immunoprecipitation Analysis in Filamentous Fungi 221
Stefan Boedi, Yazmid Reyes-Dominguez, and Joseph Strauss

17 Targeted Proteomics for Metabolic Pathway Optimization 237
Tanveer S. Batth, Jay D. Keasling, and Christopher J. Petzold

18 Investigation of In Vivo Protein Interactions in *Aspergillus* Spores. 251
Kwang Cheol Jeong and Jae-Hyuk Yu

19 Purification of a Vesicle-Vacuole (V) Fraction from *Aspergillus*. 259
Anindya Chanda, Ludmila V. Roze, and John E. Linz

20 Use of the Hollow Fiber Assay for the Discovery of Novel Anticancer Agents from Fungi. 267
Cedric J. Pearce, Daniel D. Lantvit, Qi Shen, David Jarjoura, Xiaoli Zhang, Nicholas H. Oberlies, David J. Kroll, Mansukh C. Wani, Jimmy Orjala, Djaja D. Soejarto, Norman R. Farnsworth, Esperanza J. Carcache de Blanco, James R. Fuchs, A. Douglas Kinghorn, and Steven M. Swanson

21 Molecular Analysis of Fungal Gene Expression upon Interkingdom Competition with Insects 279
Ulrike Fohgrub and Frank Kempken

Index *287*

Contributors

Francisco J. Asensio • *Fundación MEDINA, Centro de Excelencia en Investigación de Medicamentos Innovadores en Andalucía, ArmillaGranada, Spain*

Tanveer S. Batth • *Physical Biosciences Division, Lawrence Berkeley National Laboratory, Joint BioEnergy Institute (JBEI), Berkeley, CA, USA*

Özgür Bayram • *Abteilung Molekulare Mikrobiologie und Genetik, Institut für Mikrobiologie und Genetik, DFG Research Center for Molecular Physiology of the Brain (CMPB), Georg-August-Universität Göttingen, Göttingen, Germany*

Özlem Sarikaya Bayram • *Abteilung Molekulare Mikrobiologie und Genetik, Institut für Mikrobiologie und Genetik, DFG Research Center for Molecular Physiology of the Brain (CMPB), Georg-August-Universität Göttingen, Göttingen, Germany*

Randolph M. Beaudry • *Department of Horticulture, Michigan State University, East Lansing, MI, USA*

Gerald F. Bills • *Fundación MEDINA, Centro de Excelencia en Investigación de Medicamentos Innovadores en Andalucía, Armilla, Granada, Spain*

Jin Woo Bok • *Department of Medical Microbiology and Immunology, University of Wisconsin-Madison, Madison, WI, USA*

Stefan Boedi • *Fungal Genetics and Genomics Unit, Department of Applied Genetics and Cell Biology, AIT—Austrian Institute of Technology GmbH Health and Environment Department, Bioresources, University and Research Center, UFT Campus TullnTulln/Donau, Austria*

Gerhard H. Braus • *Institut für Mikrobiologie und Genetik, Abteilung Molekulare Mikrobiologie und Genetik, DFG Research Center for Molecular Physiology of the Brain (CMPB), Georg-August-Universität Göttingen, Göttingen, Germany*

Erin L. Bredeweg • *Program for Molecular and Cellular Biology, Department of Biochemistry and Biophysics, Center for Genome Research and Biocomputing (CGRB), Oregon State University, Corvallis, OR, USA*

Esperanza J. Carcache de Blanco • *College of Pharmacy, University of Illinois at Chicago, Chicago, IL, USA*

Anindya Chanda • *Department of Food Science and Human Nutrition, Michigan State University, East Lansing, MI, USA; Department of Cell Biology, Harvard Medical School, Boston, MA, USA*

Lanelle R. Connolly • *Program for Molecular and Cellular Biology, Department of Biochemistry and Biophysics, Center for Genome Research and Biocomputing (CGRB), Oregon State University, Corvallis, OR, USA*

ANNE W. DOMBROWSKI • *Natural Products Discovery Institute, Doylestown, Pennsylvania, USA*
HEATHER EDGERTON-MORGAN • *Department of Molecular Biosciences, University of Kansas, Lawrence, KS, USA*
NORMAN R. FARNSWORTH • *College of Pharmacy, University of Illinois at Chicago, Chicago, IL, USA*
NATALIE D. FEDOROVA • *The J. Craig Venter Institute, Rockville, MD, USA*
ULRIKE FOHGRUB • *Abteilung für Botanische Genetik und Molekularbiologie, Botanisches Institut und Botanischer Garten, Christian-Albrechts-Universität zu Kiel, Kiel, Germany*
RY R. FORSETH • *Department of Chemistry and Chemical Biology, Boyce Thompson Institute, Cornell University, Ithaca, NY, USA*
MICHAEL FREITAG • *Program for Molecular and Cellular Biology, Department of Biochemistry and Biophysics, Center for Genome Research and Biocomputing (CGRB), Oregon State University, Corvallis, OR, USA*
JENS C. FRISVAD • *Department of Systems Biology, Center for Microbial Biotechnology, Technical University of Denmark, Kongens Lyngby, Denmark*
JAMES R. FUCHS • *College of Pharmacy, University of Illinois at Chicago, Chicago, IL, USA*
ISAO FUJII • *School of Pharmacy, Iwate Medical University, Iwate, Japan*
MICHAEL A. GOETZ • *Natural Products Discovery Institute, Doylestown, Pennsylvania, USA*
TAKAYUKI ITOH • *Research Institute for Bioscience Products and Fine Chemicals, Ajinomoto Co., Kawasaki, Kanagawa, Japan*
DAVID JARJOURA • *College of Pharmacy, University of Illinois at Chicago, Chicago, IL, USA*
KWANG CHEOL JEONG • *Department of Animal Sciences and Emerging Pathogens Institute, University of Florida, Gainesville, FL, USA*
BASTIAN JÖHNK • *Abteilung Molekulare Mikrobiologie und Genetik, Institut für Mikrobiologie und Genetik, and DFG Research Center for Molecular Physiology of the Brain (CMPB), Georg-August-Universität Göttingen, Göttingen, Germany*
JAY D. KEASLING • *Physical Biosciences Division, Joint BioEnergy Institute (JBEI), Lawrence Berkeley National Laboratory, Berkeley, CA, USA; Department of Chemical & Biomolecular Engineering, Department of Bioengineering, Berkeley Center for Synthetic Biology (SynBERC), University of California, Berkeley, CA, USA*
NANCY P. KELLER • *Department of Medical Microbiology and Immunology, University of Wisconsin-Madison, Madison, WI, USA*
FRANK KEMPKEN • *Abteilung für Botanische Genetik und Molekularbiologie, Botanisches Institut und Botanischer Garten, Christian-Albrechts-Universität zu Kiel, Kiel, Germany*
A. DOUGLAS KINGHORN • *College of Pharmacy, University of Illinois at Chicago, Chicago, IL, USA*
DAVID J. KROLL • *College of Pharmacy, University of Illinois at Chicago, Chicago, IL, USA*
TETSUO KUSHIRO • *Department of Agricultural Chemistry, Meiji University, Kawasaki, Kanagawa, Japan*

Daniel D. Lantvit • *College of Pharmacy, University of Illinois at Chicago, Chicago, IL, USA*
Shu-Ming Li • *Philipps-Universität Marburg, Marburg, Germany*
John E. Linz • *Department of Food Science & Human Nutrition and Department of Microbiology & Molecular Genetics, National Food Safety & Toxicology Center and Center for Integrative Toxicology, Michigan State University, East Lansing, MI, USA*
Marnix H. Medema • *Groningen Bioinformatics Centre and Department of Microbial Physiology, Groningen Biomolecular Sciences and Biotechnology Institute, University of Groningen, Groningen, The Netherlands*
Venkatesh Moktali • *The J. Craig Venter Institute, Rockville, MD, USA*
Berl R. Oakley • *Department of Molecular Biosciences, University of Kansas, Lawrence, KS, USA*
C. Elizabeth Oakley • *Department of Molecular Biosciences, University of Kansas, Lawrence, KS, USA*
Nicholas H. Oberlies • *College of Pharmacy, University of Illinois at Chicago, Chicago, IL, USA*
Jimmy Orjala • *College of Pharmacy, University of Illinois at Chicago, Chicago, IL, USA*
Hee Soo Park • *Molecular and Environmental Toxicology Center, University of Wisconsin-Madison, Madison, WI, USA*
Cedric J. Pearce • *College of Pharmacy, University of Illinois at Chicago, Chicago, IL, USA*
Christopher J. Petzold • *Physical Biosciences Division, Joint BioEnergy Institute (JBEI), Lawrence Berkeley National Laboratory, Berkeley, CA, USA*
Pallavi A. Phatale • *Program for Molecular and Cellular Biology, Department of Biochemistry and Biophysics, Center for Genome Research and Biocomputing (CGRB), Oregon State University, Corvallis, OR, USA*
Kyle R. Pomraning • *Program for Molecular and Cellular Biology, Department of Biochemistry and Biophysics, Center for Genome Research and Biocomputing (CGRB), Oregon State University, Corvallis, OR, USA*
Yazmid Reyes-Dominguez • *Fungal Genetics and Genomics Unit, Department of Applied Genetics and Cell Biology, AIT—Austrian Institute of Technology GmbH—Health and Environment Department, Bioresources, University and Research Center—UFT Campus Tulln, Tulln/Donau, Austria*
Ludmila V. Roze • *Department of Food Science and Human Nutrition, Michigan State University, East Lansing, MI, USA*
James F. Sanchez • *Department of Pharmacology and Pharmaceutical Sciences, School of Pharmacy, University of Southern California, Los Angeles, CA, USA*
Frank C. Schroeder • *Department of Chemistry and Chemical Biology, Boyce Thompson Institute, Cornell University, Ithaca, NY, USA*
Qi Shen • *College of Pharmacy, University of Illinois at Chicago, Chicago, IL, USA*
Kristina M. Smith • *Program for Molecular and Cellular Biology, Department of Biochemistry and Biophysics, Center for Genome Research and Biocomputing (CGRB), Oregon State University, Corvallis, OR, USA*
Djaja D. Soejarto • *College of Pharmacy, University of Illinois at Chicago, Chicago, IL, USA*

JOSEPH STRAUSS • *Fungal Genetics and Genomics Unit, Department of Applied Genetics and Cell Biology, AIT—Austrian Institute of Technology GmbH—Health and Environment Department, Bioresources, University and Research Center—UFT Campus Tulln, Tulln/Donau, Austria*

STEVEN M. SWANSON • *College of Pharmacy, University of Illinois at Chicago, Chicago, IL, USA*

JOSÉ R. TORMO • *Fundación MEDINA, Centro de Excelencia en Investigación de Medicamentos Innovadores en Andalucía, Armilla, Granada, Spain*

OLIVER VALERIUS • *Abteilung Molekulare Mikrobiologie und Genetik, Institut für Mikrobiologie und Genetik, and DFG Research Center for Molecular Physiology of the Brain (CMPB), Georg-August-Universität Göttingen, Göttingen, Germany*

CHRISTIANE WALLWEY • *Philipps-Universität Marburg, Marburg, Germany*

CLAY C.C. WANG • *Department of Pharmacology and Pharmaceutical Sciences, School of Pharmacy, University of Southern California, Los Angeles, CA, USA; Department of Chemistry, College of Letters, Arts, and Sciences, University of Southern California, Los Angeles, CA, USA*

MANSUKH C. WANI • *College of Pharmacy, University of Illinois at Chicago, Chicago, IL, USA*

BEATE WOLLINSKY • *Philipps-Universität Marburg, Marburg, Germany*

JAE-HYUK YU • *Departments of Bacteriology and Genetics, Molecular and Environmental Toxicology Center, University of Wisconsin-Madison, Madison, WI, USA*

XIAOLI ZHANG • *College of Pharmacy, University of Illinois at Chicago, Chicago, IL, USA*

Chapter 1

Library Preparation and Data Analysis Packages for Rapid Genome Sequencing

Kyle R. Pomraning, Kristina M. Smith, Erin L. Bredeweg, Lanelle R. Connolly, Pallavi A. Phatale, and Michael Freitag

Abstract

High-throughput sequencing (HTS) has quickly become a valuable tool for comparative genetics and genomics and is now regularly carried out in laboratories that are not connected to large sequencing centers. Here we describe an updated version of our protocol for constructing single- and paired-end Illumina sequencing libraries, beginning with purified genomic DNA. The present protocol can also be used for "multiplexing," i.e. the analysis of several samples in a single flowcell lane by generating "barcoded" or "indexed" Illumina sequencing libraries in a way that is independent from Illumina-supported methods. To analyze sequencing results, we suggest several independent approaches but end users should be aware that this is a quickly evolving field and that currently many alignment (or "mapping") and counting algorithms are being developed and tested.

Key words: High-throughput sequencing, Solexa/illumina library, Reference genome, Paired-end library, De novo assembly, Barcoding

1. Introduction

High-throughput DNA sequencing (HTS) is now commonly used to determine the DNA sequence of closely related strains within one species, referred to as "re-sequencing" because a reference genome sequence had been previously determined by more traditional chain termination, or "Sanger," sequencing. Single- or paired-end re-sequencing has proven useful for identification of SNPs, indels, rearrangements and differences in copy number compared to a reference genome (1–5). Paired-end high-throughput sequencing has also been used as the basis for *de novo* assembly of previously unknown genome sequences (6–9). De novo assembly from

Nancy P. Keller and Geoffrey Turner (eds.), *Fungal Secondary Metabolism: Methods and Protocols*,
Methods in Molecular Biology, vol. 944, DOI 10.1007/978-1-62703-122-6_1, © Springer Science+Business Media, LLC 2012

short paired-end reads is particularly feasible in filamentous fungi where small genomes (<100 Mb) are common (7, 9). In these cases, the coding portion of the genome is typically well represented, especially in species where the fraction of repetitive DNA is low (9) because of excision of repeats by recombination or mutation by RIP, respectively (10). The methods to construct high-throughput sequencing libraries with genomic DNA are essentially the same as those for DNA derived from chromatin immunoprecipitation (ChIP) or cDNA generated for transcriptome analyses as long as sufficient starting material can be produced.

Prior to preparation of a sequencing library, DNA must be isolated and purified (see http://www.illumina.com/support/documentation.ilmn; "Genomic DNA Sample Preparation Guide"). Here we begin with purified genomic DNA as a starting point (Fig. 1). We previously published a detailed protocol for the isolation of high-quality genomic DNA from *Neurospora crassa* (11), which we have now also used with *Fusarium* spp., *Trichoderma* spp., and *Aspergillus nidulans*. Genomic DNA, which should first be analyzed for contaminating levels of RNA and if necessary treated with RNase A (see Note 1), must be sheared to near-homogeneous fragment sizes. Three commonly used methods to generate small DNA fragments for HTS library generation are (1) digestion, (2) nebulization, and (3) sonication. For rapid whole-genome sequencing digestion is not suitable because of bias in enzyme recognition sites and enzyme activity. Nebulization has a tendency to generate heterogeneous DNA fragments and results in substantial loss of DNA by vaporization (12). Therefore, we describe here shearing of genomic DNA or whole cells by sonication with a Bioruptor, which allows us to reproducibly sonicate 24 samples at one time. We also provide an alternative protocol that describes sonication with a standard tip sonicator. In both cases the DNA is sheared almost randomly, generating double strand breaks with blunt ends, as well as 3′- and 5′-overhangs. The ends are made flush with a mixture of enzymes. A 3′ adenine overhang is added to make the DNA library compatible for ligation with Illumina sequencing adapters (Table 1), which all have 3′ thymine overhangs. After ligation, PCR primers that are compatible with the adapter sequences are used to amplify and enrich for successfully ligated DNA (see Note 2).

To drastically reduce sequencing cost, several samples can be combined and sequenced in one lane ("multiplexing"). For a typical ~40 Mb filamentous fungal genome, we routinely combine up to four samples in one lane of the Illumina GAIIx, or eight samples in the HiSeq 2000, which generally yield more than 30 or 250 million total reads per lane, respectively. Divided among four or eight equally loaded samples, this yields more than 7–40 million reads per sample, in our experience sufficient to call statistically significant ChIP-seq peaks and even single point mutations. For *de novo*

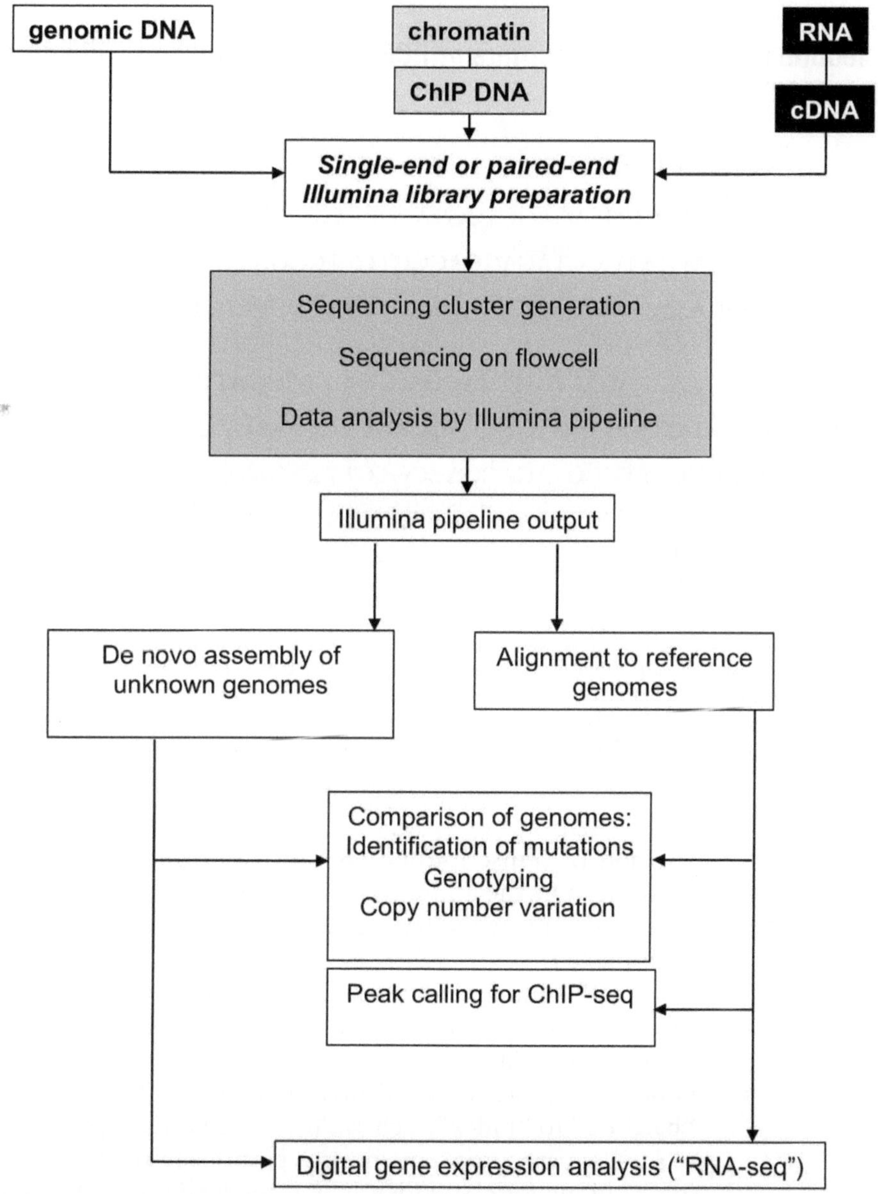

Fig. 1. Workflow for rapid genome sequencing. Genomic DNA, ChIPed DNA, or cDNA prepared from RNA is ligated to single-end or paired-end adapters to generate a sequencing library. Sequencing centers generally perform the final "wet lab" (cluster generation and sequencing) and initial computational steps (running the Illumina pipeline) indicated in the *gray box*. This generates sequence files with quality scores, the Illumina output files, which are used as input for software programs that either assemble reads *de novo* (if the genome is unknown) or map reads to a reference genome. Reference genomes are necessary for efficient analysis of ChIP-seq data and for calling mutations, genotyping strains, or other applications. RNA-seq reads can either be assembled *de novo*, or mapped to a reference set of transcripts or a reference genome. More details on various analysis software can be found in the text.

Table 1
Oligonucleotide sequences for adapters or PCR primers supplied by Illumina, Inc. (copyright, 2008)

Name	Sequence (5′–3′)
SE_P_Adaptl	Phos-**GATC**GGAAGAGCTCGTATGCCGTCTTCTGCTTG
SE_Adapt2	ACACTCTTTCCCTACACGACGCTCTTCC**GATC**T
SE_PCR1.1	AATGATACGGCGACCACCGAGATCTACACTCTTTCCCTACACGACGCT CTTCC**GATC**T
SE_PCR2.1	CAAGCAGAAGACGGCATACGAGCTCTTCC**GATC**T
PE_P_Adaptl	Phos-**GATC**GGAAGAGCGGTTCAGCAGGAATGCCGAG
PE_Adapt2	ACACTCTTTCCCTACACGACGCTCTTCC**GATC**T
PE_PCR1.1	AATGATACGGCGACCACCGAGATCTACACTCTTTCCCTACACGACGCTC TTCC**GATC**T
PE_PCR2.1	CAAGCAGAAGACGGCATACGAGATCGGTCTCGGCATTCCTGCTGAACC GCTCTTCC**GATC**T

The oligos listed here are used for single-end or paired-end library construction without barcodes. Oligos need not be highly purified (desalting is sufficient). The Adapt 1 primers need to be phosphorylated ("Phos," and note "P" in adapter name) at the 5′-most nucleotide. For more information see http://www.illumina.com/support/documentation.ilmn; "Illumina Adapter Sequences" (see Note 19). Oligonucleotide sequences © 2007–2011 Illumina, Inc. All rights reserved

assembly or transcriptome assembly we do not multiplex on this machine. On the Illumina HiSeq 2000 machine, 100–250 million reads are generated per lane, so additional samples can be multiplexed.

While Illumina advocates and supports an indexing method that does not reduce the read length (see http://www.illumina.com/support/documentation.ilmn; "Multiplexing Preparation Guide"), we often use adapters that have a four-nucleotide (nt) "barcode" to "index" each sample, based on an earlier 3-nt system (13). The barcode is sequenced as part of a normal read, thus yielding a unique four base identifier at the beginning of each read that is later used to sort or "parse" reads from the combined sample. This means that each read is reduced by 4 nt, e.g. from 58 to 54 nt on the HiSeq 2000 machine. Barcode sequences have been tested and optimized in various ways. We found problems with certain combinations of nucleotides that resulted in either higher than normal error rates or specific exclusion of some barcodes from cluster generation and/or sequencing, and thus started to balance the sequences of the adapters present in a single lane to yield even numbers of expected calls for all nucleotides in the first four cycles (one example set of four balanced barcodes would be 5′-ACGT, 5′-CGTA, 5′-GTAC, and 5′-TACG; Table 2). Because of the adapter design, the fifth base of each read is always a T.

Table 2
Adapter sequences for barcoding single-end and paired-end libraries

Name	Sequence (5′–3′)
SE_P_A1_ACGTT	**ACGT**AGATCGGAAGAGCTCGTATGCCGTCTTCTGCTTG
SE_P_A1_CGTAT	**TACG**AGATCGGAAGAGCTCGTATGCCGTCTTCTGCTTG
SE_P_A1_GTACT	**GTAC**AGATCGGAAGAGCTCGTATGCCGTCTTCTGCTTG
SE_P_A1_TACGT	**CGTA**AGATCGGAAGAGCTCGTATGCCGTCTTCTGCTTG
SE_A2_ACGTT	ACACTCTTTCCCTACACGACGCTCTTCCGATCT**ACGTT**
SE_A2_CGTAT	ACACTCTTTCCCTACACGACGCTCTTCCGATCT**CGTAT**
SE_A2_GTACT	ACACTCTTTCCCTACACGACGCTCTTCCGATCT**GTACT**
SE_A2_TACGT	ACACTCTTTCCCTACACGACGCTCTTCCGATCT**TACGT**
SE_P_A1_AGTCT	**GACT**AGATCGGAAGAGCTCGTATGCCGTCTTCTGCTTG
SE_P_A1_CTAGT	**CTAG**AGATCGGAAGAGCTCGTATGCCGTCTTCTGCTTG
SE_P_A1_GACTT	**AGTC**AGATCGGAAGAGCTCGTATGCCGTCTTCTGCTTG
SE_P_A1_TCGAT	**TCGA**AGATCGGAAGAGCTCGTATGCCGTCTTCTGCTTG
SE_A2_AGTCT	ACACTCTTTCCCTACACGACGCTCTTCCGATCT**AGTCT**
SE_A2_CTAGT	ACACTCTTTCCCTACACGACGCTCTTCCGATCT**CTAGT**
SE_A2_GACTT	ACACTCTTTCCCTACACGACGCTCTTCCGATCT**GACTT**
SE_A2_TCGAT	ACACTCTTTCCCTACACGACGCTCTTCCGATCT**TCGAT**
SE_P_A1_ATGCT	**GCAT**AGATCGGAAGAGCTCGTATGCCGTCTTCTGCTTG
SE_P_A1_CATGT	**CATG**AGATCGGAAGAGCTCGTATGCCGTCTTCTGCTTG
SE_P_A1_GCATT	**ATGC**AGATCGGAAGAGCTCGTATGCCGTCTTCTGCTTG
SE_P_A1_TGCAT	**TGCA**AGATCGGAAGAGCTCGTATGCCGTCTTCTGCTTG
SE_A2_ATGCT	ACACTCTTTCCCTACACGACGCTCTTCCGATCT**ATGCT**
SE_A2_CATGT	ACACTCTTTCCCTACACGACGCTCTTCCGATCT**CATGT**
SE_A2_GCATT	ACACTCTTTCCCTACACGACGCTCTTCCGATCT**GCATT**
SE_A2_TGCAT	ACACTCTTTCCCTACACGACGCTCTTCCGATCT**TGCAT**
SE_P_A1_ACTGT	**CAGT**AGATCGGAAGAGCTCGTATGCCGTCTTCTGCTTG
SE_P_A1_CTCAT	**TGAG**AGATCGGAAGAGCTCGTATGCCGTCTTCTGCTTG
SE_P_A1_GAGTT	**ACTC**AGATCGGAAGAGCTCGTATGCCGTCTTCTGCTTG
SE_P_A1_TGACT	**GTCA**AGATCGGAAGAGCTCGTATGCCGTCTTCTGCTTG
SE_A2_ACTGT	ACACTCTTTCCCTACACGACGCTCTTCCGATCT**ACTGT**
SE_A2_CTCAT	ACACTCTTTCCCTACACGACGCTCTTCCGATCT**CTCAT**
SE_A2_GAGTT	ACACTCTTTCCCTACACGACGCTCTTCCGATCT**GAGTT**
SE_A2_TGACT	ACACTCTTTCCCTACACGACGCTCTTCCGATCT**TGACT**

The oligos listed here are used for single-end or paired-end library construction with balanced barcodes. The same PCR primers as shown in Table 1 are used to amplify single-end or paired-end libraries. Oligos listed here have been used successfully (note that not all potential 5-nt barcodes are shown or have been tested yet). Oligos need not be highly purified (desalting is sufficient). All Adapt1 primers ("A1") need to be phosphorylated (note "P" in adapter name) at the 5′-most nucleotide. These sequences are based on and contain the oligonucleotide sequences shown in Table 1, which are copyright of Illumina, Inc.; 2008. *Note*: Illumina does not support this type of barcoding and has a different system available for purchasing. The parsing and aligning of reads obtained with the barcodes shown here will require use of in-house scripts. Oligonucleotide sequences © 2007–2011 Illumina, Inc. All rights reserved. Derivative works created by Illumina customers are authorized for use with Illumina instruments and products only. All other uses are strictly prohibited.

There are few labs with access to their own HTS machines. In most cases sequencing libraries will be processed for sequencing by dedicated staff in sequencing centers or private companies. Thus we do not elaborate on the Illumina cluster generation and sequencing steps (Fig. 1). We wish to emphasize that investment in computational hardware and personnel "know-how" is essential to successfully undertake whole-genome sequencing approaches. Here, we do not go into details with regard to the downstream handling of sequencing data, for two reasons: (1) Currently the processing pipelines supplied by Illumina change frequently as throughput increases every few months and (2) new and drastically improved analysis tools become available or are being updated almost monthly. Thus, we provide only brief outlines for how to prepare for the flood of sequencing data but we do not recommend specific software. Instead we focus on the part of the sequencing that is most influenced by the end user of this technology, the preparation and troubleshooting of Illumina sequencing libraries.

2. Materials

2.1. Equipment to Determine DNA Concentration

1. Nanodrop spectrophotometer (NanoDrop products, Wilmington, USA).
2. Qubit fluorometer (Invitrogen, Carlsbad, USA).

2.2. Equipment to Shear DNA

1. Waterbath sonicator: Bioruptor UCD-200 (Diagenode, Liège, Belgium).
2. Tip sonicator: Branson 450 Sonifier (Branson Ultrasonics, Danbury, CT).

2.3. Processing of DNA

2.3.1. Shearing of DNA by Sonication

1. Genomic DNA.
2. TE Buffer: 10 mM Tris–HCl, 1 mM EDTA in sterile distilled H_2O, pH 8.0. Stored at room temperature.
3. Components of QIAquick PCR purification kit (Qiagen, Valencia, USA) or equivalent:
   (a) PB buffer: 5.5 M guanidine hydrochloride, 20 mM Tris–HCl, pH 6.6. Store at room temperature.
   (b) PE buffer: 20 mM NaCl, 2 mM Tris–HCl in 80% ethanol, pH 7.5. Store at room temperature.
   (c) QIAquick columns.
   (d) EB buffer: 10 mM Tris–HCl, pH 8.5. Store at room temperature.

2.3.2. End Repair of Sheared DNA Fragments (See Note 3)

1. Sheared genomic DNA.
2. 10× T4 DNA ligase buffer: 500 mM Tris–HCl, 100 mM $MgCl_2$, 100 mM dithiothreitol 10 mM ATP, pH 7.5 (New England Biolabs [NEB], Ipswich, USA) (see Note 4).
3. dNTPs (20 mM of each nucleotide, i.e. mix 20 μl each of 100 mM dATP, dCTP, dGTP, dTTP, and H_2O).
4. T4 DNA polymerase, 3 units/μL (NEB).
5. DNA polymerase I, large (Klenow) fragment, 5 units/μL (NEB).
6. T4 polynucleotide kinase, 10 units/μL (NEB).
7. Components of QIAquick purification kits (see Subheading 2.3.1, items a–d; Qiagen).

2.3.3. Addition of A-Overhang

1. End-repaired, sheared DNA.
2. 10× NEBuffer 2: 500 mM NaCl, 100 mM Tris–HCl, 100 mM $MgCl_2$, 10 mM dithiothreitol, pH 7.9 (NEB).
3. 1 mM dATP.
4. DNA polymerase I, large (Klenow) fragment (3′→5′ exo^-), 5 units/μL (NEB).
5. Components of QIAquick and MinElute purification kits (see Subheading 2.3.1, items a, b, d; Qiagen) or equivalent.
6. MinElute columns (Qiagen).

2.3.4. Ligation of Sequencing Adapters

1. A-tailed, end-repaired, sheared DNA.
2. Single-end (SE) and paired-end (PE) library adapters (Table 1) or SE barcode adapters (Table 2): mix 5 μM of each of two oligonucleotides in sterile distilled H_2O. To anneal adapters, boil for 2 min (min) in a heat block and allow to cool for 30 min at room temperature. Store annealed adapters at –20°C.
3. T4 DNA ligase, 400 units/μL (NEB).
4. NuSieve GTG low melting temperature agarose (Lonza, Walkersville, USA).
5. TAE buffer: 40 mM Tris, 20 mM acetate, 1 mM EDTA. To make 50× stock, dissolve 242 g Tris base in water, add 57.1 mL glacial acetic acid, 100 mL of 500 mM EDTA (pH 8.0), and bring the final volume to 1 L. Store at 20°C.
6. Ethidium bromide solution (stocks of 10 mg/mL, use 5–10 μL stock for 100 mL of gel).
7. 5× DNA gel loading buffer: 125 mg bromophenol blue, 125 mg xylene cyanol, 25 mL of 0.2 M EDTA (pH 8.0), 15 mL glycerol, 10 mL H_2O.
8. DNA ladder able to resolve 100–1,000 bp (e.g., Qiagen GelPilot 1 kb Plus Ladder, #239095).

9. Glass microscope cover slips.
10. Components of QIAquick Gel Extraction and PCR purification kits (see Subheading 2.3.1, items b–d; Qiagen) or equivalent.
11. QG buffer: 5.5 M guanidine thiocyanate, 20 mM Tris–HCl, pH 6.6. Store at room temperature (Qiagen).
12. Isopropanol.

2.3.5. Amplification of Sequencing Library

1. DyNAzyme II DNA polymerase, 2 units/μL (Finnzymes, Vantaa, Finland).
2. 10× buffer for DyNAzyme II DNA polymerase (Finnzymes): 100 mM Tris–HCl (pH 8.8), 15 mM $MgCl_2$, 500 mM KCl, 1% Triton X-100.
3. Single-end (SE) and paired-end (PE) library PCR amplification primers (Table 1); *Dpn*II sites are indicated in bold): mix 5 μM each in H_2O. Store at −20°C.
4. 2× Phusion Flash master mix: 1 unit/μL Phusion Flash High-Fidelity polymerase, 50 mM TAPS-HCl (pH 9.3), 100 mM KCl, 3 mM $MgCl_2$, 2 mM β-mercaptoethanol, 400 μM of dATP, dCTP, dGTP, and dTTP (NEB).

2.3.6. Oligonucleotide Sequences

The two tables contain Illumina oligonucleotide sequences (Table 1; these oligonucleotide sequences are copyright of Illumina, Inc.; 2008) or barcode adapter sequences for construction of indexed single-end libraries (Table 2).

3. Methods

All procedures are carried out at room temperature unless otherwise noted.

3.1. Shearing of DNA by Sonication

1. Adjust the water bath temperature of the Bioruptor to 4°C prior to sonication (see Note 5).
2. Melt a 1% agarose gel and a 2% NuSieve low melting temperature agarose gel in TAE buffer and cool to 55°C. Add ethidium bromide to 0.5 μg/mL and pour in separate gel casting trays prior to sonication. When the gels are cool, place in separate electrophoresis chambers and cover with TAE buffer.
3. Dilute 5 μg purified genomic DNA into 100 μL TE buffer in an Eppendorf tube (recommended: VWR 87003-294; see Note 6).
4. Place the Eppendorf tube with DNA in the Bioruptor and sonicate on high for 20 min total with 30 s of sonication followed by 30 s of rest ($T_1 = 20$ min, $T_2 = 30$ s, $T_3 = 30$ s).

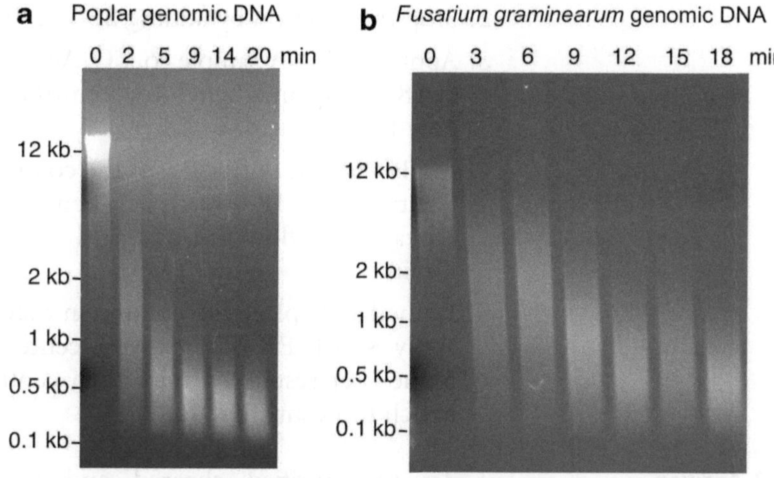

Fig. 2. Shearing of DNA in the bioruptor and by tip sonication. (**a**) Shearing of poplar (*Populus trichocarpa*) genomic DNA in the bioruptor (5 μg DNA in TE buffer, 300 μL volume) on "high" setting with 30 s on/off cycles. The total time of sonication in minutes (min) is shown above the lanes in which ~170 ng of DNA was loaded. (**b**) Shearing of *Fusarium graminearum* genomic DNA. Tissue was disrupted by bead beating for chromatin isolation, centrifuged and the supernatant sonicated in a bioruptor as described in (**a**), treated with RNAseA and run on a 1% agarose gel in TAE buffer.

5. Mix 10 μL of the sonicated DNA (0.5 μg) with 2 μL 5× DNA gel loading buffer and run on the 1% agarose gel to ensure the DNA has been sheared to less than 1,000 bp (Fig. 2). Sonicate for additional cycles if the DNA is inadequately sheared.
6. Mix the sheared DNA with five volumes of buffer PB and vortex to mix.
7. Apply the DNA sample to a QIAquick column and centrifuge at 18,000 × *g* in a bench top centrifuge for 1 min. Discard the flow-through.
8. Apply 750 μL buffer PE to the column and let rest for 5 min. Centrifuge at 18,000 × *g* in a bench top centrifuge for 1 min. Discard the flow-through and centrifuge at 18,000 × *g* for 2 min.
9. Immediately place the column in a sterile Eppendorf tube and apply 40.5 μL EB buffer to the center of the QIAquick membrane. Let rest for 1 min, then centrifuge at 18,000 × *g* in a bench top centrifuge for 1 min.

3.2. End Repair of Sheared DNA Fragments

1. To the purified sheared DNA (*from* Subheading 3.1, step 9) add 5 μL 10× T4 DNA ligase buffer with 10 mM ATP, 1 μL 20 mM dNTPs, 2.5 μL T4 DNA polymerase, 0. 5 μL large DNA polymerase I (Klenow) fragment, and 2.5 μL T4 polynucleotide kinase. Incubate at room temperature for 30 min.

2. Stop the reaction by adding 250 μL buffer PB. Vortex to mix.
3. Apply the DNA sample to a QIAquick column and centrifuge at 18,000 × *g* in a bench top centrifuge for 1 min. Discard the flow-through.
4. Apply 750 μL buffer PE to the column and let rest for 5 min. Centrifuge at 18,000 × *g* in a bench top centrifuge for 1 min. Discard the flow-through and centrifuge at 18,000 × *g* for 2 min.
5. Immediately place the column in a sterile Eppendorf tube and apply 34 μL EB buffer to the center of the QIAquick membrane. Let rest for 1 min, then centrifuge at 18,000 × *g* in a bench top centrifuge for 1 min.

3.3. Addition of A-Overhang

1. To the purified sheared and end-repaired DNA (*from* Subheading 3.2, step 5), add 5 μL 10× NEBuffer 2, 10 μL 1 mM dATP, and 3 μL DNA polymerase I Klenow fragment (3′→5′ exo⁻). Incubate at 37°C for 30 min.
2. Stop the reaction by adding 250 μL buffer PB. Vortex to mix.
3. Apply the DNA sample to a MinElute column and centrifuge at 18,000 × *g* in a bench top centrifuge for 1 min. Discard the flow-through.
4. Apply 750 μL buffer PE to the column and let rest for 5 min. Centrifuge at 18,000 × *g* in a bench top centrifuge for 1 min. Discard the flow-through and centrifuge at 18,000 × *g* for 1 min.
5. Immediately place the column in a sterile Eppendorf tube and apply 17 μL EB buffer to the center of the MinElute membrane. Let rest for 1 min, then centrifuge at 18,000 × *g* in a bench top centrifuge for 1 min.

3.4. Ligation of Sequencing Adapters

1. To the purified sheared, end-repaired, and tailed DNA (*from* Subheading 3.3, step 5), add 2 μL 10× T4 ligase buffer with 10 mM dATP, 1 μL of 5 μM library adapters (non-barcoded single- or paired-end adapters, Table 1; barcoded single-end adapters, Table 2) and 1 μL T4 DNA ligase (see Note 7). Incubate at room temperature for 20 min.
2. Stop the reaction by mixing ligated DNA with 4 μL 5× DNA gel loading buffer.
3. Load the ligated DNA on the 2% NuSieve low melting temperature agarose gel. Use DNA ladder to resolve 100–1,000 bp and run for ~45 min at 100 V (see Note 8).
4. View the gel on a UV transilluminator and excise a gel fragment between 250 and 350 bp with a clean glass cover slip (Fig. 3a; see Note 9).

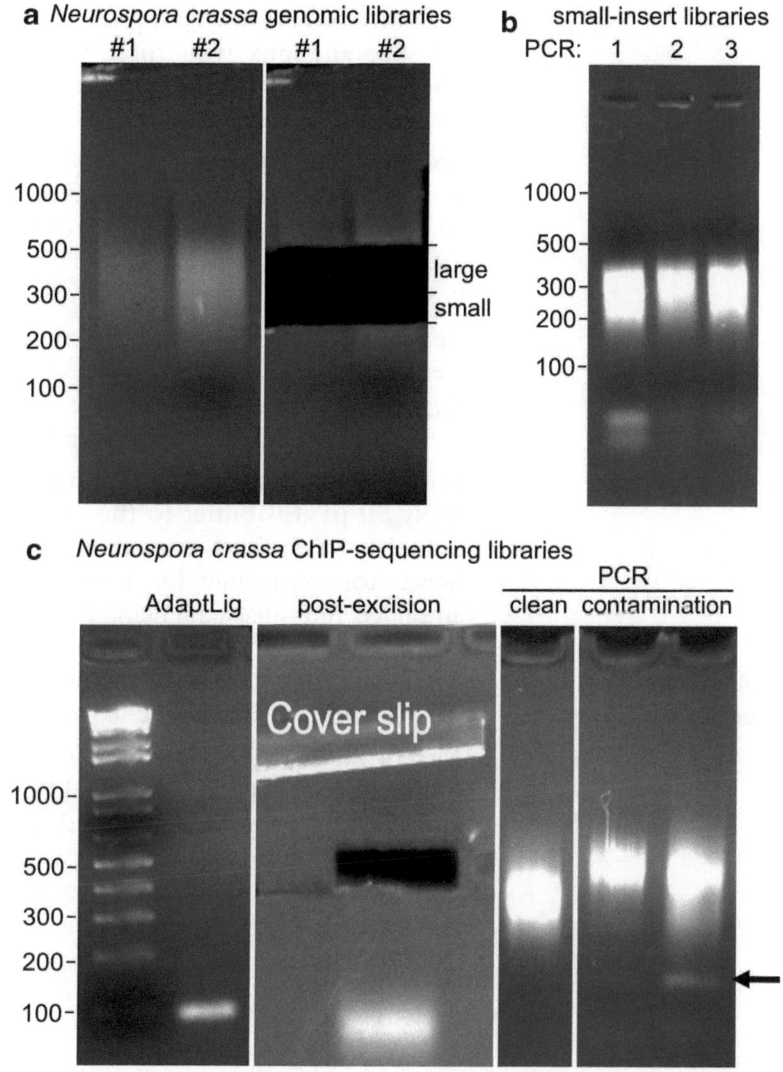

Fig. 3. Preparation of genomic and ChIP paired-end sequencing libraries. (**a**) *Neurospora crassa* genomic DNA was sonicated with a tip sonicator (see Note 5), and libraries were generated as described. Two replicates were done in parallel and loaded on a 2% NuSieve agarose gel for DNA isolation (*left panel*). Two fractions were removed from the gel (250–325 bp, "small"; 325–500 bp, "large"). We do this to insure that we have library material for future PCRs even in cases when library amplifications fail or other unforeseen difficulties arise. (**b**) Three parallel PCR amplifications of the small insert libraries shown in (**a**) (5 μL of 35 μL were loaded). Note the presence of adapter and PCR primer bands. (**c**) Example of construction of ChIP-seq libraries. After ligation of adapters ("AdaptLig") the library is invisible on the gel (but note the presence of adapter bands below 100 bp). After excision of the bands and extraction of DNA from the gel (Qiagen gel extraction kit), PCR reactions that are purified (Qiagen PCR purification kit) typically yield slightly smeary library bands ("clean") but no bands <200 bp. It is important to adjust the adapter:DNA ratio to ~10:1. In cases of great adapter excess, which can be a problem with ChIP-seq library generation, spurious bands are observed. Paired-end adapters run at ~160 bp and single-end adapters run at ~110 bp ("contamination," see *arrow*). We would not subject the library in the right-most lane to sequencing as the amount of contamination present will significantly reduce the number of useable reads, because most reads will be adapter sequence instead of insert.

5. Extract DNA with a QIAquick gel extraction kit: Weigh the gel slice and mix with three volumes of QG buffer in an Eppendorf tube. Melt the gel at 50°C for ~10 min with vortexing every 3 min (see Note 10).
6. To increase recovery of small DNA fragments, add a volume of isopropanol equivalent to one gel slice volume, vortex to mix, and apply the sample to a QIAquick column. Centrifuge at 18,000 × *g* in a bench top centrifuge for 1 min. Discard the flow-through.
7. Apply 750 μL buffer PE to the column and let rest for 5 min. Centrifuge at 18,000 × *g* in a bench top centrifuge for 1 min. Repeat the PE wash. Discard the flow-through and centrifuge at 18,000 × *g* for 2 min (see Note 11).
8. Immediately place the column in a sterile Eppendorf tube and apply 50 μL EB buffer to the center of the QIAquick membrane. Let rest for 1 min, then centrifuge at 18,000 × *g* in a bench top centrifuge for 1 min. This constitutes the non-amplified Illumina sequencing library.

3.5. Amplification of Sequencing Library

1. Test a range of cycle numbers to determine the optimum number of PCR cycles to amplify the sequencing library (see Note 12).
2. Mix 14.75 μL H_2O, 1 μL library DNA, 2 μL 10× DyNAzyme buffer, 0.25 μL of 20 mM dNTPs, 1.5 μL of 5 μM library amplification primers, and 0.5 μL DyNAzyme (see Note 13) for each number of cycles to be tested and amplify using the following protocol on a thermocycler:
   (a) 180 s at 94°C.
   (b) 30 s at 94°C.
   (c) 30 s at 60°C.
   (d) 60 s at 72°C.
   (e) Repeat (b) to (d) for 15, 18, or 21 cycles.
   (f) 300 s at 72°C.
3. Melt a 2% agarose gel in TAE buffer and cool to 55°C. Add ethidium bromide to 0.5 μg/mL and pour in a gel casting tray. When the gel cools, place in an electrophoresis chambers and cover with TAE buffer.
4. Mix the PCR reactions with 4 μL of 5× DNA gel loading buffer and load into the 2% agarose gel wells next to a DNA ladder able to resolve 100–1,000 bp and run for 45 min at 100 V.
5. View the gel on a UV transilluminator. The appropriate number of cycles results in a just visible DNA smear of ~50–100 bp greater than the size excised earlier (primers add 53 bp to the adapter ligated fragments; Notes 14 and 15).

6. After the appropriate number of cycles has been determined, amplify the DNA libraries in three separate reactions with 1–5 μL template DNA. Try to minimize the number of cycles necessary (Fig. 3b; Note 16).
7. Mix 8.5 μL template DNA and H_2O, 1.5 μL of 5 μM of each library amplification primers (SE_PCR1.1. and SE_PCR2.1. or PE_PCR1.1. and PE_PCR2.1.; Table 1) and 10 μL 2× Phusion Flash master mix. Amplify the library in three separate reactions using the following protocol on a Thermocycler (see Note 17):
   (a) 30 s at 98°C.
   (b) 10 s at 98°C.
   (c) 30 s at 65°C.
   (d) 30 s at 72°C.
   (e) Repeat steps b–d for the number of cycles determined.
   (f) 300 s at 72°C.
8. Pool the three reactions in an Eppendorf tube and add 300 μL buffer PB. Vortex to mix.
9. Apply the DNA sample to a QIAquick column and centrifuge at 18,000 × *g* in a bench top centrifuge for 1 min. Discard the flow-through.
10. Apply 750 μL buffer PE to the column and let rest for 5 min. Centrifuge at 18,000 × *g* in a bench top centrifuge for 1 min. Discard the flow-through and centrifuge at 18,000 × *g* for 2 min.
11. Immediately place the column in a sterile Eppendorf tube and apply 32 μL EB buffer to the center of the QIAquick membrane. Let rest for 1 min, then centrifuge at 18,000 × *g* in a bench top centrifuge for 1 min.
12. Melt a 2% agarose gel in TAE buffer and cool to 55°C. Add ethidium bromide to 0.5 μg/mL and pour in a gel casting tray. When the gel cools, place in an electrophoresis chamber and cover with TAE buffer.
13. Mix 5 μL of the purified PCR reactions with 5 μL of 1× DNA gel loading buffer and load into the 2% agarose gel wells next to a DNA ladder able to resolve 100–1,000 bp and run for 45 min at 100 V.
14. View the gel on a UV transilluminator. Only the library smear should be visible now, and no bands lower than 200 bp are expected (Figs. 3c and 4d).
15. Quantify the library by using a Qubit fluorometer or by quantitative PCR with the Illumina sequencing primers (see Note 18).

16. Dilute the sample in buffer EB to a concentration recommended by your core laboratory or sequencing company of choice.

3.6. Post-sequencing Data Analyses

Illumina software pipelines generate files of several formats, the most useful being "sequence.txt" files, which contain sequence and quality scores, and ELAND or bam files, which also contain the start and stop positions (or "coordinates") specific reads map to in reference genomes (see http://www.illumina.com/support/documentation.ilmn). Pre-analysis of raw data directly from the sequencer is typically run by core facilities or companies who provide the actual sequencing services, and this relies on proprietary Illumina software. File types supported by software for post-sequencing analysis vary, and so does their input and output file types. For this reason, familiarity with the program manuals is required and they should be consulted frequently for updated information.

Older pipelines of Illumina Genome Analyzers generated a different set of files than the current pipeline (CASAVA 1.8). Illumina CASAVA 1.7, for example, still used with the Illumina GAIIx machines, relies on analysis by the Illumina GERALD package to generate a comprehensive file "sequence.txt" of all the reads and quality score. Files of the format "sequence.txt" contain each read in a 4-line format: header, read, header, quality score. The header has information about the location on the flow cell from which the read was collected, the read consists of base calls made by the software analysis of the images, and the quality score has information about the probability of a base call being wrong. In this pipeline, quality scores are in a phred scale 64 Illumina-type offset format. The newer version, CASAVA 1.8, which is required for the higher throughputs on HiSeq 2000 machines, has been changed in several ways including the available output formats. Now, archival bam files and a collection of compressed fastq files are the primary source of total sequence reads. CASAVA 1.8 also uses different, Sanger-type quality scores, which are in phred scale 33 offset in ASCII format.

Many freely available programs will take Illumina "sequence.txt" files as input, such as Bowtie (14) and Velvet (15). There are tools available to convert sequence.txt files into fastq and other desired formats, when necessary. Some, such as MAQ (16), have these scripts built into a comprehensive package. Similar tools for quality score conversion are available through the BioPerl project for the savvy programmer (http://www.bioperl.org), and the Java-based Vancouver Short Read Analysis package, which also converts between popular formats (http://vancouvershortr.sourceforge.net). When samples are not barcoded for multiplexing samples, the GERALD output of CASAVA 1.7, or the compressed fastq files of CASAVA 1.8 may be used directly as input for downstream applications.

This is slightly more challenging when several distinct samples are run in one lane of a flowcell (multiplexing), as "barcoded" or "indexed" samples. In this case, several files need to be generated, grouped by the barcodes. Multiplexed samples prepared with proprietary Illumina indexing kits are supported by the CASAVA 1.8 software and thus can easily be separated into different files according to the barcode; the barcode sequence is automatically removed in this process. The barcoding scheme we mention here, however, may require one additional concatenation step of the many CASAVA 1.8 output files to combine the contents of a lane into one single huge, unwieldy file. Alternatively, and likely more convenient, the many CASAVA 1.8 output files can be used as input for sorting by barcode, followed by concatenation of the appropriate files. Both approaches require additional scripts for parsing of reads according to barcodes and removal of the barcode sequences prior to read mapping. Some institutions may have in-house software available for sorting and removal of barcodes but there are also freely available tools, such as "NovoBarCode" (novoalign package from Novocraft Technologies; http://www.novocraft.com/main/index.php). The goal of the sequencing will determine subsequent steps and file formats or conversions necessary. Prior to undertaking an analysis project, perhaps even the sequencing itself, it is recommended that users become comfortable with the command line interface and basic scripting, and read program manuals available online. For most of the software packages user groups are available that can advise on specific problems. The Seqanswers forum (http://seqanswers.com) is particularly useful for keeping up to date with constantly changing formats and software.

3.6.1. De Novo Assembly Software

A variety of open source programs for *de novo* assembly of genomes are available and are being improved regularly. Some of the more popular programs have been compared on identical datasets and are reviewed elsewhere (17). Most rely on the construction of a De Bruijn graph and differ primarily in their error correction steps. The authors have used Velvet (15) and its many updates to assemble fungal genomes (9), but other state of the art short read assemblers including ABySS (18), QSRA (19, 20), SOAPdenovo (21), Ray (22), and others described in ref. (17) promise to improve the speed and accuracy of assembly from short reads.

3.6.2. Discovery of SNPs, Insertion/Deletions (Indels), and Genome Rearrangements for Mutation Mapping

Short read sequence alignment programs are useful for identifying single nucleotide polymorphisms (SNPs), indels, and rearrangements in organisms genetically similar to a reference sequence. The programs map each read to a reference genome and allow the user to specify a variety of options including the number of mismatches allowed per read and whether reads are mapped to a single match in the genome or all possible perfect matches in the reference genome if identical kmer repeats exist; kmer refers to a specific determined length, i.e. 36mer. If the algorithm uses quality score values it is essential that the user specify the appropriate quality score scale. Popular open source programs for read alignment include SOAP (23), MAQ (16), Bowtie (14), BWA (24), CASHX (25), and Stampy (26). While some programs such as MAQ include SNP and indel analysis, others, like Bowtie, require additional analysis by a package such as SAMtools (27) to call variants. We have used these approaches to find single-point mutations in several fungal genomes, for example, see (5). Deciding which software program to use for read alignment may depend on computational power available versus what is required, which depends on the size of your genome and amount of sequence data.

3.6.3. Analysis of ChIP-Sequencing and MeDIP-Sequencing Data

The analysis of ChIP-seq data begins with aligning reads to a reference genome using one of the software packages mentioned above. How regions of enrichment are determined depends on the target of immunoprecipitation. In some cases, such as centromere proteins with localization to defined regions of each chromosome, statistical analysis of enrichment levels is not crucial (28). In the case of transcription factors, however, enriched regions must be called with software that uses statistical tools to assign a confidence level to each of the identified peaks (29). Different software packages are designed specifically to analyze localized binding, e.g. by transcription factors, versus diffuse binding, e.g. by histones and their modifications. We have used our own scripts in the past to call enrichment peaks (28, 29) but have also used MACS (30), a package which uses binomial peak identification to build a model of transcription factor binding. Diffuse binding seen in ChIP of histone modifications or other chromatin proteins requires a package such as SICER (31) or RSEG (32), which identify domains of enrichment spanning multiple nucleosomes, rather than discrete peaks. A wide variety of open source peak finding programs exist, and new additions and updates are frequent. If a genome assembly and gene annotations are available, visual confirmation of identified peaks in a browser like IGV (www.broadinstitute.org/igv/), GenomeView (http://genomeview.org/), or gbrowse (http://gmod.org/wiki/GBrowse) can be used to refine the settings and adjustable parameters of the program.

3.6.4. Analysis of Transcriptomes

Transcriptome sequencing by analysis of cDNA, or "RNA-seq," is performed to identify expressed genes from an organism with no reference genome, or in a quantitative way to identify differential expression between two or more conditions. If a reference genome is unavailable, RNA-seq reads must be assembled into contigs with Velvet (15) or SOAPdenovo (21), or any of the other *de novo* assemblers described above. If a reference genome is available, RNA-seq reads are mapped to the genome with an aligner that can find splice junctions. These include SOAP (23), Tophat (33), and

BWA (24). Other programs are designed specifically to identify novel intron/exon junctions (20). Counting cDNA tags from different samples and replicate experiments to generate comparable datasets that give meaningful results is challenging. To quantify transcript levels and call differential expression between two samples, statistical tools have been developed, including DEseq (34), Cufflinks (35, 36), and GENEcounter (37) but this field is still very much in flux.

4. Notes

1. When generating libraries from pooled DNA samples (e.g., to do bulk segregant analyses) it is essential that an equal amount of DNA from each strain be used. If different amounts of RNA are present in different samples this will affect the measured amount of nucleic acids and underestimate the amount of DNA in some samples. Imbalances in the concentration of the DNA pool are sometimes significant enough to affect mutation mapping. While presence of RNA will reduce the effective DNA concentration this is not a problem with library construction of non-pooled samples.
2. DNA shearing, end repair, A-overhang addition, ligation, and the subsequent gel extraction should be performed in a single day for best ligation efficiency. Ligated DNA can be stored at −20°C and amplified by PCR at any time prior to Illumina sequencing cluster generation. The incubation steps prior to ligation are short so plan ahead and thaw buffers for the next reaction during the incubation so that the next step can be performed immediately.
3. Several manufacturers now offer kits that contain all or most enzymes required for library preparation (e.g., NEB #E6040, contains all required enzymes; NEB Quick-Blunting kit #E1201 or Epicentre End-It end-repair kit #ER0720, contain enzymes for blunting reactions). In our hands both kits and individual enzymes work equally well. We typically use enzymes from NEB but enzymes from other suppliers can be used. Compare units/μl (e.g., Invitrogen T4 ligase is typically less concentrated than NEB T4 ligase, so we use 2.5 μl of this enzyme).
4. Ligase buffer can be prepared fresh or premade buffers from several manufacturers can be used, e.g. 10× ligase buffer from NEB or 5× ligase buffer from Invitrogen: 250 mM Tris–HCl (pH 7.6), 50 mM $MgCl_2$, 5 mM ATP, 5 mM DTT, 25% (w/v) polyethylene glycol (8000). No differences in library construction were observed when different buffers were used.

5. Instead of the Bioruptor, tip sonicators can be used, e.g. a Branson 450 sonifier equipped with a microtip. Genomic DNA is sonicated with five 10-s pulses with 30 s rest on ice between repeats (settings: output 1.2, duty cycle 80%).
6. The stiffness of the Eppendorf tube impacts shearing efficiency. Always use the same brand of tubes once a reliable Bioruptor protocol has been established. Volumes of 100–300 μL can be used in standard Eppendorf tubes in the Bioruptor.
7. One can measure the concentration of the tailed DNA with Invitrogen Picogreen kits. Typically, 1 μl of 5 μM adapters is used in a 25 μl reaction when starting with 5–10 μg of genomic DNA. In the case of chromatin immunoprecipitated (ChIPed) DNA we found that those adapter concentrations can be 1,000- to 10,000-fold too high and may contribute to spurious 160-bp bands that appear after library PCR (Fig. 3c). The goal is to have a ~10:1 adapter:insert ratio.
8. To avoid cross contamination, leave an empty lane between each sample.
9. A non-UV transilluminator will result in less DNA degradation. We routinely use a standard low- or high-range UV transilluminator but expose the DNA for as short a time as possible. The size range of DNA insert will affect the standard deviation of insert size during library assembly.
10. Instead of weighing it is sufficient to use marks on the Eppendorf tubes to judge the volume after gel slices have been centrifuged. Melting the gel closer to room temperature will increase the yield of AT-rich DNA (12).
11. Two washes with PE buffer are necessary to get rid of all the salt from the QG buffer and to allow for measurement of DNA concentration with a nanodrop spectrophotometer.
12. Typically, 15–21 cycles of PCR are optimal for generation of libraries from genomic DNA. For ChIP-seq libraries up to 24 cycles are acceptable. The best genomic DNA libraries only require as few as 12 cycles while libraries made from dilute or poorly ligated DNA can require as many as 25 cycles to yield sufficient DNA. In general fewer cycles are better because less PCR bias will be introduced, resulting in fewer "clonal" reads (i.e. large numbers of reads that have identical start and stop coordinates when mapped to the genome sequence).
13. To test the library any *Taq* DNA polymerase can be used and we have not found large differences in overall amplification behavior. For the preparation of the library used for actual sequencing a high-fidelity and highly processive enzyme is preferred but Phusion Flash polymerase is not the only such enzyme available.

14. "Tightening" of the DNA smear with more cycles indicates over-amplification. DNA greater than 100 bp but lower than the excised library size should not be present. A band at ~120 bp for single-end libraries and ~160 bp for paired-end libraries indicates adapter-primer hybrids contaminating the library (Fig. 3c). Such libraries should not be sequenced. It is possible to amplify the libraries again and to isolate and gel purify fragments above the contaminating bands but often adapter primer multimers "hide" below the library bands and still result in significant contamination of sequencing runs with these nonspecific sequences. In addition, gel purification at this stage of the protocol often results in significant loss of library DNA.

15. Complexity of libraries can be assessed by digestion of test PCR fragments with *Dpn*II, as this cleaves most of the adapter off the inserts (Fig. 4c). If predominantly genomic DNA has been cloned it will result in a smear similar to genomic DNA digested for Southern blots. Conversely, if mostly low complexity samples (i.e. adapters or small amounts of contaminants) have been cloned, sharp banding is observed instead of the smear and the library is unsuitable for further processing.

16. Ideally, doubling of the amount of template DNA is expected to reduce the number of cycles needed by one. For example, if 20 cycles with 1 μL of template is sufficient, 18 cycles with 4 μL of template should be similar and result in similar complexity of libraries. In practice this relationship is not exactly linear.

17. Increasing the initial denaturation step from 30 s to 3 min and each subsequent denaturation step from 10 to 80 s or decreasing the temperature ramp rate of the thermocycler to 2°C/s improves the yield of GC-rich amplicons. Similarly, addition of 2 M betaine also improves recovery of GC-rich amplicons. Decreasing the annealing temperature from 65 to 60°C slightly improves the yield of AT-rich amplicons (38).

18. Other quantification methods such as spectrophotometry (e.g., by Nanodrop or similar devices) are less accurate and typically result in overestimation of DNA concentration and production of significantly fewer clusters. Quantification by qPCR is most accurate since it directly measures the concentration of only DNA that is able to form clusters on the flowcell.

19. Results with in-house adapter and PCR primer mix will not be identical to results obtained with Illumina-purchased kits, as certain modifications are made to oligos in Illumina library preparation kits to improve performance.

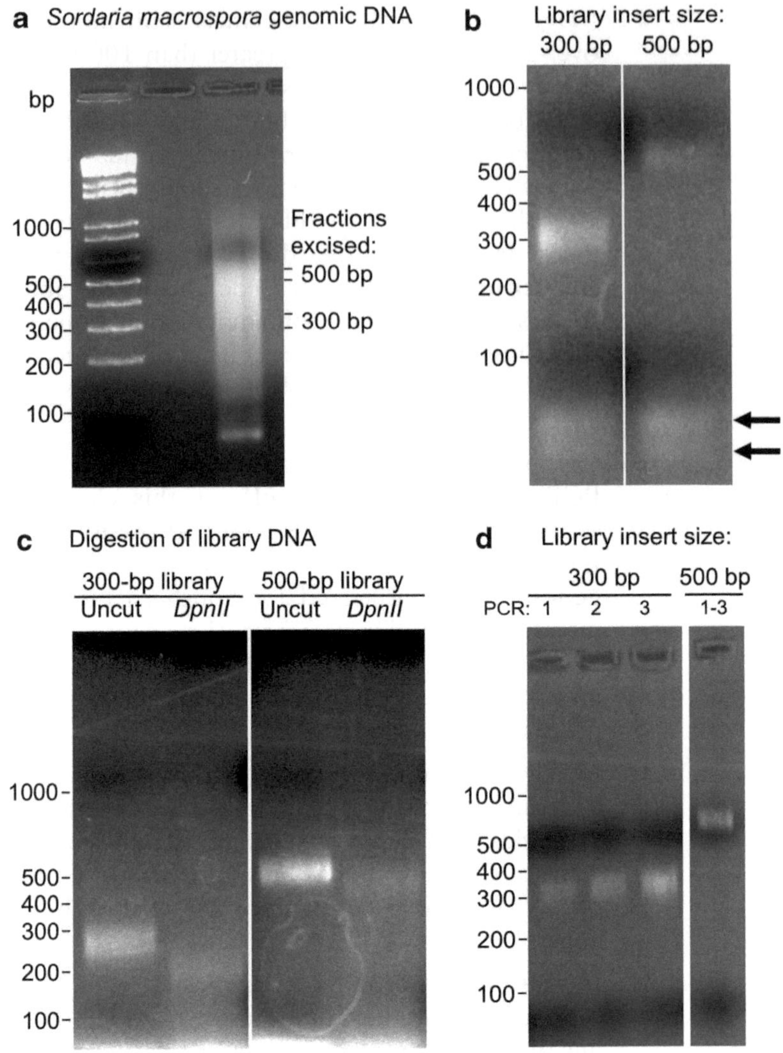

Fig. 4. Preparation of paired-end sequencing libraries from *Sordaria* macrospora. (**a**) Genomic DNA (10 μg in 450 μL TE) was prepared as described in ref. (11) and sheared by sonication with a Branson 450 Sonifier with microtip (duty cycle 80%, output 1.2; five cycles of 10 s pulses, interrupted by 30 s rest on ice). Sheared DNA was concentrated (Qiagen PCR purification kit) and separated on a 1.2% agarose gel. Gel slices of ~300 and ~500 bp fragments were isolated and purified (Qiagen Gel extraction kit, *see* Subheading 3.4, step 5). (**b**) Paired-end libraries were constructed as described in the methods from the 300- and 500-bp fractions and amplified with 12 cycles of PCR, yielding barely visible bands. Note presence of adapters and PCR primers below the 100 bp marker. (**c**) Test for library complexity. One simple test to show that libraries contain predominantly genomic DNA and are not enriched for adapter dimers or other spurious overamplified bands is to digest the DyNAzyme-amplified libraries with *Dpn*II. This cleaves most of the adapter off the inserts. If genomic DNA has been cloned it will result in a smear (similar to genomic DNA digested for Southern blots). If only low complexity samples have been cloned in the library, sharp banding is observed instead of the smear and the library is unsuitable for further processing. (**d**) Libraries were amplified with 18 cycles of PCR with Phusion Flash High-Fidelity polymerase and purified (Qiagen PCR purification kit; see three 300-bp library samples). We typically pool three independent 25 μL PCR reactions before purification and run 5 μL of 35 μL to show that only expected bands are obtained. Note the absence of the adapters or PCR primers and primer dimers (compare to (**b**)).

Acknowledgments

We thank Mark Dasenko, Chris Sullivan, Steve Drake, Matthew Peterson, and Scott Givan at the OSU CGRB core facility for assistance with Illumina sequencing, and Chris Sullivan, Jason Cumbie, Noah Fahlgren and Henry Priest for helpful discussions and sharing code. Work in our laboratory is supported by funds from the American Cancer Society (RSG-08-030-01-CCG), the National Institutes of Health (P01GM068087 and R01GM097637), and start-up funds from the OSU Computational and Genome Biology Initiative. The authors have no conflicting interests

References

1. Blumenstiel JP, Noll AC, Griffiths JA et al (2009) Identification of EMS-induced mutations in *Drosophila melanogaster* by whole-genome sequencing. Genetics 182:25–32
2. Birkeland SR, Jin N, Ozdemir AC et al (2010) Discovery of mutations in *Saccharomyces cerevisiae* by pooled linkage analysis and whole-genome sequencing. Genetics 186: 1127–1137
3. Ehrenreich IM, Torabi N, Jia Y et al (2010) Dissection of genetically complex traits with extremely large pools of yeast segregants. Nature 464:1039–1042
4. Wenger JW, Schwartz K, Sherlock G (2010) Bulk segregant analysis by high-throughput sequencing reveals a novel xylose utilization gene from *Saccharomyces cerevisiae.* PLoS Genet 6:e1000942
5. Pomraning KR, Smith KM, Freitag M (2011) Bulk segregant analysis followed by high-throughput sequencing reveals the Neurospora cell cycle gene, *ndc-1*, to be allelic with the gene for ornithine decarboxylase, *spe-1*. Eukaryot Cell 10:724–733
6. Reinhardt JA, Baltrus DA, Nishimura MT et al (2009) *De novo* assembly using low-coverage short read sequence data from the rice pathogen *Pseudomonas syringae* pv. *oryzae*. Genome Res 19:294–305
7. Diguistini S, Liao NY, Platt D et al (2009) *De novo* genome sequence assembly of a filamentous fungus using Sanger, 454 and Illumina sequence data. Genome Biol 10:R94
8. Li R, Fan W, Tian G et al (2010) The sequence and *de novo* assembly of the giant panda genome. Nature 463:311–317
9. Nowrousian M, Stajich JE, Chu M et al (2010) *De novo* assembly of a 40 Mb eukaryotic genome from short sequence reads: *Sordaria macrospora*, a model organism for fungal morphogenesis. PLoS Genet 6:e1000891
10. Pomraning KR, Connolly LR, Whalen JP et al (2011) Repeat-induced point mutation, DNA methylation and heterochromatin in *Gibberella zeae* (anamorph: *Fusarium graminearum*). In: Brown D, Proctor RH (eds) *Fusarium genomics* and molecular and cellular biology. Horizon Scientific Press, Norwich
11. Pomraning KR, Smith KM, Freitag M (2009) Genome-wide high throughput analysis of DNA methylation in eukaryotes. Methods 47:142–150
12. Quail MA, Kozarewa I, Smith F et al (2008) A large genome center's improvements to the Illumina sequencing system. Nat Methods 5:1005–1010
13. Cronn R, Liston A, Parks M et al (2008) Multiplex sequencing of plant chloroplast genomes using Solexa sequencing-by-synthesis technology. Nucleic Acids Res 36:e122
14. Langmead B, Trapnell C, Pop M et al (2009) Ultrafast and memory-efficient alignment of short DNA sequences to the human genome. Genome Biol 10:R25
15. Zerbino DR, Birney E (2008) Velvet: algorithms for de novo short read assembly using de Bruijn graphs. Genome Res 18:821–829
16. Li H, Ruan J, Durbin R (2008) Mapping short DNA sequencing reads and calling variants using mapping quality scores. Genome Res 18:1851–1858
17. Lin Y, Li J, Shen H et al (2011) Comparative studies of *de novo* assembly tools for next-generation sequencing technologies. Bioinformatics 27:2031–2037

18. Simpson JT, Wong K, Jackman SD et al (2009) ABySS: a parallel assembler for short read sequence data. Genome Res 19:1117–1123
19. Filichkin SA, Priest HD, Givan SA et al (2010) Genome-wide mapping of alternative splicing in *Arabidopsis thaliana*. Genome Res 20: 45–58
20. Bryant DW Jr, Wong WK, Mockler TC (2009) QSRA: a quality-value guided de novo short read assembler. BMC Bioinformatics 10:69
21. Li R, Zhu H, Ruan J et al (2010) *De novo* assembly of human genomes with massively parallel short read sequencing. Genome Res 20:265–272
22. Boisvert S, Laviolette F, Corbeil J (2010) Ray: simultaneous assembly of reads from a mix of high-throughput sequencing technologies. J Comput Biol 17:1519–1533
23. Li R, Li Y, Kristiansen K et al (2008) SOAP: short oligonucleotide alignment program. Bioinformatics 24:713–714
24. Li H, Durbin R (2009) Fast and accurate short read alignment with Burrows-Wheeler transform. Bioinformatics 25:1754–1760
25. Fahlgren N, Sullivan CM, Kasschau KD et al (2009) Computational and analytical framework for small RNA profiling by high-throughput sequencing. RNA 15:992–1002
26. Lunter G, Goodson M (2011) Stampy: a statistical algorithm for sensitive and fast mapping of Illumina sequence reads. Genome Res 21: 936–939
27. Li H, Handsaker B, Wysoker A et al (2009) The sequence alignment/map format and SAMtools. Bioinformatics 25:2078–2079
28. Smith KM, Phatale PA, Sullivan CM et al (2011) Heterochromatin is required for normal distribution of Neurospora CenH3. Mol Cell Biol 31:2528–2542
29. Smith KM, Sancar G, Dekhang R et al (2010) Transcription factors in light and circadian clock signaling networks revealed by genomewide mapping of direct targets for Neurospora White Collar Complex. Eukaryot Cell 9: 1549–1556
30. Zhang Y, Liu T, Meyer CA et al (2008) Model-based analysis of ChIP-Seq (MACS). Genome Biol 9:R137
31. Zang C, Schones DE, Zeng C et al (2009) A clustering approach for identification of enriched domains from histone modification ChIP-Seq data. Bioinformatics 25:1952–1958
32. Song Q, Smith AD (2011) Identifying dispersed epigenomic domains from ChIP-Seq data. Bioinformatics 27:870–871
33. Trapnell C, Pachter L, Salzberg SL (2009) TopHat: discovering splice junctions with RNA-Seq. Bioinformatics 25:1105–1111
34. Anders S, Huber W (2010) Differential expression analysis for sequence count data. Genome Biol 11:R106
35. Trapnell C, Williams BA, Pertea G et al (2010) Transcript assembly and quantification by RNA-Seq reveals unannotated transcripts and isoform switching during cell differentiation. Nat Biotechnol 28:511–515
36. Singh D, Orellana CF, Hu Y et al (2011) FDM: a graph-based statistical method to detect differential transcription using RNA-seq data. Bioinformatics 27:2633–2640
37. Cumbie JS, Kimbrel JA, Di Y et al (2011) GENE-counter: a computational pipeline for the analysis of RNA-Seq data for gene expression differences. PLoS One 6:e25279
38. Aird D, Ross MG, Chen WS et al (2011) Analyzing and minimizing PCR amplification bias in Illumina sequencing libraries. Genome Biol 12:R18

Chapter 2

Bioinformatics Approaches and Software for Detection of Secondary Metabolic Gene Clusters

Natalie D. Fedorova, Venkatesh Moktali, and Marnix H. Medema

Abstract

The accelerating pace of microbial genomics is sparking a renaissance in the field of natural products research. Researchers can now get a preview of the organism's secondary metabolome by analyzing its genomic sequence. Combined with other -omics data, this approach may provide a cost-effective alternative to industrial high-throughput screening in drug discovery. In the last few years, several computational tools have been developed to facilitate this process by identifying genes involved in secondary metabolite biosynthesis in bacterial and fungal genomes. Here, we review seven software programs that are available for this purpose, with an emphasis on *anti*biotics & *S*econdary *M*etabolite *A*nalysis *SH*ell (antiSMASH) and *S*econdary *M*etabolite *U*nknown *R*egions *F*inder (SMURF), the only tools that can comprehensively detect complete secondary metabolite biosynthesis gene clusters. We also discuss five related software packages—CLUster SEquence ANalyzer (CLUSEAN), ClustScan, *S*tructure *B*ased *S*equence Analysis of *P*oly*k*etide *S*ynthases (SBSPKS), NRPSPredictor, and *N*atural *P*roduct *searcher* (NP.searcher)—that identify secondary metabolite backbone biosynthesis genes. This chapter offers detailed protocols, suggestions, and caveats to assist researchers in using these tools most effectively.

Key words: Fungi, Genome, Gene cluster, Secondary metabolite, Mycotoxin, Polyketide, Nonribosomal peptide, Natural product, Antibiotic, Software

1. Introduction

Secondary metabolite (SM) biosynthesis pathways represent the main source of metabolic diversity in many bacterial and fungal species and have been a gold mine of new antibiotics and pharmaceuticals for decades. Although SMs have been described in plants, green algae, and protists, fungi and bacteria produce the vast majority of these natural compounds. Thus, a single bacterial or fungal genome can encode biosynthetic enzymes for over 60 SMs, such as polyketides, nonribosomal peptides, isoprenoids, flavonoids,

Nancy P. Keller and Geoffrey Turner (eds.), *Fungal Secondary Metabolism: Methods and Protocols*,
Methods in Molecular Biology, vol. 944, DOI 10.1007/978-1-62703-122-6_2,

and many others. In these organisms, genes involved in secondary metabolite biosynthesis are often co-regulated and physically linked into gene clusters on the chromosomes. This salient feature, along with sequence conservation, aids in computational discovery of new SM biosynthetic pathways encoded by microbial genomes (1, 2). Fueled by the torrent of genomic sequences and other -omics data, *in silico* approaches hold much promise for cost-effective alternatives to industrial high-throughput screening in drug discovery.

To date, several computational tools have been developed for detection of SM biosynthetic pathways in sequenced fungal or bacterial genomes. In addition to SM gene identification, some tools can predict substrate specificity of polymerization enzymes, generate approximations for products' 3D structures, and more. The usability of the tools, however, varies greatly depending on the quality of the sequence data, the user's needs, and his/her computer proficiency. Even though detailed manuals are available, funding limitations often prevent "freeware" developers from creating a perfect user-friendly interface. To assist researchers in using the software efficiently, this chapter describes approaches, software packages, and step-by-step protocols for the use of these tools. The main focus is on *S*econdary *M*etabolite *U*nknown *R*egions *F*inder (SMURF) and *anti*biotics & *S*econdary *M*etabolite *A*nalysis *SH*ell (antiSMASH), the only software packages that can detect the entire SM gene clusters in microbial genomes (3, 4). We also discuss five more specialized software tools (Table 1): CLUster SEquence ANalyzer (CLUSEAN), NRPSPredictor, ClustScan, *S*tructure *B*ased *S*equence Analysis of *P*oly*k*etide *S*ynthases (SBSPKS), and *N*atural *P*roduct *searcher* (NP.searcher) (5–10).

2. Materials

2.1. Operating System and Hardware Requirements

Except for CLUSEAN, most tools described here (antiSMASH, SMURF, ClustScan, SBSPKS, NRPSPredictor, and NP.searcher) are platform independent and work equally well on Windows, Mac, or Linux environments and are available as Web servers (Table 1). These simply require a computer with an Internet connection and a Web browser like Mozilla Firefox, Google Chrome, Apple Safari, or Microsoft Internet Explorer. antiSMASH also has a stand-alone program available, for which a computer with at least 1.0 GHz CPU and 2 GB of RAM memory is recommended, as well as a Windows XP/Vista/7, Linux Ubuntu, or Mac OS X 10.6+ operating system. Except for ClustScan, all tools are freely available and accessible to users without any programming skills.

Table 1
Software tools for detection of SM genes and gene clusters

Tools	System requirements	Input	Output	Reference
antiSMASH	– Web server: Any recent Web browser in any OS – Stand-alone version: Windows XP/Vista/7, Ubuntu Linux, or Mac OS X 10.6+	– List of GenBank/RefSeq accession numbers OR – DNA sequences in GenBank/EMBL/FASTA formats – Bacterial and fungal sequences	– Interactive output with annotation and structure for each cluster – Graphical representation – Homologous clusters – BLAST/Pfam results – Functional predictions – Backbone genes and clusters	(4)
SMURF	Any Web browser in any OS	– Protein sequences in FASTA format AND – Tab-delimited text file containing chromosomal coordinates of genes – Fungal sequences	– List of predicted backbone genes – Functional predictions for genes – List of predicted clusters – Backbone genes and clusters	(3)
CLUSEAN	– Preferred OS: Linux/Unix; except for NRPSPredictor – Compatible with MS Win 2000, XP, and VISTA – Installation required: - NCBI BLAST - HMMer - SVMlight - NRPSPredictor - Artemis Perl 5.8 or higher	– EMBL-formatted DNA files – Bacterial and fungal sequences	– Domain content – Functional predictions for genes – Substrate specificity predictions – NRPS and PKS genes only	(5)

(continued)

Table 1 (continued)

Tools	System requirements	Input	Output	Reference
ClustScan	Java Runtime Environment	– DNA sequences in FASTA or any format supported by ReadSeq – Bacterial and fungal sequences	– Annotated genes – Domain prediction – Browsing by domain searches – Backbone genes only	(6)
SBSPKS	Any Web browser in any OS	– Protein sequences in FASTA format (<10) – Sequences have to be entered manually by copying and pasting – Bacterial and fungal sequences	– Visualization of PKS domains – Prediction of docking order of PKSs – Prediction of PKS 3D structures – PKS genes only	(7)
NRPSPredictor	Any Web browser in any OS environment	– Protein sequences in FASTA format OR – Extracted domain A signatures – Bacterial and fungal sequences	– List of domains – NRPS substrate specificity – SVM scores for domains – NRPS and NRPS–PKS genes only	(9)
NP.searcher	Any Web browser in any OS	DNA sequences in FASTA format – Bacterial and fungal sequences	– Coordinates for NRPS/PKS genes – Number of PKS/NRPS modules – Backbone genes only	(8)

2.2. Input Sequence Data Formats

Correctly formatting input files is critical for getting meaningful results from any software tool. Yet for a biologist with no computational background, this step can sometimes present a serious obstacle. Also, input sequence data formats may vary depending on the software used. Below, we provide a few examples of data formats used by the tools for detection of SM genes (Fig. 1).

FASTA files. The FASTA format is a standard text-based format used in bioinformatics to represent nucleotide or peptide sequences. These sequences are usually preceded by sequence identifiers (IDs), names, and comments, collectively referred as FASTA headers. Most algorithms described here accept the NCBI FASTA specification (http://blast.ncbi.nlm.nih.gov/blastcgihelp.shtml). Files containing more than one nucleotide or protein sequence are sometimes called multiFASTA files. SMURF for example takes multiFASTA files that contain protein IDs in FASTA headers and sequences (Table 1).

GenBank and EMBL files. Both GenBank flat file and EMBL formats (Fig. 1) are rich formats used for storing nucleotide and protein sequences and meta-information. They are also the standard formats accepted by the International Nucleotide Sequence Database Collaboration, which includes the DNA DataBank of Japan (DDBJ), the European Molecular Biology Laboratory (EMBL), and GenBank at NCBI. The two formats are highly similar and contain common meta-data fields such as the accession number, source organism, literature references, as well as key genomic features (protein-coding genes, RNAs, exons, introns, etc.) and their coordinates. Detailed information about these formats is available at http://www.ebi.ac.uk/help/formats.html.

Tab-delimited TXT files. This format was designed to store data by separating the values in each row with a tab. Alternative types of delimited TXT files have other specific delimiter characters such as the comma, colon, space, and vertical bar (pipe). The lines are separated by line breaks. For example, SMURF accepts tab-separated files where data items are separated by the tab (Fig. 1). Tab-delimited TXT files can be easily generated using a spreadsheet program such as Microsoft Excel, by using the "Save as TXT file" option from the top drop-down File menu.

3. Methods

3.1. Choosing the Right Approaches and Software for the Job

Several software tools are currently available for identification of SM biosynthesis genes and gene clusters in bacterial fungal genomes (Table 1). Most tools approach the problem of SM gene detection using the LEGO principle, which states that most biological complexity is generated by using only a small number of building

FASTA Format

```
>gi|211587300|emb|AM920435.1| Penicillium chrysogenum Wisconsin 54-1255
complete genome, contig Pc00c20
TCTTCAACATATGATGTTTCCCGATGATCCATTCGGTCTGTACATCCCCCCAGACGCCAAACGAATATTC
AACTTTGGCTTACCATGAAGATAATCTGGGAAAGCGATTTCCACAAGAACAAGAACTACAATTGATGAAG
TTTGATGGGCTCCACAAAATAGTTCTATACATTTACTAAACATTACAACTCTCTTACTAGCACAGAAAAC
TTTCAATACGCATTTTAGAGAAAGCCACGGTTAATCCAATTGAATTAACGTTGTATCTTCGTCATCAGCG
TTTATTCACCGCGTGTATCTCATTTAGTGCA
...
```

GenBank Format

```
LOCUS       AM920435        3652229 bp    DNA     linear   PLN 01-NOV-2008
DEFINITION  Penicillium chrysogenum Wisconsin 54-1255 complete genome
ACCESSION   AM920435
VERSION     AM920435.1  GI:211587300
SOURCE      Penicillium chrysogenum Wisconsin 54-1255
FEATURES             Location/Qualifiers
     source          1..3652229
                     /organism="Penicillium chrysogenum Wisconsin 54-1255"
                     /mol_type="genomic DNA"
                     /db_xref="taxon:500485"
     gene            complement(<967..>2641)
                     /locus_tag="Pc20g00010"
     CDS             complement(join(967..1004,1054..1224,1273..2107,
                     2138..2641))
                     /locus_tag="Pc20g00010"
                     /protein_id="CAP85330.1"
...
```

EMBL Format

```
ID   AM920435; SV 1; linear; genomic DNA; STD; FUN; 3652229 BP.
XX
AC   AM920435;
XX
DE   Penicillium chrysogenum Wisconsin 54-1255 complete genome
XX
FH   Key             Location/Qualifiers
FH
FT   source          1..3652229
FT                   /organism="Penicillium chrysogenum Wisconsin 54-1255"
FT                   /mol_type="genomic DNA"
FT                   /db_xref="taxon:500485"
FT   CDS             complement(join(967..1004,1054..1224,1273..2107,
FT                   2138..2641))
FT                   /gene="Pc20g00010"
FT                   /locus_tag="Pc20g00010"
...
```

Fig. 1. FASTA, GenBank, and EMBL data formats. All software packages described here accept input files in at least one of these formats.

blocks. Indeed, analysis of characterized SM pathways shows that most biosynthetic reactions are performed by a limited number of enzyme classes, which are responsible for the enormous spectrum of natural products produced by bacteria and fungi (11). Consequently, known SM enzymes share highly conserved protein domains, which can be identified by sequence similarity.

Additionally, identification of new pathways is greatly facilitated by SM gene clustering and the presence of the so-called backbone genes, which are easy to detect based on their domain content. They encode enzymes that catalyze the polymerization reaction, which generates the product's backbone and is often the first reaction in an SM pathway. Backbone genes are typically surrounded by genes encoding "decorating enzymes," which catalyze subsequent reactions such as oxidation, reduction, methylation, and glycosylation of the backbone molecule. These reactions may alter the compound's hydrophilicity, stability, subcellular localization, and bioactivity. The end product is often translocated outside the cell by transporters, also encoded within a typical SM cluster. Finally, clusters may contain additional genes encoding precursor biosynthesis enzymes, additional transporters, detoxification enzymes, or pathway-specific transcriptional regulators. Most of these proteins contain one or more SM-related domains, which can be identified computationally.

Since most available tools use similar approaches, the choice of the software largely depends on the desired output, quality of the sequence data, and taxonomy of the organism of interest (Table 1). All seven software tools use protein domain searches to reliably predict backbone genes. SMURF and antiSMASH are the only tools that examine neighboring regions on the chromosome to predict the entire gene clusters. SMURF is specific to fungal genomes, and generates a quick report as a single complete and easily parsable TXT file. antiSMASH identifies clusters in both bacterial and fungal genomes. It also generates an interactive graphical output with integrated additional analyses and cross-links to other services. For analysis of fungal genomes, it is a good idea to use both tools because they rely on different algorithms for predicting cluster boundaries (see Note 2).

3.2. antibiotics and Secondary Metabolite Analysis SHell

antiSMASH is a comprehensive pipeline for identification of gene clusters encoding biosynthetic enzymes for a wide range of SM compound classes (4). It identifies polyketides, nonribosomal peptides, terpenes, aminoglycosides, aminocoumarins, indolocarbazoles, lantibiotics, bacteriocins, nucleosides, beta-lactams, butyrolactones, and others in both fungal and bacterial nucleotide sequences. The underlying algorithm is presented in Fig. 2. The antiSMASH software is available at http://antismash.secondarymetabolites.org, both as a Web server and as a stand-alone application.

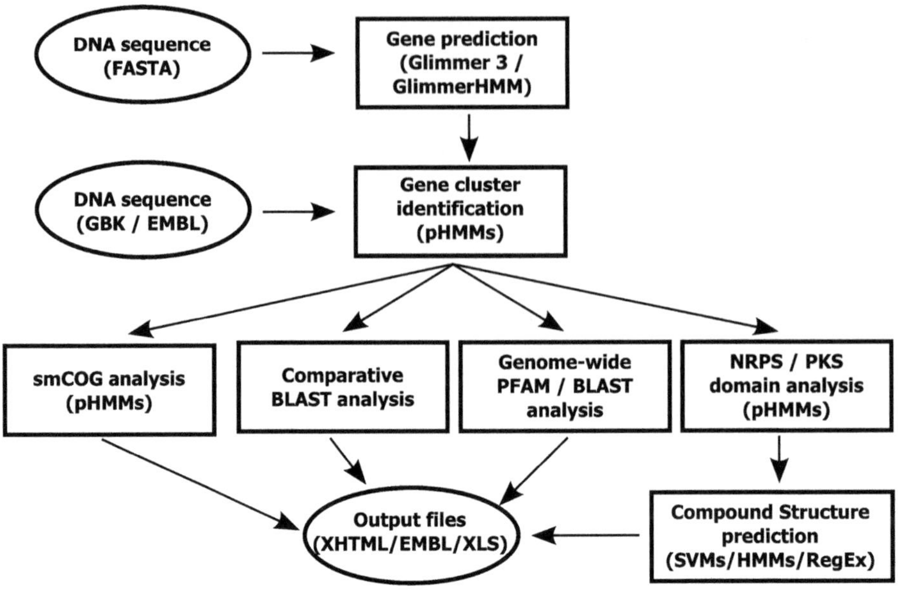

Fig. 2. Key steps in the antiSMASH algorithm. The figure shows principle steps in the algorithm from a DNA sequence to an interactive output file. Adapted from (4).

3.2.1. Preparing Input Data

Besides accepting nucleotide accession numbers from GenBank or RefSeq, antiSMASH accepts fully annotated GenBank and EMBL files as well as FASTA files that contain a single DNA sequence. MultiFASTA files are currently not accepted and should be split into single-sequence FASTA files. When preparing input files manually, it is critical to use a plain text editor like Notepad or Emacs and to save the files as "All files (*.*)" with the correct file extension (.fasta, .fas, or .fa for FASTA files; .gbk or .gb for GenBank files; .embl or .emb for EMBL files).

For a wide range of fungal and bacterial species, an annotated genome sequence is already available at NCBI. In such case, a corresponding genome accession number can simply be inserted into the third text box on the antiSMASH Web site, labeled as "input GenBank/RefSeq accession number of desired file". Accession numbers can be acquired by querying the NCBI Entrez system (http://www.ncbi.nlm.nih.gov/). The search can be performed by selecting the Bioproject (Genome Project) database from the top pull-down menu and entering the organism name in the search box as a query (e.g., "*Aspergillus fumigatus* AF293 [Organism]"). If more than one genome build is available for a particular organism, it is recommended to use the most current build. Accession numbers of any nucleotide GenBank/RefSeq entries that are smaller than full genomes can also be used.

If no genome accession number is available for the species of interest, its annotated DNA sequence can be entered as an EMBL or GBK file in the second text box (labeled "upload a file") in the antiSMASH front end. These files can be generated from a FASTA sequence file and annotation table as described below:

1. Download and install Python from http://www.python.org/download/.
2. Download the GenBank formatting script from the antiSMASH Web site http://antismash.secondarymetabolites.org/downloads.html.
3. In a spreadsheet program such as Microsoft Excel, prepare a table with the coordinate information on each exon/gene structure in the following columns: (1) name of contig FASTA file, (2) gene locus tag (must be unique), (3) 5′ exon/gene start, (4) 3′ exon/gene end, 5) gene annotation. For every exon or gene, fill in the information in a single line. Make sure that all exons from one gene have the same locus tag and that the exons/genes are listed in the same order in which they occur on the DNA sequence. Save the table as a tab-delimited TXT file "annotationtable.txt".
4. Copy 'annotationtable.txt' and the config FASTA files to the same folder where the EMBL-formatting script is located.
5. Open the Terminal/Command Prompt, and navigate to the folder, which contains the files.
6. Type in "python format_embl.py" and press enter.
7. For every config, an EMBL file will be created that can serve as an input for antiSMASH.

Finally, if only an unannotated DNA sequence is available, a FASTA file with the unannotated nucleotide sequence can be uploaded to antiSMASH. The algorithm will then run a gene prediction algorithm: GlimmerHMM (12) for eukaryotic sequences and Glimmer3 (13) for prokaryotic sequences to perform a gene annotation of the DNA sequence before running the rest of the pipeline (see Note 3). Alternatively, the genomic sequence can be annotated using standard genome annotation tools, converted into GenBank or EMBL format, and then submitted to antiSMASH. Automated genome annotation can be performed using automated Web pipelines such as the MAKER Web Annotation Service (14) or RAST (for prokaryotic sequences only) (15).

3.2.2. Choosing Parameters and Running antiSMASH

One of the main advantages of using antiSMASH is that it allows users to run other analyses (in addition to gene cluster identification) by selecting various options at the front end (Fig. 2). The first option allows user to select cluster types such as polyketides, nonribosomal peptides, terpenes, or aminoglycosides. Subsequently, the user can choose whether or not to include a number of additional, computationally intensive analyses such as the following:

- Secondary metabolite Cluster of Orthologous Groups (smCOG) analysis: This module attempts to assign all genes to one of the smCOGs, which represent SM-specific gene families. For every assigned gene, a phylogenetic tree is generated to show the relatedness of that gene to other genes in the smCOG family.
- Gene Cluster Blast: For each gene cluster, this module detects putative homologs in a global database of gene clusters from across the tree of life, by searching for homologous genes and synteny (conserved gene order) inferred from BLASTp hits among gene clusters.
- Whole genome BLAST and PFAM analyses: This module executes genome-wide analyses from the CLUSEAN pipeline. It provides PFAM domain annotations for each gene in the DNA sequence, as well as BLAST hits to a custom nonredundant database of bacterial and fungal genes. Based on these results, it also generates annotation suggestions for each gene. Finally, if the PFAM analysis option is selected, the module generates a graph that displays genomic regions enriched in SM-related PFAM domains.

To very quickly obtain gene cluster predictions for a genome, the user may choose not to include all the analyses available. However, choosing these additional options can be beneficial for more in-depth analysis of gene clusters. Domain architecture analysis of PKS and NRPS proteins is automatically included in each analysis and therefore does not need to be selected.

When an antiSMASH run is finished, it generates an interactive XHTML file containing the output. Depending on the options chosen, this file will include a functional description of each gene cluster, its domain annotations for PKS and NRPS genes, a list of homologous clusters in other species, a putative chemical structure of the cluster's product, and several downloadable output files. All sections have a question mark button at the top right to provide more information on each section. Additionally, the user can access more detailed information by clicking on elements of the page or hovering over them with the mouse. To store antiSMASH results locally, the user can click the "Download all results" button at the bottom right. Doing so will allow access to the antiSMASH results at any later point in time, while the output page on the server will be deleted eventually.

3.2.3. Viewing antiSMASH Results in a Genome Browser

For each run, antiSMASH generates an EMBL file that can be analyzed in a genome browser such as Artemis. This can especially be useful for viewing the whole-genome PFAM and BLAST analyses, which are only displayed in the EMBL file, but not in the general output XHTML file. To view the results in Artemis, follow these steps:

1. Download and extract the zip file that contains the results by clicking on the link "Download all results" at the bottom right of the antiSMASH output page.
2. Download and install Artemis from http://www.sanger.ac.uk/resources/software/artemis/.
3. Start Artemis, select "File » Open...", and open the output file ending with ".final.embl" downloaded from antiSMASH.
4. The antiSMASH results should appear in the main Artemis window (Figs. 3 and 4). NRPS and PKS domains as well as conserved motifs within these domains are saved as "CDS_motif" or "misc_feature" features. More details for every feature can be viewed by selecting it and pressing Ctrl-V. If PFAM and BLAST options were selected, detected PFAM domains get saved as "CDS_motif" features. BLASTP results for each gene can be viewed by selecting a coding sequence (CDS) icon and pressing Ctrl-B (please note that the full download of all results is required for this option to be functional). The putative function of the CDS is displayed as a note.

3.2.4. Batch Processing of Multiple Files

Some users may want to analyze a large number of input files such as multiple genome sequences or a single draft genome that contains a large number of contigs. In such cases, the user can perform a batch run using the stand-alone version of antiSMASH in combination with a batch processing script, available from the Web site. To perform such a batch run, one needs to follow these steps:

1. Download the stand-alone version of antiSMASH for your operating system from http://antismash.secondarymetabolites.org/download.html.
2. Download and install Python from http://www.python.org/download/.
3. Download the script for batch processing from http://antismash.secondarymetabolites.org/downloads.html and copy it into the antiSMASH installation folder.
4. In the folder where antiSMASH is installed, create a subfolder "batch_input" and copy the files that need to be analyzed into this folder.
5. Alternatively, create a multiGBK/multiEMBL/multiFASTA file with your GBK/EMBL/FASTA entries.
6. Open the Terminal/Command Prompt, and navigate to the antiSMASH installation folder.
7. Type in a command with the structure "python batch_antismash.py <multiple_entry_file.gbk> --options'" and press enter. The multiple entry file is optional. If not supplied, the script will search your 'batch_input' folder for single-entry files. You can add antiSMASH options to batch_antismash.py (e.g. "--clusterblast n --taxon e"), which will be forwarded to antiSMASH.)

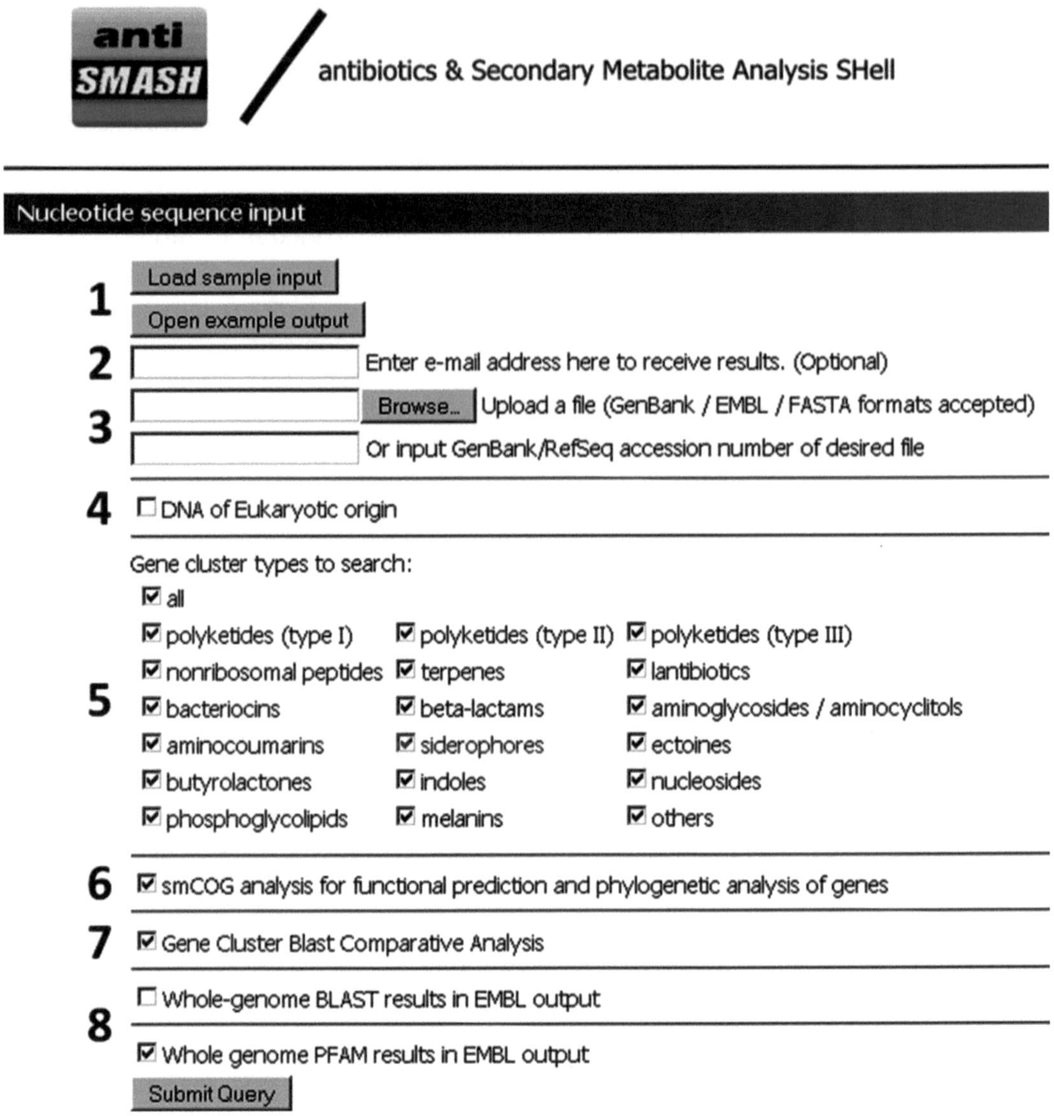

Fig. 3. antiSMASH front end. The figure shows different sections of the input screen including (1) buttons to load a sample input file or open an example output file; (2) optional input for an e-mail address, where results can be sent; (3) input file options for GBK/EMBL/FASTA files and genome accession numbers; (4) a check box for fungal or other eukaryotic inputs; (5) a check box for gene cluster type selection; (6) a check box for smCOG analysis; (7) a check box for BLAST analysis; and (8) check boxes for CLUSEAN-derived analyses.

8. The files will now be processed automatically and the outputs will be copied to a separate folder "'batch_output'" in case of directory input, and into the same folder in case of a multiple entry-file input.
9. Complete and integrated multiple-entry input functionality is planned for antiSMASH 2.0, which is currently under development.

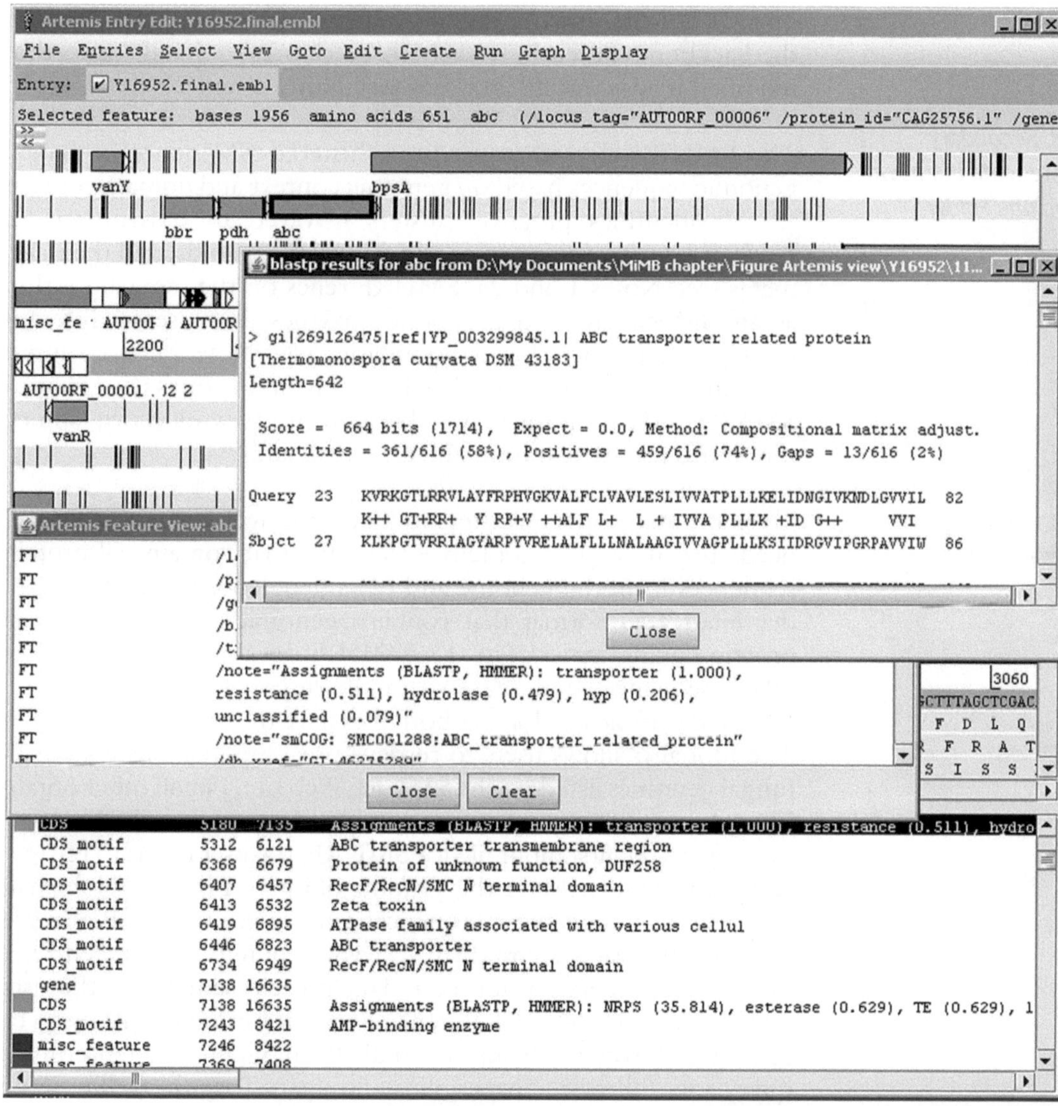

Fig. 4. Screenshot of an antiSMASH EMBL output file viewed in the Artemis genome browser. antiSMASH results such as gene clusters, NRPS/PKS domains, and conserved motifs are displayed as sequence features. Gene function predictions are also available for each gene.

3.2.5. Interpreting antiSMASH Results

antiSMASH has been shown to predict SM clusters with reasonable accuracy (see Notes 3–5) (4). However, the sensitivity and specificity of predictions vary depending on the nature of the data and the type of analyses performed. For example, in order to minimize the number of missed SM genes, it predicts cluster boundaries in a very greedy manner. The user can then modify the predictions based on smCOG and comparative BLAST results (see Note 2). antiSMASH also tends to merge adjacent putative clusters into one gene cluster if they are located within 20 kb (see Note 4). Similarly, products' core structure predictions based on the NRPS or PKS domain content should be treated with caution, because the backbone molecule is likely to be circularized and/or further modified by decorating enzymes (see Notes 1 and 2).

3.3. Secondary Metabolite Unknown Regions Finder

The SMURF algorithm predicts clustered SM genes in fungal genomic sequences based on genomic context and domain content (3). It identifies putative clusters involved in biosynthesis of polyketides, nonribosomal peptides, indole alkaloids, and their chimeras (see Notes 1 and 2). SMURF relies on the same algorithm as antiSMASH to find backbone enzymes, but uses a different principle to predict cluster boundaries (see Note 2). Both input and output file formats are also distinct from the ones used in antiSMASH. SMURF is freely available at http://www.jcvi.org/smurf; it requires user registration and an e-mail address.

3.3.1. Obtaining and Preparing Input Files

To find putative gene clusters in a given fungal genome, SMURF needs two files: (1) a protein FASTA file that contains all protein sequences from the genome and (2) a gene coordinate file in tab-delimited TXT format that contains genomic coordinates for all protein-coding genes (Fig. 1). SMURF does not require a completely sequenced genome, so it can be run on a subset of chromosomes or scaffolds, as long as both files have the same sequence IDs. The tool also allows users to retrieve precomputed clusters for 27 fungal genomes listed on the SMURF Web site. For all other organisms, the user must prepare the two input files as described below. To prepare the first input file for SMURF, the user needs to retrieve all protein sequences for the genome of interest in FASTA format. One way to obtain these sequences is to query the NCBI Entrez general retrieval system (http://www.ncbi.nlm.nih.gov/entrez). To limit the query to nonredundant protein sequences, the user needs to use the NCBI Reference Sequence database (RefSeq) by selecting "Protein" from the pull-down menu and entering a query as follows: "Aspergillus fumigatus[orgn] AND srcdb refseq[properties]". Alternatively, he/she can use the NCBI Whole Genome Sequences (WGS) sequencing project database available at http://www.ncbi.nlm.nih.gov/projects/WGS/WGSprojectlist.cgi. As with antiSMASH, it is best to use the most current genome build if possible. Protein sequences can be retrieved by clicking first on a species name and then on the Protein link in the Entrez records table. Another alternative would be to use fungal specific sequence data warehouses such as FungiDB (www.fungidb.org), the Broad's Genome Sequencing Platform (http://www.broadinstitute.org), or the Comparative Fungal Genomics Platform (CFGP; http://cfgp.snu.ac.kr).

Although many scientific journals require sequence data submission to NCBI prior to publication, some sequenced genomes may not yet be available at the NCBI Web site. For these fungal species, preliminary sequence data may be available directly from sequencing centers such as the Broad Institute, Joint Genome Institute (JGI), Beijing Genomics Institute (BGI), Welcome Trust Sanger Institute, or J. Craig Venter Institute (JCVI; former TIGR). To avoid potential data-ownership conflicts, it is recommended to first obtain permission to use the data. If the sequence data is generated by the user or obtained from a sequencing center, the files' format may deviate from the standards. The format can be adjusted using the FASTX-Toolkit (http://hannonlab.cshl.edu/fastx_toolkit) or similar tools for formatting sequence data.

To prepare the second input file for SMURF, the user needs to generate a tab-delimited TXT file that has the following elements (values) in each line, separated by tabs: (1) protein ID; (2) chromosome or scaffold ID; (3) gene start coordinate; (4) gene stop coordinate; and (5) protein description if available. The protein IDs must be nonredundant in the dataset. The gene start and end coordinates correspond to the starting (5′ end) and ending (3′ end) position of each gene on the chromosome/scaffold. The last column is optional and can include the protein name, functional description, or any other functional information.

The data necessary to create the second input file can be obtained from the NCBI ftp site (ftp.ncbi.nih.gov/gene/DATA/ASN_BINARY/Fungi/All_Fungi.ags.gz). The downloaded files can be converted to xml and parsed using the gene2xml tool from NCBI. The tool is available at ftp.ncbi.nih.gov/toolbox/ncbi_tools/cmdline/. If not available at NCBI, the coordinates' information can sometimes be obtained from sequencing center Web sites. The coordinate data are often stored in GFF format, which can be parsed using standard scripts or tools such as the text manipulation option in Galaxy (http://galaxy.psu.edu).

3.3.2. Running SMURF

Once the two input files are obtained, they need to undergo a quality control test. FASTA files obtained from non-NCBI sources sometimes need to be "cleaned" by reformatting the file and/or by removing extra characters from the header or the sequences themselves. The coordinates file format needs to be validated manually by checking the order in which values are presented and the delimiter used. The user also needs to make sure that the same protein IDs are used in both the protein FASTA file and the gene coordinate file and that no duplicates are present.

Once both the protein FASTA file and the gene coordinates file are confirmed to be in the required file format, they can be uploaded on the SMURF Web site (Figs. 5 and 6). This can be done by simply copying and pasting the content of the files in the text boxes or by clicking on the "Browse" button and selecting the files from the respective folder. SMURF output files with results are sent to the user by e-mail.

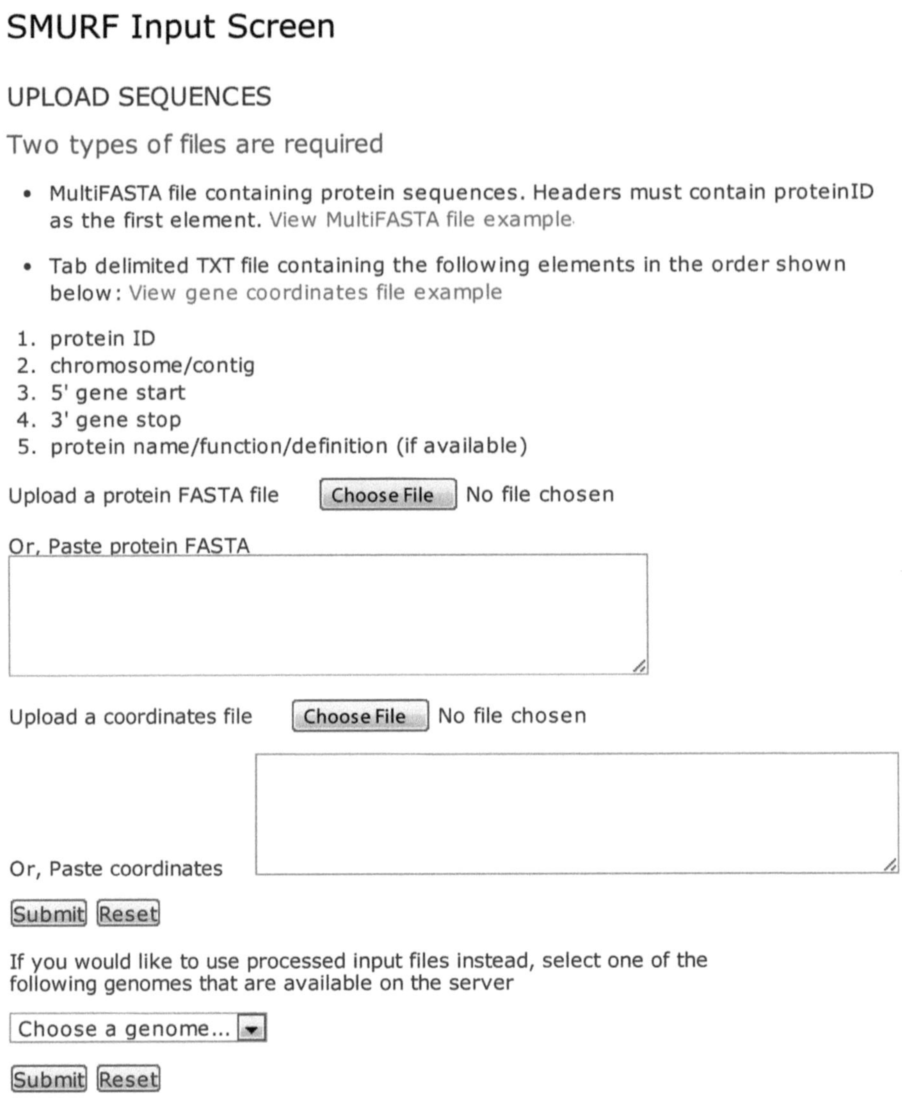

Fig. 5. SMURF front end. The figure shows different sections of the input screen including (1) links to view sample input multiFASTA and coordinate files; (2) a button to upload a protein FASTA file; (3) a text box to paste a protein FASTA file; (4) a text box to paste coordinate files; (5) a Submit button to run SMURF; and (6) a drop-down menu to choose and download precomputed results for 27 fungal genomes.

3.3.3. Interpreting SMURF Results

For each genome, SMURF generates two files: one including the list of putative backbone enzymes and another including the list of putative SM clusters. These files can be used to estimate the number of backbone genes and gene clusters in the genome. The two numbers are not the same, because many backbone genes do not appear to be part of any gene cluster. As with antiSMASH results, caution must be taken when interpreting gene cluster boundaries predicted by SMURF (see Note 2). The accuracy of the predictions can be validated by using gene expression profiling data, if available. Nonetheless, SMURF generates a good preliminary view

SMURF Output Files

File 1: Secondary Metabolism Gene Clusters

Cluster:1

Backbone_gene_id	Gene_id	Gene_positions	Chromosome-Contig	Gene_order	5'end	3'end	Gene_distance	Domain_score	Annotated_gene_function
AFUA_1G01010	AFUA_1G01010	0	98	67	352044	358955	0	1	"polyketide synthase, putative"
AFUA_1G01010	AFUA_1G01000	1	98	66	351003	349732	1041	1	"oxidoreductase, 2OG-Fe(II) oxygenase family"
AFUA_1G01010	AFUA_1G00990	2	98	65	348512	347632	1220	1	short chain dehydrogenase/reductase family protein
AFUA_1G01010	AFUA_1G00980	3	98	64	346714	344953	918	1	"FAD-dependent oxidase, putative"
AFUA_1G01010	AFUA_1G00970	4	98	63	344153	342453	800	1	"MFS monocarboxylate transporter, putative"

Cluster:2

Backbone_gene_id	Gene_id	Gene_positions	Chromosome-Contig	Gene_order	5'end	3'end	Gene_distance	Domain_score	Annotated_gene_function
AFUA_1G10380	AFUA_1G10380	0	95	162	447716	428528	0	0	nonribosomal peptide synthase Pes1
AFUA_1G10380	AFUA_1G10370	1	95	161	426122	428041	487	1	"MFS multidrug transporter, putative"
AFUA_1G10380	AFUA_1G10360	2	95	160	422521	425058	1064	0	conserved hypothetical protein
AFUA_1G10380	AFUA_1G10355	3	95	159	421515	421046	1006	1	"26 proteasome complex subunit Sem1, putative"
AFUA_1G10380	AFUA_1G10350	4	95	158	419083	420498	548	0	"phosphoglycerate kinase PgkA, putative"
AFUA_1G10380	AFUA_1G10340	5	95	157	417419	418337	746	0	"integral membrane protein, Mpv17/PMP22 family, putative"
AFUA_1G10380	AFUA_1G10330	6	95	156	414576	416503	916	0	conserved hypothetical protein
AFUA_1G10380	AFUA_1G10320	7	95	155	414256	411705	320	0	"formin binding protein (FNB3), putative"
AFUA_1G10380	AFUA_1G10310	8	95	154	408473	410949	756	1	"RNase L inhibitor of the ABC superfamily, putative"

File 2: Backbone genes

Backbone_gene_id	Annotated_gene_function	Chromosome-Contig	Gene_order	5'_end	3'_end	SMURF_backbone_gene_prediction
AFUA_1G01010	"polyketide synthase, putative"	98	67	352044	358955	PKS
AFUA_1G10380	nonribosomal peptide synthase Pes1	95	162	447716	428528	NRPS
AFUA_1G17200	nonribosomal siderophore peptide synthase SidC	95	843	2441629	2455970	NRPS
AFUA_1G17740	"polyketide synthase, putative"	95	894	2655126	2663085	PKS
AFUA_2G01290	"polyketide synthase, putative"	57	120	314706	306145	PKS
AFUA_2G05760	"beta-ketoacyl synthase (Cem1), putative"	57	553	1621987	1623626	PKS-Like
AFUA_2G17600	conidial pigment polyketide synthase PksP/Alb1	92	967	2817817	2824478	PKS
AFUA_2G17990	dimethylallyl tryptophan synthase FgaPT1	92	1005	2913852	2912488	DMAT
AFUA_2G18040	dimethylallyl tryptophan synthase FgaPT2	92	1010	2925193	2923679	DMAT
AFUA_3G01410	"polyketide synthase, putative"	100	126	360985	352856	PKS
AFUA_3G02530	"PKS-like enzyme, putative"	100	230	634289	632049	PKS-Like
AFUA_3G02570	"polyketide synthase, putative"	100	234	653701	646906	PKS
AFUA_3G02670	"NRPS-like enzyme, putative"	100	244	694320	697670	NRPS-Like
AFUA_3G03350	nonribosomal peptide synthase SidE	100	307	892438	898767	NRPS
AFUA_3G03420	nonribosomal peptide synthase SidD	100	313	908168	914474	NRPS
AFUA_3G12920	"nonribosomal peptide synthase GliP-like, putative"	93	238	648974	642245	NRPS
AFUA_3G12930	"dimethylallyl tryptophan synthase SirD-like, putative"	93	237	641551	640080	DMAT
AFUA_3G13730	"nonribosomal peptide synthase, putative"	93	159	459847	455975	NRPS
AFUA_3G14700	"polyketide synthase, putative"	93	64	181917	174183	PKS
AFUA_3G15270	"nonribosomal peptide synthase, putative"	93	11	68646	61531	NRPS
AFUA_4G00210	"polyketide synthase, putative"	109	134	384731	390228	PKS
AFUA_4G14560	"polyketide synthase, putative"	96	30	81245	86775	PKS
AFUA_5G10120	"NRPS-like enzyme, putative"	94	473	1345356	1341532	NRPS-Like
AFUA_5G12730	"nonribosomal peptide synthase, putative"	94	225	633905	608358	NRPS
AFUA_6G03480	"NRPS-like enzyme, putative"	101	186	545229	540113	NRPS-Like
AFUA_6G08560	"NRPS-like enzyme, putative"	108	225	704656	707884	NRPS-Like
AFUA_6G09610	"nonribosomal peptide synthase, putative"	108	327	1024733	1020915	NRPS
AFUA_6G09660	nonribosomal peptide synthase GliP	108	332	1040501	1033997	NRPS
AFUA_6G12050	"nonribosomal peptide synthase, putative"	108	556	1694970	1698884	NRPS
AFUA_6G12080	"nonribosomal peptide synthase, putative"	108	559	1716682	1704693	NRPS
AFUA_6G13930	"polyketide synthase LovB-like, putative"	108	737	2239872	2232481	PKS

Fig. 6. SMURF output. Typical SMURF results are presented: a TXT file with gene clusters and a TXT file with backbone genes and their functional descriptions.

of the species' secondary metabolome. If desired, further predictions can be made of the substrate specificity and core structure based on domain architecture of the backbone enzymes such as NRPSs and PKSs using other tools described in this chapter. To visualize the genomic context of the identified clusters, the user can use available browsers such as the CFGP (http://cfgp.snu.ac.kr) or FungiDB (http://www.fungidb.org). Batch processing of multiple genomes is currently not available in SMURF.

3.3.4. SMURF Algorithm

To better interpret SMURF output, it may be helpful to understand the underlying algorithm. At the heart of SMURF lie hidden

Markov models (HMM) for each PFAM and TIGRFAM domain that have been associated with backbone and decorating proteins (16, 17). The HMMER program (http://hmmer.janelia.org) is employed to search for these domains in protein dataset (16). NRPS enzymes are identified as enzymes with at least one module composed of an amino acid adenylation domain (A), a thiolation domain (PCP), and a condensation domain (C). PKS enzymes are identified as enzymes with at least one acyl transferase domain (AT), a beta-ketoacyl synthase C-terminal domain (BKS-C), and a beta-ketoacyl synthase N-terminal domain (BKS-N). Hybrid PKS–NRPS enzymes are identified as enzymes with at least one instance from each set of three domains listed above.

NRPS-like enzymes are identified with a combination of at least two domains from any of those in the NRPS enzyme module; a combination of an A domain and a NAD_binding_4 domain; or a combination of an A domain and short chain dehydrogenase domain. PKS-like enzymes were identified with a combination of at least two domains from any of those in the PKS enzyme module. To eliminate false positives among PKS-like enzymes, they were defined as proteins with AT, BKS-C, and BKS-N domains that scored below a trusted HMM cutoff. To identify false positives such as alpha-aminoadipate reductase among NRPSs, candidate NRPSs are screened for the presence of the C-terminal domain of L-aminoadipate-semialdehyde dehydrogenase alpha subunit with the cutoff score set above the default cutoff. Those that score below the cutoff are removed from the final list of putative NRPSs. Prenyltransferase enzymes are identified as enzymes with at least one DMATS-type prenyltransferase domain (DMATS). The corresponding de novo HMM model for this domain (TIGR03429) was created in this study from the seed alignment generated using *A. fumigatus* dimethylallyl tryptophan synthase FtmPT2 as a seed sequence as previously described (16).

Putative decorating enzymes are predicted by checking for the presence of one of the 27 SM-defining domains, as well as by their genomic neighborhood, i.e., the distance to the nearest backbone genes. The algorithm first identifies SM-defining domains in the window of ±20 genes on each side of a backbone gene. The algorithm uses two fixed parameters: d and y to define the cluster's boundaries. d is the maximum intergenic distance permitted between two adjacent genes in the same cluster, while y is the maximum number of SM domain-negative genes. These parameters have been set as $d = 3{,}814$ bp and $y = 10$ genes, based on optimization with experimental data. Finally, additional genes are trimmed at both ends of the cluster until the algorithm reaches the first gene that contains an SM-defining domain. If SMURF predicts overlapping clusters, they are merged into one (see Note 4).

3.4. NRPSPredictor

NRPSPredictor ((9, 10) is a widely used tool for prediction of substrate specificities of NRPS enzymes. Predictions are based on the sequence of the adenylation (A) domain, which is responsible for the recruitment of the amino acid monomers by NRPS enzymes. NRPSPredictor is a Web-based tool, available at http://nrps.informatik.uni-tuebingen.de. A new version, NRPSPredictor2, based on updated training data, has recently been published (10). Unlike the original version, NRPSPredictor2 contains a separate fungal module. The input consists either of a full NRPS protein sequence in FASTA format or of a set of extracted A domain signatures based on amino acid residues surrounding the A domain active site (Table 1). Input sequences can be pasted in the text box called "Sequence to analyze") or uploaded using the "Browse" button. The output contains (1) a list of A domains identified in the submitted NRPS sequences, (2) their predicted substrate specificities, (3) support-vector machine scores for domains, and (4) substrate specificity of their nearest homolog in the NRPSPredictor signature database.

The tool relies on two algorithms: (1) a machine-learning technique called transductive support-vector machines to predict the substrate specificity of an unknown sequence and (2) a sequence similarity search using amino acid residues surrounding the A domain active site (active site code) as a query against a set of codes from experimentally characterized A domains. It also provides support-vector machine predictions at four hierarchical levels, from predictions of gross physicochemical properties of the amino acid to predictions of single amino acid specificity. If substrate specificity cannot be predicted based on training data, this module can still predict the general nature of the substrate (e.g., hydrophilic or hydrophobic). Since antiSMASH includes the NRPSPredictor2 module, this pipeline can also be used to predict substrate specificities of NRPS enzymes. Yet, for a quick analysis of single NRPS genes, the NRPSPredictor2 Web server is most efficient.

3.5. CLUster SEquence ANalyzer

CLUSEAN (5) is a comprehensive SM analysis pipeline for bacterial genomes. The latest version has been adapted to also process fungal genomes, and has been merged into the antiSMASH pipeline (represented by part of the NRPS/PKS domain detection module, as well as the genome-wide PFAM and BLAST options) (4). A stand-alone version of CLUSEAN is available at http://redmine.secondarymetabolites.org/projects/clusean. It is an open-source stand-alone tool, which requires installation. The preferred environment is Linux or Unix, although it is compatible with MS Win 2000, XP, and Vista. Installation of additional tools is also required, including NCBI BLAST, HMMer, SVMlight, NRPSPredictor (described above), Artemis, and Perl 5.8 or higher. CLUSEAN takes DNA sequences in EMBL format as input and generates functional predictions for NRPS and type I PKS genes as well as substrate specificity predictions for NRPS genes.

3.6. Structure Based Sequence Analysis of Polyketide Synthases

Designed mainly for analysis of bacterial type I modular PKS genes, the SBSPKS Web server offers some interesting features, which complement those offered in SMURF and antiSMASH (7). SBSPKS input consists of protein sequences in FASTA format, to a maximum of ten sequences (Table 1). Input sequences can be entered in a text box manually by copying and pasting. In addition to domain architecture and docking analysis (also present in antiSMASH), SBSPKS hosts a database of PKS genes associated with known compounds. This feature can be used to link PKS domain architecture with the polyketide structure and to generate a structural homology model of the entire PKS module. The model can be used to view putative molecular interactions in 3D as well as to make inferences on the probable functions of highly conserved residues. SBSPKS can be accessed at http://www.nii.ac.in/sbspks.html.

3.7. Natural Product searcher

NP.searcher (8) is a tool for detecting genes encoding PKSs, NRPSs, and some terpene synthases in bacterial and fungal genomes. Although a recent benchmark (4) showed that the tool has limited sensitivity, it has the interesting functionality to generate putative chemical structures for polyketide or nonribosomal peptide compounds, including circularization and glycosylation variants. It is a Web-based tool that takes DNA sequences in FASTA format as input (Table 1). Input sequences can be pasted in a text box or uploaded using the "Browse" button. The output contains chromosomal coordinates for putative NRPS and PKS genes and the number of detected PKS and NRPS modules. NP.searcher is available at http://dna.sherman.lsi.umich.edu.

3.8. ClustScan

One of the earliest programs for detection of SM gene clusters is ClustScan (6), a commercial tool that runs on an external server using a stand-alone client program that can be downloaded from their Web site. As input, the program takes DNA sequences in FASTA format or any other format supported by ReadSeq, a biosequence conversion tool (http://www-bimas.cit.nih.gov/molbio/readseq/). Output includes annotated backbone genes and domain prediction. Interestingly, ClustScan can also use custom-built HMM to identify SM genes in bacterial genomes (see Note 3). The program output is interactive, which allows users to edit annotations based on their preference. Recently, an algorithm for detection of recombination events between gene clusters has also been added to the tool (18). In order to use ClustScan, registration is required, although the program can be used free of charge on a trial basis for 30 days. It is available from http://bioserv.pbf.hr/cms/.

4. Notes

1. Necessity of experimental conformation of gene cluster prediction validity

   Further genetic and structural elucidation studies are needed to experimentally confirm putative gene clusters detected by these tools, and to characterize the structures of the compounds that they encode. Although some tools, such as antiSMASH and NP.searcher, can give an approximation of the core structure of some classes of compounds, the circularization and tailoring steps remain difficult to predict.

2. Specificity and sensitivity of gene cluster boundary predictions

   Both antiSMASH and SMURF predict genes encoding backbone enzymes with almost 100% accuracy. However, other genes in a cluster are more difficult to predict. Some of them contain unknown domains, which also contribute to poor predictions. Therefore, predicting the exact gene cluster boundaries remains a challenge, which antiSMASH and SMURF currently solve in a somewhat different manner. SMURF utilizes a list of 27 SM-specific domains to try and give an exact prediction of the gene cluster boundaries. antiSMASH, on the other hand, uses a safe margin of 5, 10, 15, or 20 kb around the backbone genes, depending on the gene cluster type. The user can then choose cluster boundaries based on additional output from smCOG gene family predictions and cluster–cluster alignments with related gene clusters. By design, both tools tend to overpredict cluster boundaries. SMURF, for example, over-predicted the number of genes included in each cluster by 8.4 on average (3). Presently, further tool refinement is limited by the number of experimentally characterized clusters and related transcriptomics, proteomics, and metabolomics datasets.

3. Problems associated with genome annotation

   Some of the tools described here (notably, ClustScan and antiSMASH) can annotate a DNA sequence using gene predictors, Glimmer, GlimmerHMM, or Genemark. While bacterial genome annotation tools may work reasonably well, gene predictors mis-annotate a large portion of genes in eukaryotic genomes, including some model organism genomes (19). This is especially true for fungal genomes, which often have small introns, tightly spaced genes, and other unusual features (20). Furthermore, fungal NRPS and PKS genes are extremely long and tend to be fragmented in poorly annotated genomes. This in turn may cause problems in domain identification and subsequent predictions. To obtain more reliable gene predictions, it is best to use consensus or post-processing methods (21).

4. Artificially merged and split gene clusters

   Experimental validation is necessary to determine whether all these genes are indeed part of one "supercluster" with multiple backbone genes, or whether two or more separate gene clusters happen to be located very closely together. Close inspection of the "homologous gene clusters" section in antiSMASH may aid in obtaining a better prediction in such case. If close homologs of the supercluster include multiple separate gene clusters, most likely the supercluster needs to be split into two or more clusters.

5. One pathway represented by two or more subclusters

   Some pathways may be encoded by two or more subclusters found at different locations on chromosomes (e.g., (22)). These types of clusters are more difficult to predict, because some of them may lack a backbone gene and most programs rely on identification of backbone genes to "anchor" clusters. In some cases, manual examination of the "homologous gene clusters" section in antiSMASH may be helpful.

Acknowledgements

We thank Suman Pakala and Bill Nierman at JCVI for critical suggestions and comments. The work of MHM was supported by the Dutch Technology Foundation, which is the applied-science division of The Netherlands Organisation for Scientific Research and the Technology Programme of the Ministry of Economic Affairs under STW grant number 10463. This project has been funded in part with federal funds to NDF from the National Institute of Allergy and Infectious Diseases, National Institutes of Health, Department of Health and Human Services, under contract numbers N01-AI-30071 and HHSN272200900007C.

References

1. Winter JM, Behnken S, Hertweck C (2011) Genomics-inspired discovery of natural products. Curr Opin Chem Biol 15(1):22–31
2. Keller NP, Hohn TM (1997) Metabolic pathway gene clusters in filamentous fungi. Fungal Genet Biol 21(1):17–29
3. Khaldi N, Seifuddin FT, Turner G, Haft D, Nierman WC, Wolfe KH, Fedorova ND (2010) SMURF: Genomic mapping of fungal secondary metabolite clusters. Fungal Genet Biol 47(9):736–741
4. Medema MH, Blin K, Cimermancic P, de Jager V, Zakrzewski P, Fischbach MA, Weber T, Takano E, Breitling R (2011) antiSMASH: rapid identification, annotation and analysis of secondary metabolite biosynthesis gene clusters in bacterial and fungal genome sequences. Nucleic Acids Res 39:W339–W346
5. Weber T, Rausch C, Lopez P, Hoof I, Gaykova V, Huson DH, Wohlleben W (2009) CLUSEAN: a computer-based framework for the automated analysis of bacterial secondary metabolite biosynthetic gene clusters. J Biotechnol 140(1–2):13–17
6. Starcevic A, Zucko J, Simunkovic J, Long PF, Cullum J, Hranueli D (2008) ClustScan: an integrated program package for the semi-automatic annotation of modular biosynthetic gene

clusters and in silico prediction of novel chemical structures. Nucleic Acids Res 36(21): 6882–6892

7. Anand S, Prasad MV, Yadav G, Kumar N, Shehara J, Ansari MZ, Mohanty D (2010) SBSPKS: structure based sequence analysis of polyketide synthases. Nucleic Acids Res 38:W487–W496
8. Li MH, Ung PM, Zajkowski J, Garneau-Tsodikova S, Sherman DH (2009) Automated genome mining for natural products. BMC Bioinformatics 10:185
9. Rausch C, Weber T, Kohlbacher O, Wohlleben W, Huson DH (2005) Specificity prediction of adenylation domains in nonribosomal peptide synthetases (NRPS) using transductive support vector machines (TSVMs). Nucleic Acids Res 33(18):5799–5808
10. Röttig M, Medema MH, Blin K, Weber T, Rausch C, Kohlbacher O (2011) NRPSpredictor2—a web server for predicting NRPS adenylation domain specificity. Nucleic Acids Res 39(2):W362–W367
11. Lansini G, Demain AL (1999) Biology of the prokaryotes. Georg Thieme, Stuttgart
12. Majoros WH, Pertea M, Salzberg SL (2004) TigrScan and GlimmerHMM: two open source ab initio eukaryotic gene-finders. Bioinformatics 20(16):2878–2879
13. Delcher AL, Bratke KA, Powers EC, Salzberg SL (2007) Identifying bacterial genes and endosymbiont DNA with Glimmer. Bioinformatics 23(6):673–679
14. Cantarel BL, Korf I, Robb SM, Parra G, Ross E, Moore B, Holt C, Sanchez Alvarado A, Yandell M (2008) MAKER: an easy-to-use annotation pipeline designed for emerging model organism genomes. Genome Res 18(1):188–196
15. Aziz RK, Bartels D, Best AA, DeJongh M, Disz T, Edwards RA, Formsma K, Gerdes S, Glass EM, Kubal M et al (2008) The RAST Server: rapid annotations using subsystems technology. BMC Genomics 9:75
16. Sonnhammer EL, Eddy SR, Birney E, Bateman A, Durbin R (1998) Pfam: multiple sequence alignments and HMM-profiles of protein domains. Nucleic Acids Res 26(1): 320–322
17. Haft DH, Selengut JD, White O (2003) The TIGRFAMs database of protein families. Nucleic Acids Res 31(1):371–373
18. Starcevic A, Diminic J, Zucko J, Elbekali M, Schlosser T, Lisfi M, Vukelic A, Long PF, Hranueli D, Cullum J (2011) A novel docking domain interface model predicting recombination between homoeologous modular biosynthetic gene clusters. J Ind Microbiol Biotechnol 38(9):1295–1304. doi:10.1007/s10295-10010-10909-10290
19. Wortman JR, Gilsenan JM, Joardar V, Deegan J, Clutterbuck J, Andersen MR, Archer D, Bencina M, Braus G, Coutinho P et al (2009) The 2008 update of the Aspergillus nidulans genome annotation: a community effort. Fungal Genet Biol 46(Suppl 1):S2–S13
20. Ma L-J, Fedorova ND (2010) A practical guide to fungal genome projects. Mycol Int J Fungal Biol 1(1):9–24
21. Pati A, Ivanova NN, Mikhailova N, Ovchinnikova G, Hooper SD, Lykidis A, Kyrpides NC (2010) GenePRIMP: a gene prediction improvement pipeline for prokaryotic genomes. Nat Methods 7(6):455–457
22. Nicholson MJ, Koulman A, Monahan BJ, Pritchard BL, Payne GA, Scott B (2009) Identification of two aflatrem biosynthesis gene loci in Aspergillus flavus and metabolic engineering of Penicillium paxilli to elucidate their function. Appl Environ Microbiol 75(23): 7469–7481

Chapter 3

Media and Growth Conditions for Induction of Secondary Metabolite Production

Jens C. Frisvad

Abstract

Growth media and incubation conditions have a very strong influence of secondary metabolite production. There is no consensus on which media are the optimal for metabolite production, but a series of useful and effective media and incubation conditions have been listed here. Chemically well-defined media are suited for biochemical studies, but in order to get chemical diversity expressed in filamentous fungi, sources rich in amino acids, vitamins, and trace metals have to be added, such as yeast extract and oatmeal. A battery of solid agar media is recommended for exploration of chemical diversity as agar plug samples are easily analyzed to get an optimal representation of the qualitative secondary metabolome. Standard incubation for a week at 25°C in darkness is recommended, but optimal conditions have to be modified depending on the ecology and physiology of different filamentous fungi.

Key words: Secondary metabolites, Growth media, Agar plugs, Yeast extract, HPLC

1. Introduction

Filamentous fungi can produce many different secondary metabolites. Profiles of secondary metabolites were introduced by Frisvad et al. (1–6) to show that a single strain of filamentous fungi can produce a series of secondary metabolites in a species-specific manner. The fact that many secondary metabolites can be produced by one single strain was taken up by Bode et al. (7) and called one strain many compounds (OSMAC). In recent years it has been shown that histone methylation or acetylation inhibitors, other microbes, plant extracts and other growth medium ingredients, water activity, pH, temperature, light, stress, and other factors can help express usually silent secondary metabolite gene clusters (8–26). Of the rather few tested, most secondary metabolites from

Nancy P. Keller and Geoffrey Turner (eds.), *Fungal Secondary Metabolism: Methods and Protocols*, Methods in Molecular Biology, vol. 944, DOI 10.1007/978-1-62703-122-6_3,

species of fungi have been shown to have highly remarkable effects on other species. It is therefore of interest to find laboratory and industrial conditions where all these secondary metabolites can be produced in order to screen for new antibiotics, quorum sensing inhibitors, and drug lead candidates and also in general to elucidate the chemical language of organisms. It is often not known which conditions are optimal for production of any one secondary metabolite, as the ecology of the producers is usually not well known. For this reason media and growth conditions for production of secondary metabolites have been based on trial and error or tradition.

In Japan rice has been used traditionally as a substrate for secondary metabolite production and many other natural product chemists prefer this substrate (e.g., (27, 28)), while wheat has been used in Europe and corn or wheat in the USA (e.g., (29, 30)). In other studies scientists have preferred chemically well-defined media or well-defined media with added yeast extract, malt extract, corn-steep liquor, peptone, or potato extract. Secondary metabolites from marine-derived fungi are often produced on media-added seawater rather than pure water (e.g., (31)).

A major problem with chemically poorly defined media is that different brands of yeast extract (32, 33), peptone (34), malt extract (35), etc. will occasionally give different morphologies and secondary metabolite profiles. On Yeast extract sucrose (YES) agar, profiles of secondary metabolites were remarkably different depending on the brand of yeast extract (32). At least Difco yeast extract gives good results concerning secondary metabolites.

It has long been known that all filamentous fungi produce a large number of secondary metabolites, biosynthetically often produced as polyketides, terpenes, non-ribosomal peptides, and alkaloids or mixtures of those, or derived from shikimate or nucleotides. Each species produce a specific profile of secondary metabolites (2, 3), and so the concept OSMAC (7) could just as well have been an abbreviation of "one species many compounds." Some research groups have used this approach in using nutritional arrays in a small scale (9), while others have suggested a large number of media in order to increase the chance of having as many secondary metabolite families expressed as possible (36).

The medium components have a pronounced effect on secondary metabolite production. For example lactate or starch as the sole carbon source only resulted in low production of secondary metabolites in *Aspergillus niger*, but the combination of the two gave a synergistic effect, so much larger amounts of fumonisins, ochratoxins, kotanins, and naphtho-γ-pyrones were produced (23). This indicates that there are many possibilities to optimize medium composition. C/N ratios and combinations of C and N sources have not often been systematically explored regarding profiles of secondary metabolites. Experimental design is a good tool for optimizing single secondary metabolite production in general (37), but multicriterium optimization may be needed in order to optimize the profile of secondary metabolites.

Addition of suboptimal concentrations of fungicides, preservatives, and histone effectors to the media may also enhance secondary metabolite production (10, 18, 21, 22, 38). Plant secondary metabolites may trigger secondary metabolite production by epigenetic remodeling (13, 26) and thus the addition of plant extracts to many media often boost secondary metabolite production. Here a general procedure for exploring the secondary metabolome using a simple agar plug extraction procedure with different kinds of media combined with (UPLC)-DAD-MS is presented.

2. Materials

2.1. Fungal Inoculation

1. Media for production of conidia:
   (a) CYA (Czapek yeast autolysate agar) for *Penicillium*, *Aspergillus*, and *Paecilomyces*: $NaNO_3$ (3 g/L), yeast extract (5 g/L), sucrose (30 g/L), $K_2HPO_4{\cdot}3H_2O$ (1.3 g/L), Czapek concentrate (10 mL/L), agar (15 g/L) in 1 L ddH_2O. Autoclave for 15 min, at 121°C on liquid cycle.
   (b) Czapek concentrate: KCl (5 g/100 mL), $MgSO_4{\cdot}7H_2O$ (5 g/100 mL), $FeSO_4{\cdot}5H_2O$ (0.1 g/100 mL), $ZnSO_4{\cdot}7H_2O$ (0.1 g/100 mL), $CuSO_4{\cdot}5H_2O$ (0.05 g/100 mL) in 100 mL ddH_2O.
   (c) Malt extract agar (MEA) for most filamentous fungi: Malt extract (20 g/L), glucose (20 g/L), peptone (1 g/L), $ZnSO_4{\cdot}7H_2O$ (0.01 g/L), $CuSO_4{\cdot}5H_2O$ (0.005 g/L), agar (25 g/L) in 1 L ddH_2O, pH approximately 4.7, so it is advisable to autoclave 500 mL water with the malt extract and 500 mL with the remaining ingredients, then mix after autoclaving. Autoclave for 15 min, at 121°C on liquid cycle (see Note 1).
2. Spore Suspension Medium (SSM): Tween 80: 0.5 g/L, agar 2 g/L, ddH_2O to 1 L, distributed in 4 mL vials, autoclaved, and stored in a refrigerator (39).

2.2. Media for Secondary Metabolite Production

1. YES agar: Sucrose (150 g/L), yeast extract (20 g/L), $MgSO_4{\cdot}7H_2O$ (0.5 g/L), $ZnSO_4{\cdot}7H_2O$ (0.01 g/L), $CuSO_4{\cdot}5H_2O$ (0.005 g/L), agar (15 g/L) in 1 L ddH_2O. Autoclave for 15 min, at 121°C on liquid cycle (see Note 2) (1, 2, 40, 41).
2. CYA agar (see above) (1, 2, 41, 42) (see Note 3).

3. Potato dextrose agar (PDA): 200 g sliced potatoes are boiled in 1 L of ddH_2O and sieved, add glucose (20 g/L), agar (20 g/L), $ZnSO_4 \cdot 7H_2O$ (0.01 g/L), $CuSO_4 \cdot 5H_2O$ (0.005 g/L). pH is approximately 5.6. Autoclave for 15 min, at 121°C on liquid cycle (1, 2, 41).
4. Mercks Malt Extract (MME) agar: Malt extract (30 g/L), soy peptone (3 g/L), $ZnSO_4 \cdot 7H_2O$ (0.01 g/L), $CuSO_4 \cdot 5H_2O$ (0.005 g/L), agar (20 g/L) in 1 L ddH_2O. Autoclave for 15 min, at 121°C on liquid cycle (43).
5. Wickerhams Antibiotic Test Medium (WATM): $NaNO_3$ (2 g/L), glucose (2 g/L), saccharose (30 g/L), yeast extract (Difco) (2 g/L), peptone (3 g/L), corn steep solids (5 g/L), $KH_2PO_4 \cdot 3H_2O$ (1 g/L), KCl (0.2 g/L), $MgSO_4 \cdot 7H_2O$ (0.5 g/L), $FeSO_4 \cdot 7H_2O$ (0.01 g/L), $ZnSO_4 \cdot 7H_2O$ (0.01 g/L), $CuSO_4 \cdot 5H_2O$ (0.005 g/L), agar (20 g/L) in 1 L ddH_2O. Autoclave for 15 min, at 121°C on liquid cycle (44).
6. Raulin Thom Oatmeal (RTO) agar: $(NH_4)H_2PO_4$ (0.4 g/L), K_2CO_3 (0.4 g), $ZnSO_4 \cdot 7H_2O$ (0.07 g/L), $CuSO_4 \cdot 5H_2O$ (0.005 g/L), $FeSO_4 \cdot 7H_2O$ (0.06 g/L), $(NH_4)_2SO_4$ (0.16 g/L), $MgCO_3$ (0.25 g), tartaric acid (2.6 g/L), di-ammonium tartrate (2.6 g/L), glucose (50 g/L), oatmeal (30 g/L), agar (20 g/L) in 1 L ddH_2O. pH adjusted to 6.5 with 1 N NaOH before autoclaving. Autoclave for 15 min, at 121°C on liquid cycle (36).
7. Rice Corn steep (RC) agar: Rice meal (50 g/L), corn steep solids (40 g/L), $ZnSO_4 \cdot 7H_2O$ (0.01 g/L), $CuSO_4 \cdot 5H_2O$ (0.005 g/L), agar (20 g/L) in 1 L ddH_2O. Autoclave for 15 min, at 121°C on liquid cycle (45).
8. Oatmeal Agar (OA): Oatmeal (30 g/L), $ZnSO_4 \cdot 7H_2O$ (0.01 g/L), $CuSO_4 \cdot 5H_2O$ (0.005 g/L), agar (20 g/L) in 1 L ddH_2O. Autoclave for 15 min, at 121°C on liquid cycle (5, 41).
9. MEA (see above) (5).
10. Czapek Yeast Autolysate Salt (CYAS) agar: CYA agar with 50 g/L NaCl (46) (see Note 4).
11. CY20 or CY20S (Czapek Yeast autolysate 20% Saccharose agar): CYA agar with 170 g/L saccharose added, but ddH_2O only to 1 L (46) (see Note 4).
12. Pharma Medium (PM): Maltose (30 g/L), polypeptone (10 g/L), $(NH_4)_2SO_4$ (2 g/L), Cotton seed flour (Pharmamedia) (40 g/L), KH_2PO_4 (2.5 g/L), K_2HPO_4 (7.5 g/L), $MgSO_4 \cdot 7H_2O$ (0.5 g/L), $ZnSO_4 \cdot 7H_2O$ (0.01 g/L), $CuSO_4 \cdot 5H_2O$ (0.005 g/L), agar (15 g/L) in 1 L ddH_2O. Autoclave for 15 min, at 121°C on liquid cycle (see Note 5) (47).

13. Yeast Extract (YE) agar: Yeast extract (20 g/L), $MgSO_4 \cdot 7H_2O$ (0.5 g/L), $ZnSO_4 \cdot 7H_2O$ (0.01 g/L), $CuSO_4 \cdot 5H_2O$ (0.005 g/L), agar (15 g/L) in 1 L ddH_2O. Autoclave for 15 min, at 121°C on liquid cycle (see Note 6) (36).
14. LB (Lysogeny Broth, "Luria Broth," "Lennox Broth," "Luria-Bertani medium") agar: Tryptone (10 g/L), yeast extract (5 g/L), NaCl (10 g/L), agar (15 g/L) in 1 L ddH_2O. pH 7.0. Autoclave for 15 min, at 121°C on liquid cycle (see Note 6) (48, 49).
15. Dichloran Rose bengal Yeast Extract Sucrose (DRYES) agar: Sucrose (150 g/L), yeast extract (20 g/L), dichloran (0.002 g/L), rose bengal (0.025 g/L), chloramphenicol (0.05 g/L), $MgSO_4 \cdot 7H_2O$ (0.5 g/L), $ZnSO_4 \cdot 7H_2O$ (0.01 g/L), $CuSO_4 \cdot 5H_2O$ (0.005 g/L), agar (15 g/L) in 1 L ddH_2O. Autoclave for 15 min, at 121°C on liquid cycle. After autoclaving add 0.05 g/L chlortetracycline aseptically. pH 5.6 (41) (see Note 7).

2.3. Extraction of Fungal Cultures

1. Extraction liquid: Ethylacetate:dichloromethane:methanol = 3:2:1, this mixture containing 1% formic acid (see Note 8).
2. Organic solvent for redissolving fungal plug extract: Methanol.

2.4. Ultrahigh-Performance Liquid Chromatography Procedure

1. Elution liquid I: Acetonitrile with 50 ppm trifluoro acetic acid (TFA).
2. Elution liquid II: Water with 50 ppm TFA.
3. Alkylphenone retention standard mixture: Acetophenone, propiophenone, butyrophenone, valerophenone, hexanophenone, octanophenone, decanophenone, 0.1% of each dissolved in acetonitrile. Keep in −18°C freezer until use.
4. If available dissolve secondary metabolite authentic standards in methanol, and keep in −18°C freezer until use.

3. Methods

Qualitative and semiquantitative screening for secondary metabolites in filamentous fungi can be done efficiently using a microscale agar plug method. In combination with ultrahigh-performance liquid chromatography (UPLC), this method will give very reproducible separation in 10 min for each sample, and many analyses can be done in a short time. It is recommendable to take plugs from agar cultures, as it is easily seen if there are any contaminants in the cultures, and such 9 cm agar plates are used for fungal identification anyway. Furthermore it can be determined if different parts of the fungal colony have different chemical profiles and which metabolites are secreted into the medium. Some of these advantages are lost when using either small-scale 96-well plates (9) or at the other extreme, large fermentations. However, other screening methods have been suggested, for example one based on still rice cultures and a shaken liquid medium (28, 50).

Regarding growth conditions, usually filamentous fungi are grown at 25–30°C and will produce their secondary metabolites at their growth optimum, and often less when they are approaching their growth maximum or minimum. Of course thermophilic fungi will produce secondary metabolites at high temperatures. Chu et al. (51) found that *Talaromyces thermophilus* produced talathermophilins A and B at temperatures from 35 to 65°C. Likewise psychrotolerant fungi produce secondary metabolites optimally at 15–20°C (52). Concerning light, most fungi produce most of their secondary metabolites in darkness, but there are some exceptions. For example *Talaromyces stipitatus* produces some additional anthraquinones only when exposed to light (24), and in general the perfect and imperfect states of filamentous fungi may contain different secondary metabolites. As some of these fungi produce the imperfect state mostly when exposed to light and the perfect state preferentially when grown in darkness, secondary metabolite profiles have to be different in light and darkness (8). Furthermore fungi have a circadian regulation of secondary metabolite production and different kinds of light (blue light, yellow to green) have a pronounced effect on secondary metabolism (22). Furthermore light has a direct influence on carotene production (14, 15), and if these compounds are to be considered, incubation in light may be important. Low water activity will usually mean a lower production of secondary metabolites (17, 53), but Schmidt-Heydt et al. (21) have shown that while large amounts of secondary metabolites are produced close to the growth optimum, there is another smaller production optimal peak of secondary metabolite production at more extreme conditions. This was also the case for temperature and pH. Oxygen and carbon dioxide can also have a positive effect on mycotoxin production, but usually secondary metabolite-producing fungi produce most mycotoxins at high oxygen and low carbon dioxide levels (54). At high CO_2 levels or anaerobic conditions fungi will turn their metabolism to production of alcohol, esters, and other small volatile molecules as is known in yeasts.

3.1. Fungal Inoculation

Obtain fungal isolates directly from a substrate or via direct or dilution plating from the same substrate or order cultures from culture collections. Purify the fungal isolates by streaking on CYA, or similar agar media, or by single spore inoculation. Identify the fungal isolates to species level, using a polyphasic approach, using morphology, physiology, and molecular methods, and accession the cultures in one or preferably two international culture collections (i.e., CBS, the Netherlands; ATCC, the USA; NRRL, the USA;

IMI, Great Britain; IHEM, Belgium; CCF, the Czech Republic; CECT, Spain; VKM, Russia; DSMZ, Germany; IBT, Denmark; ITEM, Italy; IFO, Japan; IITA, Nigeria). Most of these culture collections also have a fungal identification service. Make a literature search on the known compounds from the species identified. For each fungal isolate make a conidium suspension (2×10^6 conidia per mL) in 0.2% agar in water with 0.05% Tween 80 and three point inoculate (39) onto the agar media chosen. Incubate at 25°C for 7 days in darkness.

3.2. Media for Secondary Metabolite Production

Three point inoculate the media YES, CYA, MEA, and OA (for *Penicillium, Aspergillus, Talaromyces, Rasamsonia, Hamigera*, and *Paecilomyces*); DRYES and PDA for *Alternaria, Cladosporium*, and other dematiaceous hyphomycetes; and YES and PDA for *Fusarium, Trichoderma, Verticillium, Stachybotrys*, and other genera with spores produced in slime. For a greater secondary metabolite profile, inoculate each fungal isolate on YE, LB, MME, WATM, RTO, RC, CYAS, CY20, and PM.

3.3. Extraction of Fungal Cultures

1. Take three 6 mm diam. agar plugs from one of the colonies from each agar medium used (one plug in the center of the colony and one from the periphery closest and furthest away from the other colonies) (55) and place in a 1.5 mm autosampler dram vial with a screw cap.
2. Extract secondary metabolites with ethyl acetate/methanol/dichloromethane (3:2:1) with 1% formic acid in 1.5 mL autosampler screw-cap vials.
3. Ultrasonicate for 50 min and transfer extract to a new 1.5 mL vial. Evaporate the contents and redissolve in 400 μl methanol by ultrasonication in 10 min, filter through a PTFE 0.45 μm filter, and keep at −18°C prior to analysis.

3.4. UPLC

1. Arrange all samples in a rack (for the autosampler of the liquid chromatograph) and analyze in the following order: (a) a blank sample, (b) the alkylphenone mixture, (c) any standards, (d) plugs extracts of the media without fungi as controls, (e) the actual samples with fungal secondary metabolites, and (f) the alkylphenone mixture once again.
2. Inject 1 μl extract or standard into a liquid chromatograph with two pumps and with diode array detection, fluorescence detection, and electrospray MS detection (56). Analyze by UPLC using alkylphenone retention indices and diode array UV–VIS detection as described by Frisvad and Thrane (57, 58).
3. Separate the secondary metabolites on a C_{18} (150×2.1 mm) column with 2.6 μm particles. Hold column temperature at 60°C. Inject 1 μl of fungal extract into the UPLC. Use a

gradient based on water with 50 ppm TFA and acetonitrile (AcCN) with 50 ppm TFA, at a flow rate of 0.8 mL/min: Start conditions 85% H_2O with TFA and 15% AcCN with TFA. Gradient: 15–25% AcCN in ½ min, from 25 to 65% AcCN in 5½ min, from 60 to 100% AcCN in 1 min, at 100% AcCN in 1 min, back to 15% AcCN in 1 min, and then 1 min constant at 15% AcCN. The run-time is therefore 10 min.

4. Record UV chromatograms at 210 and 280 nm, and record the full diode array spectra in a range from 190 to 600 nm.
5. Print or examine chromatograms, DAD spectra, and MS spectra on the screen. Identify secondary by comparison of the UV-Visible DAD spectra, MS spectra, and retention times of the secondary metabolites with authentic standards (56).
6. Record the peak area of each chromatographic peak for semi-quantitative measurements. In that case make biological replicates of the extracts of the fungal cultures to calculate the average and standard deviation of the results.

3.5. Conclusions

No single medium under standardized growth conditions can secure expression of the full potential for producing secondary metabolites (the secondary metabolome) of a fungal culture. Despite a plethora of media and growth conditions for fungal secondary metabolite production used by natural product scientists, secondary metabolite profiling can be done using a simple compromise method. YES agar, PDA (or equivalent media such as MEA or oatmeal agar), and a carbohydrate-poor medium such as Lysogeny Broth agar incubated at 25°C for 1 week in darkness will give a relatively good representation of major secondary metabolites of a fungal species. Extracted agar plugs from the fungal agar cultures can then be analyzed by HPLC-DAD-MS in order to identify as many secondary metabolites as possible. However there are many possibilities to have fungal cultures express even more biosynthetic families of secondary metabolites, including adding histone methylation inhibitors, adding plant parts in the medium, adding other microorganisms, using light, using longer incubation periods, using other temperatures, using low or high pH, etc.

4. Notes

1. Many other media are excellent for production of large numbers of fungal conidia: PDA, Potato Carrot Agar (PCA), Spezieller Nährstoffarmer Agar (SNA) agar, and several other media (see (41) for medium composition of those). For genera like *Fusarium*, *Trichoderma*, and *Alternaria*, an alternate light cycle will promote sporulation (41).

2. YES agar appears to be optimal when made with Difco yeast extract, but can also be made from Sigma yeast extract and many other yeast extracts. The addition of magnesium sulfate will make YES agars made from different yeast extracts less variable concerning secondary metabolite production (32). Sometimes a modification of YES agar, DRYES (59), will give even better secondary metabolite production (60).
3. This medium is especially suited for non-ribosomal peptides, but alternative media, such as ALK and ABE, have also been proposed for alkaloids (61).
4. CYAS and CY20 are especially suited for secondary metabolite production by osmophilic and halophilic fungi such as *Eurotium* (*Aspergillus* section *Aspergillus*), but have also been shown to be optimal for certain secondary metabolites in less osmophilic and halophilic species, for example fumonisin production by *Aspergillus niger* (46).
5. PM is especially suited for Zygomycetes, such as *Rhizopus* (47).
6. These two media contain no or very little direct carbon source and are supportive of production of certain non-ribosomal peptides such as penicillin, gliotoxin, and aspergillic acids. Other media that have been used for this purpose are Tryptone Glucose Yeast (TGY) extract agar (62) and Aspergillus Flavus Parasiticus Agar (AFPA) (63). There have as yet not been any systematic comparisons of these media for their efficiency in metabolite profiling, but many known secondary metabolites, such as aflatoxins, are not produced on these low-carbon-source media.
7. Secondary metabolites can be produced in still or shaken broth cultures, in agar media, on other kinds of solid support, for example Lightweight Expanded Clay Aggregates (LECA) "nuts" (64), and on other gelling agents such a pluronic polyol F-127 (65, 66), but solid agar media have given consistently good results.
8. Other alternative extraction liquids are, for example, ethyl acetate with 1% formic acid, isopropanol, 75% methanol (excellent for fumonisins (46)), and hexane, for apolar sclerotial metabolites (30).

Acknowledgements

The project was supported by the Danish Research Council for Technology and Production.

References

1. Frisvad JC (1981) Physiological criteria and mycotoxin production as aids in identification of common asymmetric Penicillia. Appl Environ Microbiol 41:568–579
2. Frisvad JC, Filtenborg O (1983) Classification of terverticillate Penicillia based on profiles of mycotoxins and other secondary metabolites. Appl Environ Microbiol 46:1301–1310
3. Frisvad JC, Filtenborg O (1989) Terverticillate penicillia: chemotaxonomy and mycotoxin production. Mycologia 81:836–861
4. Frisvad JC, Samson RA (2004) Polyphasic taxonomy of *Penicillium* subgenus *Penicillium*. A guide to identification of the food and airborne terverticillate Penicillia and their mycotoxins. Stud Mycol 49:1–173
5. Frisvad JC, Filtenborg O, Samson RA, Stolk AC (1990) Chemotaxonomy of the genus *Talaromyces*. Antonie Van Leeuwenhoek 57: 179–189
6. Frisvad JC, Andersen B, Thrane U (2008) The use of secondary metabolite profiling in fungal taxonomy. Mycol Res 112:231–240
7. Bode HB, Bethe B, Höfs R, Zeeck A (2002) Big effects from small changes: possible ways to explore nature's chemical diversity. Chembiochem 3:619–627
8. Bayram Ö, Krappmann S, Ni M, Bok JW, Helmstaedt K, Valerius O, Braus-Stromeyer S, Kwon N-J, Keller NP, Yu J-H, Braus GH (2008) VelB/VeA/LaeA complex coordinates light signal with fungal development and secondary metabolism. Science 320:1504–1506
9. Bills GF, Platas G, Fillola A, Jimenez MR, Collado J, Vicente F, Martin J, Gonzalez A, Bur-Zimmermann J, Tormo JR, Pelaez F (2008) Enhancement of antibiotic and secondary metabolite detection from filamentous fungi by growth on nutritional arrays. J Appl Microbiol 104:1644–1658
10. Bok JW, Chiang YM, Szewczyk E, Reyes-Domingez Y, Davidson AD, Sanchez JF, Lo HC, Watanabe K, Strauss J, Oakley BR, Wang CCC, Keller NP (2009) Chromatin-level regulation of biosynthetic gene clusters. Nat Chem Biol 5:462–464
11. Chiang Y-M, Oakley BR, Keller NP, Wang CCC (2010) Unraveling polyketide synthesis in members of the genus *Aspergillus*. Appl Microbiol Biotechnol 86:1719–1736
12. Chiang YM, Chang SL, Oakley BR, Wang CCC (2011) Recent advances in awakening silent biosynthetic gene clusters to natural products in microorganisms. Curr Opin Chem Biol 15:137–143
13. Cichewicz RH (2010) Epigenome manipulation as a pathway to new natural product scaffolds and their congeners. Nat Prod Rep 27:11–22
14. Corrochano LM, Cerda Olmedo E (1992) Sex, light and carotenes - the development of *Phycomyces*. Trends Genet 8:268–274
15. Graafman WD (1974) Metabolism in *Penicillium isariiforme* on exposure to light, with special reference to citric acid synthesis. J Gen Microbiol 82:247–252
16. Knight V, Sanglier J-J, Tullio D, Braccili D, Bonner P, Waters J, Hughes P, Zhang L (2003) Diversifying microbial natural products for drug discovery. Appl Microbiol Biotechnol 62:446–458
17. Nielsen PV, Beuchat LR, Frisvad JC (1988) Growth and fumitremorgin production by *Neosartorya fischeri* as affected by temperature, light, and water activity. Appl Environ Microbiol 54:1504–1510
18. Pettit PK (2011) Small molecule elicitation of microbial secondary metabolites. Microb Biotechnol 4:471–478
19. Scherlach K, Hertweck C (2009) Triggering cryptic natural product biosynthesis in microorganisms. Org Biomol Chem 7: 1753–1760
20. Schmidt-Heydt M, Baxter E, Geisen R, Magan N (2007) Physiological relationships between food preservatives, environmental factors, ochratoxin and *otapksPv* gene expression by *Penicillium verrucosum*. Int J Food Microbiol 119:277–283
21. Schmidt-Heydt M, Magan N, Geisen R (2008) Stress induction of mycotoxin biosynthesis genes by abiotic factors. FEMS Microbiol Lett 284:142–149
22. Schmidt-Heydt M, Rüfer C, Raupp F, Bruchmann A, Perrone G, Geisen R (2011) Influence of light on food relevant fungi with emphasis on ochratoxin producing species. Int J Food Microbiol 145:229–237
23. Sørensen LM, Lametsch R, Andersen MR, Nielsen PV, Frisvad JC (2009) Proteome analysis of *Aspergillus niger*: Lactate added in starch-containing medium can increase production of the mycotoxin fumonisin B_2 by modifying acetyl-CoA metabolism. BMC Microbiol 9:255
24. Van Eijk GW (1973) Anthraquinones in fungus *Talaromyces stipitatus*. Experientia 29:522–523
25. Wang FZ, Wei HJ, Zhu TJ, Li DH, Linz J, Gu QQ (2011) Three new cytochalasins from the marine-derived fungus *Spicaria elegans* KLA03

by supplementing the cultures with L- and D-tryptophan. Chem Biodivers 8:887–894

26. Williams RB, Henrikson JC, Hoover AR, Lee AE, Cichewicz RH (2008) Epigenetic remodeling of the fungal secondary metabolome. Org Biomol Chem 6:1895–1897
27. Ito T, Odake T, Katoh H, Yamaguchi Y, Aoki M (2011) High-throughput profiling of microbial extracts. J Nat Prod 74:983–988
28. Kjer J, Debbab A, Proksch P (2010) Methods for isolation of marine-derived endophytic fungi and their bioactive secondary metabolites. Nat Protoc 5:479–490
29. MacGeorge KM, Mantle PG (1990) Nephrotoxicity of *Penicillium aurantiogriseum* and *P. commune* from an endemic nephropathy area of Yugoslavia. Mycopathology 112: 139–145
30. Wicklow DT, Dowd PF, TePaske MR, Gloer JB (1988) Sclerotial metabolites of *Aspergillus flavus* toxic to a detrivorous maize insect (*Carpophilus hemiterus*, Nitidulidae). Trans Br Mycol Soc 91:433–438
31. Liu J, Li F, Kim EL, Li JL, Hong J, Bae KS, Chung HY, Kim HS, Jung JH (2011) Antibacterial polyketides from the jellyfish-derived fungus *Paecilomyces variotii*. J Nat Prod 74:1826–1829
32. Filtenborg O, Frisvad JC, Thrane U (1990) The significance of yeast extract composition on metabolite production in *Penicillium*. In: Samson RA, Pitt JI (eds) *Modern concepts in* Penicillium *and* Aspergillus *classification*. Plenum, New York, pp 433–441
33. Gaudreau H, Champagne CP, Conway J, Degré R (1999) Effect of ultrafiltration on yeast extracts on their ability to promote lactic acid bacteria growth. Can J Microbiol 45: 891–897
34. Odds FC, Hall CA, Abbott AB (1978) Peptones and mycological reproducibility. Sabouraudia 16:237–246
35. Frisvad JC, Filtenborg O, Thrane U, Samson RA (2000) Collaborative study on stipe roughness and conidium form in some terverticillate penicillia. In: Samson RA, Pitt JI (eds) Integration of modern taxonomic methods for Aspergillus and Penicillium classification. Harwood Scientific Publishers, Reading, MA, pp 113–125
36. Nielsen ML, Nielsen JB, Rank C, Klejnstrup ML, Holm DMK, Brogaard KH, Hansen BG, Frisvad JC, Larsen TO, Mortensen UH (2011) A genome-wide polyketide synthase deletion library uncovers novel genetic links to polyketides and meroterpenoids in *Aspergillus nidulans*. FEMS Microbiol Lett 321:157–166
37. Pimenta EF, Vita-Marques AM, Tininis A, Seleghim MHR, Sette LD, Veloso K, Fereirra AG, Williams DE, Patrick BO, Dalisay DS, Andersen RJ, Berkinck RGS (2010) Use of experimental design for the optimization of new secondary metabolites in two *Penicillium* species. J Nat Prod 73:1827–1832
38. Ghosh J, Häggblom P (1985) Effect of sublethal concentrations of propionic acid or butyric acid on growth and aflatoxin production by *Aspergillus flavus*. Int J Food Microbiol 2:323–330
39. Pitt JI (1979) The genus Penicillium and its teleomorphic states Eupenicillium and Talaromyces. Academic, New York
40. Davis ND, Diener UL, Eldridge DW (1966) Production of aflatoxins B_1 and G_1 by *Aspergillus flavus* in a semisynthetic medium. Appl Microbiol 14:378–380
41. Samson RA, Houbraken J, Thrane U, Frisvad JC, Andersen B (2010) Food and indoor fungi. CBS laboratory manual Series 2. CBS KNAW Fungal Biodiversity Center, Utrecht, NL, 381 pp
42. Pitt JI (1973) An appraisal of identification methods for *Penicillium* species: novel taxonomic criteria based on temperature and water relations. Mycologia 65:1135–1157
43. Leistner L, Pitt JI (1977) Miscellaneous *Penicillium* toxins. In: Rodricks JV, Hesseltine CW, Mehlmann MA (eds) Mycotoxins in human and animal health. Pathotox Publishers, Park Forest South, IL, pp 639–653
44. Raper KB, Thom C (1949) Manual of the Penicillia. Williams & Wilkins, Baltimore, MD
45. Bullerman LB (1974) A screening medium and method to detect several mycotoxins in mold cultures. J Milk Food Technol 37:1–3
46. Frisvad JC, Smedsgaard J, Samson RA, Larsen TO, Thrane U (2007) Fumonisin B_2 production by *Aspergillus niger*. J Agric Food Chem 55:9727–9732
47. Jennesen J, Nielsen KF, Houbraken J, Lyhne EK, Schnürer J, Frisvad JC, Samson RA (2005) Secondary metabolite and mycotoxin production by the *Rhizopus microsporus* group. J Agric Food Chem 53:1833–1840
48. Bertani G (1951) Studies on lysogenesis. I. The mode of phage liberation by lysogenic Escherichia coli. J Bacteriol 62:293–300
49. Bertani G (2004) Lysogeny at mid-twentieth century: P1, P2, and other experimental systems. J Bacteriol 186:595–600
50. Aly AH, Debbab A, Proksch P (2011) Fungal endophytes: unique plant inhabitants with great promises. Appl Microbiol Biotechnol 90:1829–1845

51. Chu Y-S, Niu X-M, Wang Y-L, Guo J-P, Pan W-Z, Huang X-W, Zhang K-Q (2010) Isolation of putative biosynthetic intermediates of prenylated indole alkaloids from a thermophilic fungus *Talaromyces thermophilus.* Org Lett 12:4356–4359
52. Frisvad JC, Larsen TO, Dalsgaard PW, Seifert KA, Louis-Seize G, Lyhne EK, Jarvis BB, Fettinger JC, Overy DP (2006) Four psychrotolerant species with high chemical diversity consistently producing cycloaspeptide A, *P. jamesonlandense* sp. nov., *P. ribium* sp. nov., *P. soppii* and *P. lanosum.* Int J Syst Evol Microbiol 56:1427–1437
53. Baxter CJ, Magan N, Lane B, Wildman HG (1998) Influence of water activity and temperature on *in vitro* growth of surface culture of a *Phoma* sp. Appl Microbiol Biotechnol 49:328–332
54. Jayashree T, Subramanyam C (2000) Oxidative stress as a prerequisite for aflatoxin production by *Aspergilus parasiticus.* Free Radic Biol Med 29:981–985
55. Smedsgaard J (1997) Micro-scale extraction procedure for standardized screening of frungal metabolite production in cultures. J Chromatogr A 760:264–270
56. Nielsen KF, Månsson M, Rank C, Frisvad JC, Larsen TO (2011) Dereplication of microbial natural products by LC-DAD-TOFMS. Statically analysis of adduct formation, in-source fragmentation and chromatographic retention as an in-house standard database of 719 microbial metabolites and mycotoxins. J Nat Prod 74(11):2338–2348
57. Frisvad JC, Thrane U (1987) Standardized High-Performance Liquid Chromatography of 182 mycotoxins and other fungal metabolites based on alkylphenone indices and UV-VIS spectra (diode-array detection). J Chromatogr 404:195–214
58. Frisvad JC, Thrane U (1993) Liquid column chromatography of mycotoxins. In: Betina V (ed) Chromatography of mycotoxins: techniques and applications. Elsevier, Amsterdam, pp 253–372
59. Frisvad JC (1983) A selective and indicative medium for groups of *Penicillium viridicatum* producing different mycotoxins in cereals. J Appl Bacteriol 54:409–416
60. Andersen B, Sørensen JL, Nielsen KF, van den Ende BG, de Hoog S (2009) A polyphasic approach to the taxonomy of the *Alternaria infectoria* species-group. Fungal Genet Biol 46:642–656
61. Reshetilova TA, Solov'eva TF, Baskunov BP, Kozlovskii AG (1992) Investigation of alkaloid formation by certain species of fungi of the *Penicillium* genus. Mikrobiologiya 61:873–879
62. Frisvad JC, Filtenborg O, Wicklow DT (1987) Terverticillate penicillia isolated from underground seed caches and cheek pouches of banner-tailed kangaroo rats (*Dipodomys spectabilis*). Can J Bot 65:765–773
63. Pitt JI, Hocking AD, Glenn DR (1983) An improved medium for the detection of *Aspergillus flavus* and *A. parasiticus.* J Appl Bacteriol 54:109–114
64. Nielsen KF, Larsen TO, Frisvad JC (2004) Lightweight expanded clay aggregates (LECA), a new up-scaleable matrix for production of microfungal metabolites. J Antibiot 57:29–36
65. Olson LW, Lange L (1989) The use of BASF pluronic polyol F-127, low temperature liquefying polyol, for the isolation of microbial antagonists. Opera Bot 100:197–199
66. Reeslev M, Kjøller A (1995) Comparison of biomass dry weights and radial growth rates of fungal colonies on media solidified with different gelling compounds. Appl Environ Microbiol 61:4236–4239

Chapter 4

Manipulating Filamentous Fungus Chemical Phenotypes by Growth on Nutritional Arrays

José R. Tormo, Francisco J. Asensio, and Gerald F. Bills

Abstract

Methods for manipulating and fermenting microorganisms in multi-well plates offer unlimited possibilities for high-throughput parallel experimentation. Furthermore, bar-coded data tracking and downstream processing with modern liquid handling equipment reduce handling errors and are able to format microbial products for autosampler-equipped analytical instruments, e.g., HPLCs, mass spectrometers, and plate readers. An integrated system for high-throughput culturing of filamentous fungi replicating strains across many fermentation parameters, called nutritional arrays, was developed. It takes advantage of this equipment while addressing the age-old dilemma of how to manipulate fungal phenotypes to express a more complete spectrum of their secondary metabolites. Growth of any given strain in a well-designed nutritional array increases the chances of detecting a biologically active metabolite while reducing the manpower and materials needed for preparing individual fermentations and extracts. Fungi fermented in nutritional arrays are directly processed in a semi-automated fashion and the extracts prepared for bioassays and analytical chemistry. The necessary equipment, custom tools, and protocols to grow fungi in nutritional arrays are described along with examples of bioactive secondary metabolites discovered using this system.

Key words: Automation, Fungi, OSMAC, Secondary metabolites, Microfermentation, Microextraction, LIMS, Data mining, Drug discovery

1. Introduction

For decades, scientists analyzing data from high-throughput screening of many fungal and bacterial strains for bioactive secondary metabolites (SMs) have been aware of the unpredictability of fermentation-dependent biosynthesis of SMs. Maximizing the qualitative and quantitative metabolic output of a prospective SM-producing strain has long been a major objective of industrial screening programs searching for new applications of bioactive

Nancy P. Keller and Geoffrey Turner (eds.), *Fungal Secondary Metabolism: Methods and Protocols*, Methods in Molecular Biology, vol. 944, DOI 10.1007/978-1-62703-122-6_4, © Springer Science+Business Media, LLC 2012

small molecules in medicine and agriculture. However, only during the last decade have the genetics underlying the secondary metabolome and the evolutionary processes driving natural product synthesis been substantially clarified. It is now clear that the number of presumptive SM gene clusters in every filamentous fungal genome examined so far exceeds the number of detectable metabolites from a given species (1–3) indicating the existence of numerous unexpressed pathways possibly or likely encoding the biosynthesis of both known and heretofore undescribed (novel) secondary metabolites. Referred to as "cryptic" or "silent" gene clusters, unexpressed gene clusters might be better called "orphan" gene clusters, because their "silence" is probably an artifact caused by our inability to simulate the environmental conditions or chemical signals required for their expression (4).

The value of empirical strategies for manipulating nutritional and environmental parameters to stimulate the metabolic diversity of a microorganism has long been recognized. In industrial microbial screening laboratories, e.g., at Merck Research Laboratories, these strategies were called "media screens." Media screens, empirical comparisons of an organism's metabolic response to a large number of medium formulations, over a range of temperatures and aerations, were often the first-line approach for improving product formation and associated bioactivity from a newly discovered fungus or actinomycete. The impact of media screens was also evident during the initial fermentation of large number of new strains and testing extracts across multiple biological assays (5). Review of screening data often recognized that interesting active metabolites were formed or produced in larger quantities in only one of several media. Countless hours were spent debating the hypothetical number of media needed for getting a new metabolite from a newly isolated strain, resulting in the general acknowledgement that several growth conditions were always better than one. However, when faced with the task of manually preparing, labeling, inoculating, and extracting many flasks, the practical reality set in that the number of possible conditions for growing each organism was limited. These empirical methods were later formalized in a paradigm designated, one strain, many compounds (OSMAC) (6–8). OSMAC articulated a strategy for modulating the metabolism of poorly characterized microorganisms, often with unknown nutrition, genetics, and taxonomy. Statistical models of the effects of increased number of media on the detection of antifungal signals from extracts have confirmed that the OSMAC approach indeed increases the probability of detecting antimicrobial products from a given strain (7, 9). The major difficulty with OSMAC is that open-ended testing of fermentation media, temperatures, agitation rates, and other parameters and subsequent product analysis can be laborious and expensive. However interpretation of annotated genomic data can now inform what types and numbers of metabolites can be expected, and miniaturized fermentations and robotics can facilitate the work of finding optimal conditions for their expression.

By the early 2000s, parallel experimentation with microfermentations of bacteria and yeasts in multi-well plates (MWPs) had become widely accepted (10, 11). Effective MWP fermentation systems, pin tool inoculators, and colony picking robots became available for use with heterogeneous strain collections of aerobic bacteria and actinomycetes, and clone libraries in *Escherichia coli* and yeasts. However, when our laboratory considered implementing a new high-throughput strategy to prescreen filamentous fungi under many nutrient regimes and amplifying their ability to produce antifungal or antibacterial metabolites (Fig. 5), we found few precedents that suggested that filamentous fungi could be effectively handled in MWPs, or much less that they would produce their SMs in microplate wells. When first challenged with the idea, several complications were evident. Most fungi in agar culture lack spores and grow only as multicellular filaments and therefore could not be transferred as homogeneous microliter volumes. Mixed populations of fungi vary widely in their rates and patterns of hyphal growth. Aggressively growing species may produce enough biomass to overgrow and escape from microwells. When transferring inoculum or pipetting extracted mycelia, hyphal masses clog and interfere with pipette tips. Furthermore, orbital shaking to aerate microwell cultures would be futile because mycelia quickly grow into solid masses. Therefore, we improvised and adapted existing protocols for MWP fermentations of bacteria and actinomycetes to overcome and accommodate these peculiar behaviors of filamentous fungi, particularly those fungi that only grow as hyphae in culture (9, 10).

We adapted a commercially available spring-loaded replicator, and sandwich covers for deep well MWPs for inoculation and growth of filamentous fungi in 96-well and 24-well formats (www.enzyscreen.com; www.kuhner.com). A technique was developed for efficiently generating inoculum from fungal hyphae and transferring inocula among microwells as hyphal suspensions (Figs. 1, 2, and 3) (9). A simple screening system was established using heterogeneous groups of 80 fungi grown in 96-well format replicating hyphal suspensions across multiple media. Fungi were incubated statically to simulate solid-state fermentations (Fig. 4). Fungi could be efficiently extracted from MWPs, and metabolite production was adequate for detection of biological activity (Fig. 5). Liquid handling robots had many different incompatibilities with direct extraction in MWPs and presented significant obstacles achieving the desired throughput. These obstacles were overcome by designing a new system that allowed solvent dispensing and direct pipetting from MWPs and the use of disposable tips for extract handling. A custom tool (Figs. 8 and 9) was built to compress

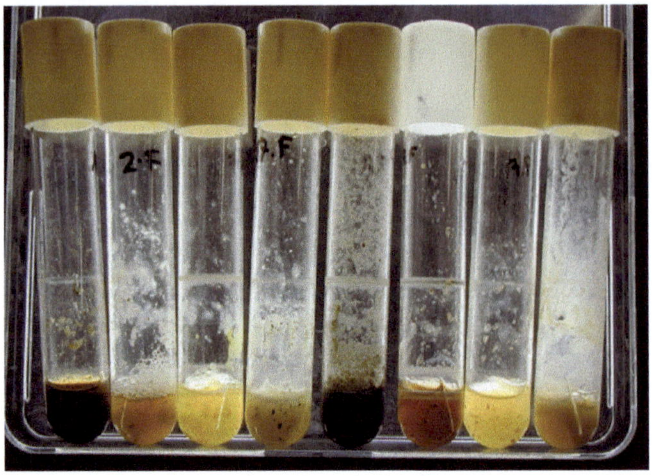

Fig. 1. Preparation of fungal nutritional arrays. Fungal hyphal suspensions prepared in tubes with cover glasses.

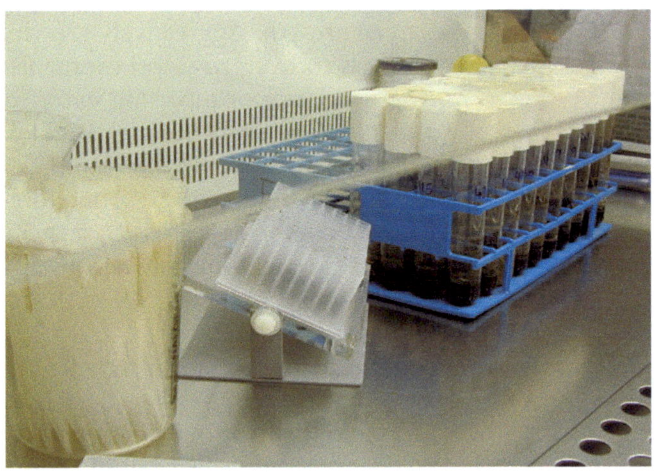

Fig. 2. Preparation of fungal nutritional arrays. Rack of hyphal suspensions organized in the same pattern as the master plate.

fungal cells to the bottom of deep well microplates enabling clean pipetting of extracts (10). Growth in 96-well plates could yield 500–1,000 μl of mycelial extract and supernatant, enough to carry out several assays, repeat them if necessary, and evaluate the extract via HPLC-MS. Active strains discovered by high-throughput screening of microplate fermentations usually can be scaled up in shaken flask or static flask cultures in order to isolate and characterize the active components.

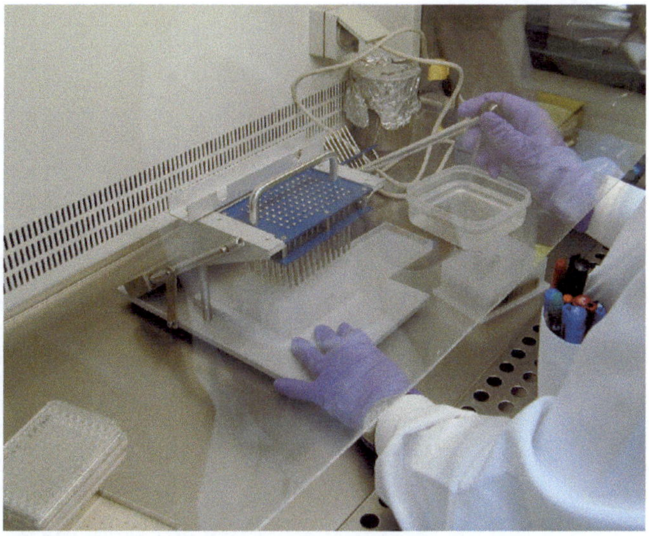

Fig. 3. Preparation of fungal nutritional arrays. Inoculating a medium plate with 80 fungi.

Fig. 4. Preparation of fungal nutritional arrays. Incubation of nutritional arrays in an incubator. Plates on *upper shelf* are closed with carpenter clamps. Plates on *bottom shelf* are closed with custom-built supports.

Because of the potential to generate masses of screening data, a suitable data-mining environment is recommended for getting the most out of the nutritional array methodology, especially for making comparisons in assay responses among fungus × medium combinations (e.g., Fig. 5). Automation of laboratory procedures effectively increases the chances of successfully implementing innovative solutions. Computer-assisted follow-up of integrated

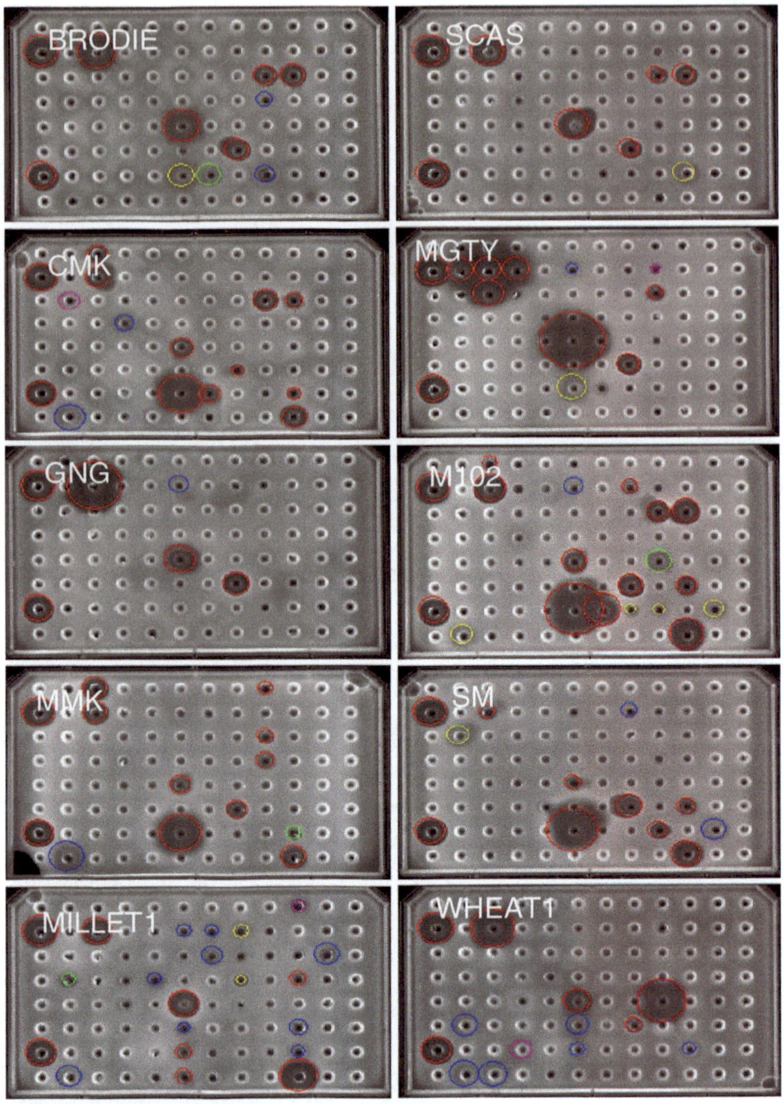

Fig. 5. Prescreening of 800 extracts from a fungal nutritional array with a zone of inhibition assay against *Staphylococcus aureus*. Eighty fungi were replicated and grown in ten different media in the central 80 wells of 96-well plates. Extracts were applied to preformed wells in Omnitray plates (Nunc) seeded with *S. aureus*. Positions second and seventh from the first columns per plate are positive controls, kanamycin and vancomycin, respectively. Images were captured by a custom image analyzer, and zones of inhibition were outlined and measured by the analysis program. Media designations are in *white* letters.

processes, given the number of strains, media, fermentations, and extracts that grow factorially with nutritional array methodology, substantially improves data tracking and interpretation. An internal computer network environment for managing users and instruments with access control and method monitoring will help to get the most of these programs. Connecting liquid handling robots and assay readers enables both storage of valuable data and direct access to processed data. These operations can be managed by determining the application needed for basic operational use of each piece of laboratory equipment, plus the design and development of specific procedures needed for each of the nutritional array protocol steps. Correct configuration of each instrument and installation of integrated data transfer solutions enable a productive integration of the different steps and the automated storage of the data for its processing and decision making.

Storage of data regarding strains, fermentation media, solvent extracts, well positions, plates, volumes, biological results, chemical results, and standard operating procedures (SOPs) that mirror each step in large-scale research programs is commonly approached by the use of databases. Our data tracking employed Oracle Database because of its robustness, simplicity, and ability to create distributive networks that facilitate integration of each component of the system and sharing of information with other research environments (users, laboratories, departments, companies, and institutes). The connectivity offered by Oracle DB allows the use of a large series of client/Web-server applications for processing, visualization, or sharing of data from the database that, in coordination with a file server, will allow effective distribution of data between the different pieces of equipment, laboratory technicians, and scientists. Security of data is also recommended with specific SOPs that ensure duplication and integrity of each data point generated from every fungus × medium combination.

Below we describe the necessary equipment, custom tools, protocols to grow fungi in nutritional arrays, and data management for recording and analyzing results. First-time users should understand that our system evolved in a high-throughput pharmaceutical screening environment; however most of the liquid handling steps and software solutions can be accomplished with simpler tools, e.g., multichannel pipettes, peristaltic pumps coupled to manifolds, and common small-business data-mining applications. We encourage anyone interested in maximizing the full potential of fungi to express their range of chemical phenotypes to experiment with nutritional arrays and design their own microfermentation experiments.

2. Materials

2.1. Labware Infrastructure and Equipment

1. Autoclave for sterilizing tools, media, and plates.
2. Climatic chamber or controlled temperature room for maintaining stable fermentation conditions. Mesophilic fungi are normally grown between 20 and 30°C. If possible, relative

humidity should be maintained at 60% or higher, especially in arid climates.

3. Freezers (–80°C or liquid N_2) for cryopreservation of fungal strains.
4. Stereo- and bright field microscopes for checking purity and identity of strains.
5. System Duetz (www.enzyscreen.com; www.kuhner.com), a set of tools developed for miniaturized parallel preservation and aerated cultivation of bacterial strains in microplates with 96-, 48-, or 24-well formats.
6. Rotary shaker (e.g., New Brunswick Scientific, USA; Kühner AG, Switzerland; or other manufacturers) with clamps for aerated culturing of inoculum and shaking platforms (5 cm displacement) with inclinable (approximately 75° angle) steel racks (25 mm diameter, 16 tubes per rack) for incubating and agitating seed inoculum tubes.
7. Level P2 biological safety cabinet for culture and media manipulations.
8. Ventilated chemical safety cabinet for the handling of volatile organic solvents, extraction of fermentations, and solvent evaporation.
9. Regulatory approved disposal system for biohazards and used chemicals.
10. Small liquid handling tools, e.g., multichannel pipettes, Thermo Multidrop Combi.
11. Liquid handling station (e.g., Tecan Genesis, Biomek FX) for the distribution of extracts in 96-well assay plates.
12. Barcode printers and readers for plate tracking.
13. Low-speed vacuum centrifuges, e.g., Genevac HT-24 (Genevac Inc.) computer-controlled vacuum centrifuge, or Savant Speedvac Plus (SC2010A) for solvent evaporation in MWPs.
14. A custom aluminum plunger that filters and squeezes the mycelia from the walls into the bottom of the plate. Alternatively, a high-speed centrifuge for MWPs can be used to partially separate fungal cells from extracts.
15. Database or spreadsheet programs for barcode labeling and tracking of strains, strain × medium combinations, plates, plate coordinates, extracts, and assay results.

2.2. Information Technologies Infrastructure and Equipment

1. Data mining hardware: Different options for servers are available, with a range of quality-to-cost ratios. We selected a specific solution (HP, www.hp.com) because of our long experience and confidence in its reliability, e.g., 4× ProLiant DL360 G5 with 2 Xeon E5405 2 GHz processors, 64 Gb of RAM

memory, 2×146 Gb hard drives for operative systems (OS) and virtualization software, plus two hard drive cabinets with 11 450 Gb discs for a total of 4 Tb scalable up to 21.6 Tb with two additional hard drive cabinets.

2. Data-mining configuration: Modern IT hardware allows the distribution of computer flows virtually among physical servers in a balanced way. Moreover, hardware power can be divided in isolated virtual servers for better performance. HTS databases should be placed, if possible, in one of these virtual divisions to reduce chances of failure and data corruption.
3. Storage of data from databases in hard drives can also follow the same paradigm of dividing physical space (e.g., 2× 450 Gb hard drives) into virtual units with redundancy of data in case of technical failure. The combination of virtual servers and storage of data results in a scalable and highly versatile programming environment that can evolve with the research projects as they evolve.

2.3. Applications and Software Solutions

1. Databases: Two databases, production and development, are commonly needed to allow the implementation of software applications and laboratory procedures while laboratory operations evolve in response to experimental results; the two databases ensure security and functional structure for the data.
2. Database Software: Oracle Database (www.oracle.com) has been our preferred solution. Oracle on Linux ensures robustness, good design, easy learning curve, and administration, and has the largest capabilities of online maintenance and interaction with other external or previous historical databases.
3. Laboratory Information Management System (LIMS): LIMS are central applications essential for tracking and recording experiments and related manipulations (sample preps, equipment quality control, etc.) involved in all steps of a research process (Fig. 6). We created an internal LIMS (Culture Tracking System or CTS) with a nutritional array-specific module for supporting this complex fermentation scheme and other programs related to the fermentation of strains. Other commercial options, e.g., ActivityBase (IDBS, www.idbs.com) or Nautilus (Thermo Scientific, www.thermo.com), oriented, respectively, to chemical or biological laboratories, also include a complete series of modules for tracking, calculations, and statistics.
4. Physical Tracking and Documentation: Electronic databases efficiently maintain information, but data input can be a time-consuming task and becomes an important handicap for the implementation of any high-throughput system. Barcode labeling software (e.g., Bartender, Seagull Scientific, www.seagullscientific.com) is recommended for the correct

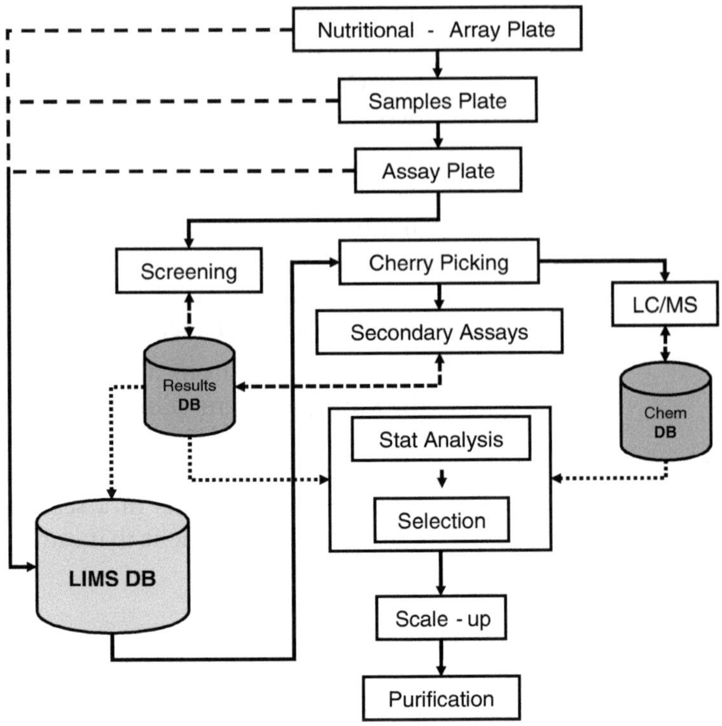

Fig. 6. LIMS system for centralizing data, workflow tracking, and registering inocula, fermentations, and extracts for high-throughput generation of nutritional arrays. The configuration enables sample tracking, hit selection and comparison, cherry-picking of actives for confirmation and follow-up, and scale-up selection.

identification (human and robot reading) of inocula, fermentations, extractions, and assay plates created by the LIMS system.

5. Visualization tools and statistics: Data-mining applications are needed to summarize and extract information from the extensive raw data. Statistical packages can find patterns among serendipitous data and can discern and control experimental errors. Visualization of data (e.g., Tibco Spotfire, www.tibco.com) can aid in comparisons between fermentation media for a given population of strains, identification of unique media for the production of new SMs, or simple evaluation of the number and type of strains, fermentations, extracts, or actives from a nutritional-array program.

2.4. Ingredients and Reagents

1. Petri dishes and agar media appropriate for fungi, e.g., malt extract agar, potato-dextrose agar. Omnitray plates (Nunc) for sterility controls.
2. Yeast extract medium (YM): Malt extract 10 g; yeast extract 2 g; bacteriological agar 20 g; distilled H_2O 1 l.

3. Sabouraud's maltose yeast extract medium: Maltose 40 g; yeast extract 10 g; neopeptone 10 g; agar 3.5 g; distilled H_2O 1 l.
4. Cryovials with sterile 20% glycerol and storage boxes for culture preservation.
5. Automatic and disposable pipettes, Transfer tubes (Spectrum Laboratories, Rancho Dominguez, CA).
6. Adhesive labels and marking pens.
7. 25 × 150 mm glass tubes with closures.
8. 22 mm long glass covers.
9. Deep-well polypropylene 96-well (2.4 ml total volume) or 24-well plates. Steel lids, breathable membranes, and silicone closures for plates (see System Duetz).
10. 250-ml Erlenmeyer flasks for medium aliquots.
11. Ingredients for media: Media selection will depend on the kinds of fungi grown and the investigator's experience and preferences. Because one of the goals of the technique is to take advantage of automated liquid handling, generally media with defined or soluble complex ingredients are used to facilitate medium dispensing.
12. Solvents for extraction: Historically acetone has been used, but other water-miscible solvents could work as well. DMSO for suspension of dried extracts.

3. Methods

3.1. Inoculum Preparation, Replication, and Growth of 80 Fungi in MWPs

1. Select sets of 90–100 strains of newly isolated fungi or fungi from a culture collection and grown in YM agar for 2–3 weeks in 60 mm Petri dishes. Avoid heavily sporulating species with dry airborne spores (e.g., *Aspergillus* spp.) or species with extremely fast radial growth (e.g., *Trichoderma* spp.) to minimize the possibility of cross-well contamination (see Note 1).
2. Prepare seed medium in 25 × 150 mm glass tubes. Add two 22-mm-long cover glasses to each tube before adding medium. Agitation on an orbital shaker causes the cover glasses to continually shear hyphae and mycelial pellets resulting in a homogeneous hyphal suspension (Fig. 1).
3. Dispense 8 ml of melted and homogenously stirred Sabouraud's maltose yeast extract medium with 0.35% agar into each tube.
4. Cap tubes and autoclave.
5. Inoculate the tube with two 5-mm diameter agar discs cut from each culture with a Transfer Tube (Spectrum Laboratories). The Transfer Tube's plunger is pushed against the bottom of

the tube to mash and extrude the agar discs against the glass; this increases the number of mycelial growing points.

6. Agitate tubes at 200 rpm, 5 cm displacement, for 3–6 days, where most strains will exhibit moderate to dense growth. However, tubes that appear to the unaided eye to have little growth, will, upon microscopic inspection, often reveal a dense suspension of fine hyphal fragments.

3.2. Preparation of Master Plate for Replication

1. Select 80 strains from among sets of 100 tubes, with adequate growth for the preparation of the master plates for replication (Figs. 1, 2, and 3).
2. Use a sterilized master plate for holding the hyphal inoculum for replication. Use the Sterilized Transfer Tubes to transfer 1 ml of hyphal suspensions to the wells (see Note 1).
3. Additional aliquots of each inoculum tube in 20% glycerol in 1.8 ml cryovials at −80°C can be frozen after preparation of the master inoculum plate for easy recovery of screening actives.
4. Close the master plate after loading, until replication in fermentation plates is needed.

3.3. Preparation of Fermentation Media in MWPs

1. Prepare the fermentation media in advance, in aliquots of 100 ml in Erlenmeyer flasks.
2. Use 80 ml of the aliquots to fill the 80 central wells of each deep-well growth plate at 1 ml per well (see Note 2).
3. The first and last columns of the plates are left empty, so those positions can be used for evaluation of assay-positive and -negative controls. Media formulations usually consist of soluble components. See Note 3 for solid media recommendations.
4. Close fermentation media plates with the sandwich lids.
5. Autoclave the assembled plates with media and then store them in a refrigerator until use.

3.4. Replication of Master Plates Across the Nutritional Array

1. Sterilize the cryoreplicator pin tool by immersing the pins in 70% ethanol for 1 min and then evaporating the ethanol on a hot plate.
2. Once thoroughly cooled insert the cryoreplicator pins several mm into the hyphal suspensions of the master plate and then raise them.
3. Switch the master plate for a medium-filled growth plate, and lower the inoculum-covered pins to the bottom of the liquid medium.
4. Switch the growth plate again for the master plate, and repeat the inoculation three to five times. Repeated inoculation from the master plate ensures that sufficient fungal cells are transferred to initiate growth.

5. Repeat the process for each different medium plate. Sterilization of the cryoreplicator during multiple transfers from the same master plate is unnecessary because the cryoreplicator press prevents accidental lateral movements, and the inoculum source for each well remains constant.
6. Change cryoreplicators and sterilize them between consecutive master plate inoculations.

3.5. Incubation

1. Place plates statically at about an 85° angle in a custom-built steel support (Fig. 3). See Note 4 for alternative incubation methods. The support's thumbscrews seal the silicone layer preventing cross-well contaminations. Inclined incubation increases the surface area of the liquid media and improves aeration.
2. Incubate plates for two- to three-week growth cycles at 22°C.

3.6. Contamination and Viability Check

1. Fill Omnitray plates with YM inoculum agar and inoculate them from the same master plate incubated at 22°C.
2. Fill a second Omnitray plate with Luria broth agar (from Sigma) and inoculate it.
3. Incubate both plates at 28°C to check for bacterial contamination. Replication of master plates onto agar is used in this case to verify that each well contains viable inoculum and that strains are not contaminated.

3.7. Extract Preparation

1. After growth, open plates and inspect them for contamination.
2. Replace the sandwich cover's 96-hole silicone mat with a solid silicone mat, removing the breathable cotton layer.
3. Dispense a volume of 850 μl of acetone (or other water-miscible solvent) into each fungal culture (see Note 5).
4. Gently dislodge mycelia adhering to the well walls. Crush and mix mycelial masses with the help of the cryoreplicator tool.
5. Reassemble and clamp Sandwich covers with the solid silicone mats to a shaker board and agitate for 1 h. After about 30 min, reverse the shaker orientation to change the direction of agitation.
6. Without the Sandwich covers, reduce the culture-solvent mixture up to its 50% of original volume by vacuum evaporation in a Genevac HT-24 for 1 h (Fig. 7).
7. After solvent removal, approximately 500 μl of aqueous extract remains per well. Plates are centrifuged to settle the mycelia (3,800 rpm, 4036 g, 20 min).
8. Add 3–4 h in a chemical hood after vacuum centrifuge evaporation of organic solvent to achieve a more homogeneous and complete solvent evaporation.

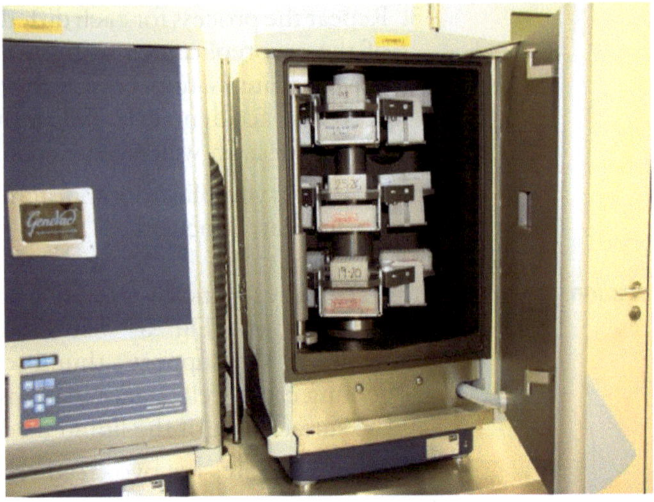

Fig. 7. Extracts' preparation from fungal nutritional arrays. Plates with solvent and mycelia supernatants in GeneVac vacuum evaporator.

9. To facilitate extract pipetting from growth plates, a custom-built aluminum 96-well hollow plunger (Figs. 8 and 9) can be used to compress mycelium to the bottom of wells while the extract and supernatant rise in the plunger's hollow columns. The hollow center of the plunger columns and open windows at the sides prevent spilling and overflow of the extracts and eliminate mycelial interference during pipetting with a liquid handling robot.
10. Transfer a total of 500 μl of the aqueous-based supernatant from each well of the original 2-ml fermentation MWPs to a 500–800 μl well plate with a Beckman 96 head Biomek FX robot (Fig. 9). This machine was selected for its ability to handle with a conveyor belt the 2-ml heavy plates with the metal plungers from a Cytomat Carrousel to the pipetting platform. Fly-by barcode readers, one for the source plates and a different one for the destination plates, are convenient for cross-checking of source and destination plates (see Note 6 for alternative extraction methods for growth media with insoluble components).
11. Store the samples at −20°C until assay by creating the library of potential sources of new SMs for screening.
12. Before assaying, thaw sample plates (Fig. 10) and briefly shake them for 20 min on a plate mixer.

3.8. Structure of the Data

Data mining of nutritional array experimentation can be a limiting step depending on the dimensions of the experimental data. Small projects can be handled with a calculus spreadsheet and a picture

Fig. 8. Extracts' preparation from fungal nutritional arrays. Aluminum plungers used to compress mycelium in deep well plates.

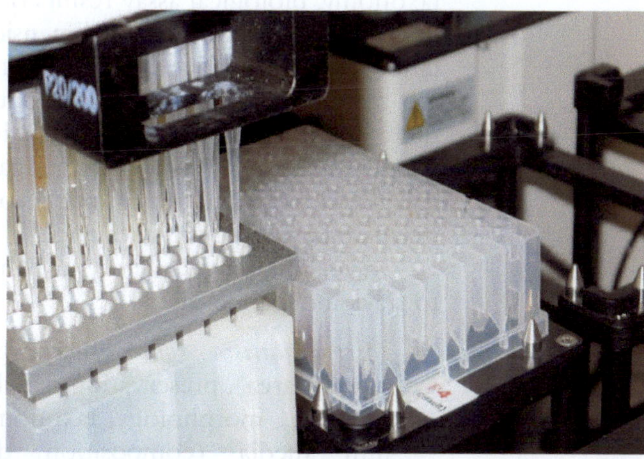

Fig. 9. Extracts' preparation from fungal nutritional arrays. Pipetting the aqueous phase from inside the metal plunger holes into assay blocks.

of the assay plates (e.g., Fig. 5); larger projects will need an IT database development environment as described in the infrastructure section. Resources in each lab will need to evaluate and determine the quantity of data that will be useful to confirm an experimental hypothesis.

In practice, development of nutritional array methodologies would resemble expanding the secondary metabolite production dimension from fermentations of a given group of strains. The SM production will be evaluated by different methodologies such as

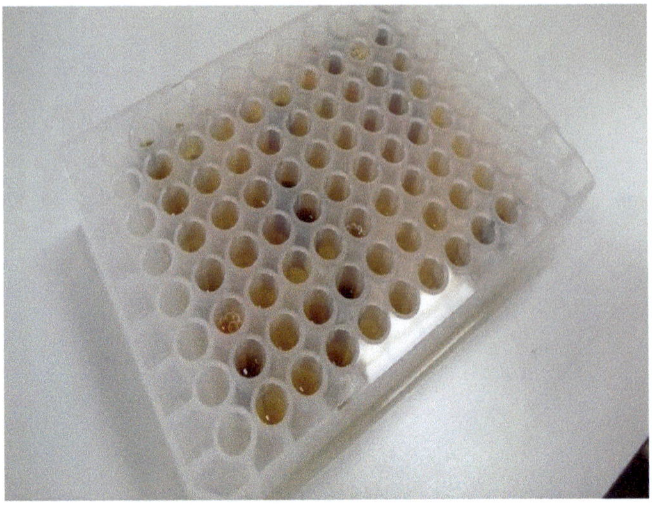

Fig. 10. Extracts' preparation from fungal nutritional arrays. Extract sample plate ready for bioassay. Columns 1 and 12 are used for assay controls.

taxonomy, biological assay results or chemical evaluation (Fig. 6); in reality data becomes multidimensional as the number of fermentations becomes wider and a number of analyses become available. No matter whether the infrastructure available is small or large, there are several pieces of information that should be kept independent and open for later interrogation of the data. Experiments will afford more and better conclusions if the raw data is kept organized and available for querying.

In general, for each fermentation batch, we suggest, if possible, to track independently:

1. *Microorganism code*, microorganism substratum (source, county, area), preservation details, microorganism characteristics (e.g., morphology, taxonomy, taxonomy details), inoculum media (composition details), physical inoculum fermentation parameters (e.g., temperature, days, shaken, light, etc.), fermentation media (composition details) and physical parameters (e.g., temperature, length of cycle, agitation rate, light, etc.), fermentation SOP.
2. *Extraction code*, part of the fermentation extracted (e.g., crude, mycelium, pellet, etc.), kind of extraction (e.g., liquid/liquid partition, solid/liquid extraction, solid-phase extraction, etc.), extraction solvent, final sample characteristics (solvent, concentration, aliquot number, volume), extraction SOP.
3. *Plate code*, format (e.g., 24 wells, 96 wells, 384 wells, etc.), plate layout (e.g., 80 central positions with extracts leaving the first and last column empty for biological controls), coordinates

of the extract in the plate (row, column), volume per well of extract, storage history (freeze–thaw cycles), sample management SOP.

4. *Biological data per each assay performed*: Assay test, replicate, response (e.g., 19 mm halos), response compared to controls (e.g., % vs. controls), characteristics of the response (e.g., turbidity, definition), biological evaluation SOP.
5. *Chemical data for each analysis*: Chemical test (e.g., LC, MS, LC/MS), characteristics of the test, detailed result in a database format, chemical evaluation SOP.
6. *Other* evaluation dimensions possible.

Depending on the questions that need to be answered from the nutritional array fermentations, the structure of the database will need to be elaborated. Simple questions only will need simple data tracking, but with the situation of fermenting 80 strains with several fermentation media, with several biological or chemical evaluations, the better the data mining tools, the more answers we will be able to extract from the same experimental effort.

3.9. Conclusion

During the 2 years that the nutritional array program was fully operative in our laboratory, more than 163,000 extract samples were prepared from fungi; from those crude extracts 3,900 fungal fermentations were selected for scale-up due to their suitable antibiotic and metabolite profiles (12). The resulting collection of prescreened bioactive samples constituted one of the most valuable collections of natural product extracts generated within an industrial environment (12, 13) and copies of those extracts remain available at Fundación MEDINA (www.medinaandalucia.es).

In summary, we find that growth performed in 96 or 24 well plates allow for creative and high-throughput experimentation with fungi that otherwise would be resource-limited and cumbersome with shake-flask systems. Growth among phylogenetically related strains can be quickly compared under identical conditions for their production of bioactive metabolites (12, 14–16). Nutritional arrays enabled the prescreening of thousands of fungal strains for medium conditions producing antifungal and antibacterial metabolites; identification of the optimal culture medium facilitated scale-up for chemical genetic profiling in genome-wide *Candida albicans* and *Staphylococcus fitness* tests (12, 13). Profiling of some of these extracts, first generated at 1-ml volumes, led to the identification of new fungal metabolites with novel mechanisms of bacterial and fungal antibiosis (13, 17, 18). A convincing example of the potential of nutritional arrays was the discovery of fellutamides C and D from an unknown *Metulocladosporiella* sp. (13, 19). The *Metulocladosporiella* strain's antifungal phenotype was revealed during a validation experiment comparing the antifungal activity from a set of 320 fungi grown in typical vial fermentations on three media to that from the same fungi grown in a 12-media nutritional array. From the entire set of 15 media, only one extract from a sucrose and yeast extract medium in the nutritional array was antifungal. A larger set of samples was generated by growing multiple wells of the fungus; components of its organic extract were determined to affect genes for the 20S subunit of the *C. albicans* proteasome, a novel mode of action for an antifungal agent. The biosynthesis of fellutamides C and D was so specific to the medium that it was necessary to mimic the 1-ml microfermentation by growing the fungus statically in a thin layer of the same medium in Fernbach flasks for the production of the compounds (19).

4. Notes

1. Screening applications that need to use heavily sporulating species with dry airborne spores or with very fast growth should consider inoculation patterns that skip alternate rows or wells for avoiding cross-growth contamination.
2. Adding in the medium more agar than indicated reduces mycelial pelleting and aids in adherence of mycelial fragments to the replicator tips.
3. Growth media with insoluble components or precipitates (e.g., $CaCO_3$ or starch) can be used with this system with a magnetic stirrer during dispensing. Seed-based solid media can be prepared by filling an MWP with 200-μl wells with seeds (wheat, rice, millet, cracked corn), leveling the seed depth to the tops of the wells. The open wells of an inverted growth plate can be aligned on top of the seed-containing microtiter plate, and the two can be quickly flipped, so that a measured seed volume drops into each well. Finally add a liquid basal medium (700 μl) to the seeds. See Note 6 for appropriate extraction.
4. An alternative method for plate incubation can be improvised by holding the sandwich covers in place on growth plates with heavy rubber bands or with carpenter's wood clamps that clamp lids securely to growth plates, and then carefully tilting the plates on the edge of a 4-mm high edge (e.g., two wooden tongue depressors).
5. DMSO can be added during the extraction process to minimize metabolite precipitation during solvent evaporation. A final 20% of DMSO in the sample vehicle (water) is compatible with most of the cell-based assays.
6. Description of extracting protocol covers most of the fungi fermentations. A variant of the protocol was developed in

response to the presence of excess of starch in the media when processing solid seed-based fermentations (2.5% of our total extracts). In the case of solid fermentations (see Note 3), it is more convenient to add solvents, shake, centrifuge, pipette the supernatant, and evaporate the extracts in an additional barcoded 2-ml deep-well plate before their final transfer to the 500–800 μl extract well plate rather than evaporating solvents in the presence of the mycelia.

Acknowledgements

We would like to express our gratitude to all our former coworkers and colleagues who participated in the development, validation, and improvement of the nutritional array method.

References

1. Khaldi N, Seifuddin FT, Turner G, Haft D, Nierman WC, Wolfe KH, Fedorova ND (2010) SMURF: genomic mapping of fungal secondary metabolite clusters. Fungal Genet Biol 47:736–741
2. Brakhage AA, Schroeckh V (2010) Fungal secondary metabolites - strategies to activate silent gene clusters. Fungal Genet Biol 48:15–22
3. Hoffmeister D, Keller NP (2007) Natural products of filamentous fungi: enzymes, genes, and their regulation. Nat Prod Rep 24:393–516
4. Gross H (2007) Strategies to unravel the function of orphan biosynthesis pathways: recent examples and future prospects. Appl Microbiol Biotechnol 75:267–277
5. Yarbrough GG, Taylor DP, Rowlands RT, Crawford MS, Lasure LL (1993) Screening microbial metabolites for new drugs - theoretical and practical issues. J Antibiot 46:535–544
6. Höfs R, Walker M, Zeeck A (2000) Hexacyclinic acid, a polyketide from *Streptomyces* with a novel carbon skeleton. Angew Chem 39:3258–3261
7. Bode HB, Bethe B, Höfs R, Zeeck A (2002) Big effects from small changes: possible ways to explore nature's chemical diversity. Chembiochem 3:619–627
8. Scherlach K, Hertweck C (2006) Discovery of aspoquinolones A-D, prenylated quinoline-2-one alkaloids from *Aspergillus nidulans*, motivated by genome mining. Org Biomol Chem 4:3517–3520
9. Bills G, Platas G, Fillola A, Jiménez MR, Collado J, Vicente F, Martín J, González A, Bur-Zimmermann J, Tormo JR, Peláez F (2008) Enhancement of antibiotic and secondary metabolite detection from filamentous fungi by growth on nutritional arrays. J Appl Microbiol 104:1644–1658
10. Duetz W, Chase M, Bills G (2010) Miniaturization of fermentations. In: Demain A, Davies J, Baltz R (eds) Manual of industrial microbiology and biotechnology, 3rd edn. ASM, Washington, DC, pp 99–116
11. Duetz WA (2007) Microtiter plates as minibioreactors: miniaturization of fermentation methods. Trends Microbiol 15:469–475
12. Bills GF, Martín J, Collado J, Platas G, Overy D, Tormo JR, Vicente F, Verkleij G, Crous P (2009) Measuring the distribution and diversity of antibiosis and secondary metabolites in the filamentous fungi. Soc Ind Microbiol News 59:133–147
13. Roemer T, Xu DM, Singh SB, Parish CA, Harris G, Wang H, Davies JE, Bills GF (2011) Confronting the challenges of natural product-based antifungal discovery. Chem Biol 18: 148–164
14. Vicente F, Basilio A, Platas G, Collado J, Bills GF, Del Val AG, Martin J, Tormo JR, Harris GH, Zink DL, Justice M, Kahn JN, Pelaez F (2009) Distribution of the antifungal agents sordarins across filamentous fungi. Mycol Res 113:754–770
15. Bills GF, Platas G, Overy D, Collado J, Fillola A, Jiménez MR, Martín J, del Val AG, Vicente F, Tormo JR, Peláez F, Calati K, Harris G, Parish C, Xu D, Roemer T (2009) Discovery of the parnafungins, antifungal metabolites

that inhibit mRNA poly-adenylation, from the *Fusarium larvarum* complex and other hypocrealean fungi. Mycologia 101:445–469

16. Peláez F, Collado J, Platas G, Overy DP, Martín J, Vicente F, González del Val A, Basilio A, de la Cruz M, Tormo JR, Fillola A, Arenal F, Villareal M, Rubio V, Baral HO, Galán R, Bills GF (2010) The phylogeny and intercontinental distribution of the pneumocandin-producing anamorphic fungus *Glarea lozoyensis*. Mycology 2:1–17
17. Ondeyka J, Harris G, Zink D, Basilio A, Vicente F, Bills G, Platas G, Collado J, González A, de la Cruz M, Martin J, Kahn JN, Galuska S, Giacobbe R, Abruzzo G, Hickey E, Liberator P, Jiang B, Xu DM, Roemer T, Singh SB (2009) Isolation, structure elucidation, and biological activity of virgineone from *Lachnum virgineum* using the genome-wide *Candida albicans* fitness fest. J Nat Prod 72:136–141
18. Herath K, Harris G, Jayasuriya H, Zink D, Smith S, Vicente F, Bills G, Collado J, Gonzalez A, Jiang B, Kahn JN, Galuska S, Giacobbe R, Abruzzo G, Hickey E, Liberator P, Xu DM, Roemer T, Singh SB (2009) Isolation, structure and biological activity of phomafungin, a cyclic lipodepsipeptide from a widespread tropical *Phoma* sp. Bioorg Med Chem 17:1361–1369
19. Xu D, Ondeyka J, Harris GH, Zink D, Nielsen-Kahn J, Wang H, Bills G, Platas G, Wang W, Szewczak AA, Liberator P, Roemer T, Singh SB (2011) Isolation, structure and biological activities of fellutamides C and D from an undescribed *Metulocladosporiella* (Chaetothyriales) using the genome-wide *Candida albicans* fitness test. J Nat Prod 74(8):1721–1730. doi:10.1021/np2001573

Chapter 5

The "FERMEX" Method for Metabolite-Enriched Fungal Extracts

Gerald F. Bills, Anne W. Dombrowski, and Michael A. Goetz

Abstract

The FERMEX (*Ferm*entation *Ex*tract) program was a highly successful source of microbial natural product molecules for pharmaceutical lead discovery. The program was based on the observation that solid fermentations of fungi generally exhibited more complex metabolite profiles than when the same strains were grown on liquid medium. To produce interference-free fermentations and improve organic product recovery, fungi colonized homogeneous media-saturated vermiculite thus promoting cellular and metabolic differentiation. Secondary metabolites in fungal cells were extracted from the substratum and medium with methyethylketone to generate metabolite-enriched screening samples. The necessary equipment, protocol, and media recipes are described along with examples of bioactive fungal metabolites produced in this system.

Key words: Fungi, Secondary metabolites, Solid-state fermentation, Drug discovery

1. Introduction

1.1. The FERMEX Method

The FERMEX (*Ferm*entation *Ex*tract) program was one of Merck Research Laboratories (MRL) most successful sources of natural product lead molecules when measured in terms of its manpower input per successful chemistry isolation project and its ability to consistently obtain and reproduce novel compounds. FERMEX operated at MRL in Rahway, New Jersey during the late 1980s and throughout the 1990s and produced fungal fermentations generated exclusively in solid media for primary screening in assays at Rahway, New Jersey and West Point, Pennsylvania. The program was based on the observation that solid fermentations of fungi generally exhibited more complex

Nancy P. Keller and Geoffrey Turner (eds.), *Fungal Secondary Metabolism: Methods and Protocols*,
Methods in Molecular Biology, vol. 944, DOI 10.1007/978-1-62703-122-6_5, © Springer Science+Business Media, LLC 2012

Fig. 1. The FERMEX method. Coarse particle vermiculite.

metabolite profiles than when the same strains were grown on liquid medium. Intuitively this makes sense because fungi are terrestrial organisms that penetrate, hydrolyze, and consume their substrata. However, a common problem associated with fungi grown on solid media, like agar, rice, and corn, was that often large amounts of solid material and plant metabolites would be extracted along with the fungal metabolites, diluting potency, complicating purification of fungal metabolites, and sometimes causing assay interference.

To produce "clean" solid-state fermentations (SSFs) and improve product recovery, natural matrices (rice, corn, wheat) were replaced by vermiculite (Fig. 1), an inert expanded clay mineral. The vermiculite was saturated with a defined or semidefined liquid nutrient medium to provide nutrients. To accelerate the penetration of the fungus and ensure its thorough colonization of the vermiculite, the medium–vermiculite mixture was incubated in glass tissue culture roller bottles that were rotated on cell culture roller bottle machines (Fig. 2). When the liquid to vermiculite ratio is optimized, the vermiculite forms a moist layer around the inside of the bottles. As the bottle rolls, excess liquid medium is converted to fungal biomass that penetrates and binds together the vermiculate particles (Figs. 3, 4, and 5). The fungus forms a solid matrix around a large interior surface open to the air in the bottle's central chamber stimulating sporulation (for those fungi that sporulate); a high degree of cellular differentiation is achieved corresponding to differential expression of secondary metabolite pathways within the same culture. The hyphal mass produced in roller bottle SSFs is easily broken apart and extracted to produce

Fig. 2. The FERMEX method. Roller bottles (2 l) on roller bottle machine. Note cellulose closures fastened in place with gauze and rubber bands.

Fig. 3. Autoclaved vermiculite bottle (1 l) before inoculation and growth of *Emericella parvathecia* (14 days, 22°C).

Fig. 4. The FERMEX method. Mycelium and vermiculite matrix formed by *Trichoderma virens* on roller bottle wall.

Fig. 5. The FERMEX method. Bottom view of roller bottles showing hollow central chamber. *Emericella navahoensis* (*left*) and *E. parvathecia* (*right*).

about 200–500 ml of organic extract from which small amounts are aliquoted into screening plates and distributed to assays; the rest is banked as a backup for initial analysis of biologically active components.

Below we describe how the FERMEX method was implemented to support MRL drug discovery programs and the necessary equipment, protocols, and media recipes. We list some examples of newly discovered fungal metabolites or known fungal metabolites that were prepared for various biology and chemistry programs at MRL.

1.2. History and Implementation

By the mid-1970s, the use of microbial fermentations in the discovery of bioactive natural products at Merck Research Laboratories and other pharmaceutical companies had focused primarily on antiparasitic, antibacterial, and antifungal targets assaying aqueous extracts or whole broth samples. These techniques were quite successful and remain perfectly valid to this day. With the advent of laboratory automation and high throughput mode-of-action assays, e.g., enzyme inhibition, receptor binding, and whole cell functional measures addressing disease states at the molecular level, traditional types of fermentation samples were explored in these more "delicate" systems. As was anticipated, and indeed borne out with experimentation, the raw microbial samples used for assay would often be too crude and resulted in false positives and uninterpretable results caused by salts, proteins, and spent media components.

Therefore, a fermentation sample preparation method was sought that circumvented these problems. It was decided to use the methods on fermentations of carefully selected microorganisms to ensure distinctive chemistry, taking into consideration that the high throughput of that period in the 1980s was on the order of 20–40 microorganisms per week rather than the screening of hundreds of microbes in the more classical programs.

Early successes were striking: mevinolin and dihydromevinolin, the first HMG CoA reductase inhibitors and which spawned the first generation of statins for the management of hypercholesterolemia, appeared literally in the first month of this new "Fermentation Products for Screening" program (1); shortly later came asperlicin, the first non-peptide antagonist to a peptide hormone discovered intentionally (2, 3).

This combined approach of preselecting distinct microorganisms and growing them to produce a cleaner organic solvent extract compatible with biochemical assays eventually evolved, as described in the following sections, into the FERMEX program, one of the most successful sources of natural product lead molecules at Merck Research Laboratories.

1.3. Biological and Chemical Basis of FERMEX Method

The advantages and benefits of using SSF for producing fungal biomass, fungal spores, enzymes, and pigments are well known. SSF processes are equally important in transforming solid materials (grains, milk products) into foodstuffs or decolorizing, bleaching, or detoxifying complex polymeric materials. SSF has been emphasized as a low technology method for converting biological and agricultural waste into high value chemical products. The design of SSF fermentors and their applications and the advantages of immobilizing cells in inert support materials for SSF have been reviewed extensively (4–6). The FERMEX method is based on a simple rotating drum bioreactor design using standard animal cell culture roller bottles and rolling machines (Fig. 2, Bellco, Vineland, NJ, USA). Unlike SSF on natural materials, composition of production media could be carefully controlled and reproduced because all the media components were known and could be analyzed.

The advantages of SSF for fungal secondary metabolite production are intuitive. Many mycotoxins are produced in solid substrata, including aflatoxin, ochratoxin, and fumonisins. The use of SSF for antibiotic production traces its roots all the way back to some processes implemented during the development of penicillins and can be used at the production scale (4, 7), but as a discovery tool in the pharmaceutical industry, the idea never gained much traction. Likely, the reasons were needs to develop economically viable large-scale processes in stirred tank fermentors where extraction and purification of active ingredients could be carried out with liquid processes. Results from the FERMEX method started emerging in the early 1990s (Table 1). Subsequently, other research groups independently devised similar SSF methods with inert carriers, e.g., light expanded clay aggregates (8) and membrane immobilized mycelia (9), and have reaffirmed the profound effects of solid phase growth on secondary metabolite expression relative to liquid cultures. The genetic basis for differential expression is only now starting to be understood. Transcriptional analysis of *Aspergillus oryzae* demonstrated that SSF significantly upregulated the ratios of expression of genes from non-syntenic blocks relative to those from syntenic blocks (10). These non-syntenic blocks are enriched in genes for hydrolytic enzymes and secondary metabolism. Characteristics of gluconeogenesis in SSF were observed in the presence of a nutrient-rich medium, and it was proposed that lower expression of genes in the glycolytic and tricarboxylic pathway might avoid catabolite repression (11).

Early efforts explored a variety of alternatives to find the most suitable processing protocol for this new screening program. A first approach was to proceed via adsorption of whole broth onto reverse-phase resins, e.g. Amberlite XAD-2, followed by washing and solvent elution of screenable mixtures. A second period saw the rather extreme extraction with only ethyl acetate. These methods proved to be either too cumbersome for any reasonable weekly throughput or too incomplete in extractive efficiency. Further experimentation included extraction with *n*-butanol for liquid fermentations, methanol or various solvent mixtures for solid substrate fermentations. Based on its effectiveness at extracting a wide range of polarities from both liquids and solids, its convenience of use and reasonable cost, methyethylketone (MEK) eventually became the standard extraction solvent for the FERMEX program.

Table 1
Some significant fungal secondary metabolites produced by the FERMEX method

Metabolite(s)	Fungus	Application(s)	Reference(s)
Thysanone	*Thysanophora penicilloides*	Human rhinovirus 3C-protease inhibitor	(22)
Variculanol	*Emericella variecolor*	Toxicity to *Eimeria tenella*	(23)
Ophioblin K and 6-epiophiobolin K	*Aspergillus ustus*	Toxicity to *Caenorhabditis elegans*	(24)
L-696,474 (18-dehydroxy cytochalasin H)	*Hypoxylon fragiforme*	HIV-1 protease inhibitor	(18, 25–27)
3-(Dimethylaminomethyl)-1-(1,1-dimethyl-2-propenyl)indole	*Penicillium daleae*	5-HT_{1D} receptor agonists	(28, 29)
Calbistrin A	*Aspergillus versicolor*	Broad spectrum antifungal, increased human LDL receptors	(30, 31)
Maxikdiol	Unidentified coelomycete MF5717	Agonist of maxi-K (large-conductance calcium-activated potassium) channels	(32)
Apicidins	*Fusarium* spp.	Broad spectrum antiparasitic agents, histone deacetylase inhibitors	(33–35)
Ascosteroside	*Mycoleptodiscus atromaculans*	Antifungal, β-1,3-glucan synthase inhibitor	(36)
Sporandol	*Chrysosporium merdarium*	Insecticide and antiparasitic agent	(37, 38)
Equisetin	*Fusarium heterosporum*	HIV-1 integrase inhibitor	(39, 40)
Phomasetin	*Phoma* sp.	HIV-1 integrase inhibitor	(39, 40)
Cytosporic acid	*Cytospora* sp.	HIV-1 integrase inhibitor	(41)
Nodulisporic acids	*Nodulisporium* sp.	Insecticide, invertebrate-specific glutamate-gated chloride ion channels	(42–44)
Hyalodendrosides A and B	*Hyalodendron* sp.	Antifungal, β-1,3-glucan synthase inhibitor	(45)

(continued)

Table 1
(continued)

Metabolite(s)	Fungus	Application(s)	Reference(s)
Nalanthalide	*Chaunopycnis pustulata*	Blockage of voltage-gated potassium (Kv1.3) channels and L-type Ca^{2+} channels	(46)
Candelalides	*Sesquicillium candelabrum*	Blockage of voltage-gated potassium (Kv1.3) channels	(47)
Dihydrotubingensin A, B	*Aspergillus tubingensis*	No activity	(48)
Sordarins	*Rosellinia subiculata*	Antifungal, elongation factor 2 inhibitor	(19)
Integramides A and B	*Dendrodochium* sp.	HIV-1 integrase inhibitors	(49)
Xylarenal A and B	*Xylaria liquidambar*	Selective binding to neuropeptide Y receptor Y5	(20)
Xanthonol	Non-sporulating fungus MF6460	Insecticidal activities against larvae of *Lucilia sericata*, *Aedes aegypti*, and anthelmintic against *Haemonchus contortus*	(50)
Coccidiostatin A	*Penicillium rugulosum*	Antiparasitic activity against *Besnoitia jellisoni*, *Eimeria tenella*, and *E. mitis*	(51)
Mellamide	*Aspergillus melleus*	Insecticidal activity in bioassays against larvae of *Lucilia sericata* and *Aedes aegypti*	(52)

2. Materials

2.1. Source of Strains

To obtain strains for screening, fungi were isolated at Rahway, New Jersey from living plants (12, 13), plant litter (14, 15), herbivore dung (16), soils (17), spores from field-collected fungi (18–20), and other materials (21). These sets of fungi were supplemented with strains deposited in the Merck Culture Collection and strains supplied from collaborating laboratories. Strains spanned most of the major groups of cultivable filamentous Ascomycota and Basidiomycota.

2.2. Infrastructure and Equipment

1. Autoclave with capacity to autoclave large tissue culture bottles or large media bottles.
2. Climatic chamber or controlled temperature room for maintaining stable fermentation conditions. Mesophilic fungi are normally grown at between 20 and 30°C. If possible, relative humidity should be maintained at 60% or higher, especially in arid climates.
3. Freezers (–80°C or liquid N_2) for cryopreservation of fungal strains.
4. Stereo- and bright field microscopes for checking purity and identity of strains.
5. Tissue culture roller machine (e.g., Fig. 2, Bellco bottom drive roller apparatus, Bellco Glass, Vineland, NJ, USA). The number of positions of the machine depends on the intended throughput. Our machines have 45 positions.
6. Rotary shaker with clamps for 250-ml Erlenmeyer flasks (e.g., New Brunswick Scientific, USA, Kühner AG, Switzerland or other manufacturer) for aerated culturing of inoculum.
7. Level P2 biological safety cabinet for culture and media manipulations.
8. Ventilated chemical safety cabinet for handling of volatile organic solvents, extraction of fermentations, and solvent evaporation.
9. Regulatory approved disposal system for biohazards and used chemicals.
10. Liquid handling station (e.g., Tecan Genesis, Biomex FX) for distribution of extracts in 96-well assay plates.

2.3. Glassware, Ingredients and Reagents

1. Petri dishes and agar media appropriate for fungi, e.g. malt extract agar, potato-dextrose agar.
2. Cryovials with sterile 20% glycerol and storage boxes for culture preservation.
3. Automatic and disposable pipettes, Transfertubes (Spectrum Laboratories, Rancho Dominguez, CA).

4. Adhesive labels or marking pens.
5. Autoclavable glass roller bottles. Several sizes are available ranging from 1 to 4 l. In FERMEX, 2 l tissue culture bottles (Wheaton, Bellco) were used for fermenting strains for the extract collection (Fig. 1), and 4 l tissue bottles were used for scale-up of fermentation for chemistry purification projects. However, recently to conserve resources, we have scaled down the primary extract fermentations using 1 l media bottles (e.g. Fisher, VWR) which are large enough to roll on the roller apparatus (Fig. 2). Glass bottles are reusable and are inert to the organic solvent extract.
6. Closures for roller bottles. Roller bottles are closed with a breathable lid during fungal growth. We have used reusable silicone closures (Silicone Sponge Closure for Flasks with 38 mm OD Neck), cellulose stoppers (Figs. 2 and 3), or cotton. The latter two need to be taped to the bottles so they don't fall out while rolling. During the solvent extraction, the breathable closure is replaced with the bottle's normal screw-cap closure.
7. Vermiculite. Vermiculite is readily available in the USA and parts of Europe. Extra coarse particle size is necessary, e.g., as Grade V4 (Fig. 1). The fine particle sizes used in horticultural is unsatisfactory because the particles clog and do not form proper pore spaces. Before use, the vermiculite must be sieved to remove dust and fine particles. This is best done outdoors, but if done inside, then use adequate ventilation and respiratory protection against the dust.
8. 250-ml Erlenmeyer flasks.
9. Ingredients for media. Media selection will depend on the kinds of fungi grown and the investigator's experience and preferences. Because one of the goals of the technique is to produce clean interference-free extracts, generally media with defined or soluble complex ingredients are used. See references in Table 1 for examples and formulations of different media.
10. Solvents for extraction. Historically MEK or acetone has been used, but other solvents could be used. DMSO for suspension of dried extracts.

3. Methods

1. Storage of frozen strains. An agar slant or Petri plate of each strain is used to prepare frozen vegetative mycelia, by aseptically transferring a portion of the growth into seed medium and incubating on a gyratory shaker (220 rpm) at 22–25°C and until sufficient amount of biomass is obtained. The length

of time required to obtain sufficient biomass for each culture is noted, and cultures are designated fast (1–4 days seed growth) or slow (5–10 days seed growth) for subsequent experimentation. If pellets or clumps form, the culture flask can be shaken at 150 rpm with small sterile ceramic balls and cylinders for 30–60 min to macerate the mycelial mass and facilitate its pipetting. Several vials (2 or 4 ml each) of each culture are stored in 10–20% glycerol at −80°C until used. Cultures are routinely stored at least 6 months or longer before being used in screening. After assay and chemistry analysis, repetition and scale up of fermentations may not be expected for a year or more after the culture was initially grown. The 6-month storage period prior to screening eliminates those few organisms unable to survive long-term ultracold freezing.

2. Organization and scheduling. The FERMEX program operated on a continuous weekly schedule and accumulated a large collection of screening extracts over time. Each strain is usually grown on three different production media that vary significantly in their composition and levels of primary and trace nutrients in order to express a large range of secondary metabolism. Over the course of several weeks, the schedule is rotated among 10–12 different production media (see examples in references). Approximately 240–300 extracts are generated per month.

   The strains were inoculated from frozen vials into seed media and then transferred to production media when sufficient biomass had grown. Strains were selected and segregated into weekly sets according to whether they grow slowly or quickly. A typical weekly schedule during a 6 week period is illustrated in Table 2.

3. Inoculum media. To minimize mycelium from pelleting and therefore accelerating growth, polymeric agents, e.g., agar, carboxymethyl cellulose or polyacrylic acid, are added to the inocula (53, 54). Three media typically useful are Sabouraud's maltose broth emended with yeast extract and agar (SMYA; neopeptone 10 g, maltose 40 g, yeast extract 10 g, agar 4 g, distilled H_2O 1,000 ml), yeast malt extract with Junlon (polyacrylic acid) (YMEJ; yeast extract 4 g, malt extract 8 g, glucose 4 g, Junlon 1.5 g, distilled H_2O 1,000 ml), and KFA (corn steep powder 2.5 g, tomato paste 40 g, glucose 10 g, oat flour 10 g, agar 2 g, trace element solution 10 ml, distilled H_2O 990 ml, with the trace element solution containing $FeSO_4 \cdot 7H_2O$ 1.0 g, $MnSO_4 \cdot H_2O$ 1.0 g, $CuCl_2 \cdot 2H_2O$ 0.025 g, $CaCl_2$ 0.1 g, H_3BO_3 0.056 g, $(NH_4)_6Mo_7O_{24} \cdot 4H_2O$ 0.019 g, $ZnSO_4 \cdot 7H_2O$ 0.2 g in 1,000 ml distilled H_2O).

4. Inoculum production. To revive frozen strains, vials are thawed at room temperature and a portion of the mycelial discs or

Table 2
Weekly schedule for inoculation, growth, and extraction of three sets of fungi using the FERMEX method

Week	Day 1	Day 2	Day 3	Day 4	Day 5
1	Set A-Start inocula (1–4 days fast growers)	A inocula to production (1 day)	A inocula to production (2 days)	A inocula to production (3 days)	A inocula to production (4 days) Set B-Start inocula (5–7 days slow growers)
2			B inocula to production (5 days)	B inocula to production (6 days)	B inocula to production (7 days)
3	Set C-Start inocula (1–4 days fast growers)	C inocula to production (1 day)	C inocula to production (2 days)	C inocula to production (3 days)	C inocula to production (4 days)
4	Set A Extraction (17–20 days)				
5	Set B Extraction (17–19 days)				
6	Set C Extraction (17–20 days)				

mycelial suspension is aseptically transferred to a 250-ml flask with 50 ml of SMYA, KFA, or YMEJ medium. The rest of the contents of the cryovial are transferred to a Petri plate with malt extract agar or potato-dextrose agar. Flask are incubated on a gyratory shaker (200–220 rpm, 22–25°C) 3–10 days or until abundant cells are produced.

5. Culture quality control. Mycelium cultured from cryovials should be compared to inoculum plated onto the same medium. The plates can be saved and compared to growth on vermiculite to check for possible contamination.
6. Preparation of vermiculite and roller bottles. Pre-sifted vermiculite is added to roller bottles and sealed with breathable closures (see Table 3 for relative proportions). The bottles are autoclaved with the autoclave's dry cycle twice, with at least one full day between each autoclaving (Fig. 3). A large supply of vermiculite bottles can be prepared in advance and stored for subsequent use.
7. Media preparation. The liquid media are weighed and mixed in large batches and then aliquoted and autoclaved in single

Table 3
Relative proportions of inoculum, medium, vermiculite, and solvent extract for different roller bottle dimensions

Container and dimensions	Volume inoculum (ml)	Volume liquid medium (ml)	Volume vermiculite (cm^3)	Solvent extraction with methyethyl ketone (ml)
Media bottle (1 l capacity)	7	120	340	150–165 to yield 125 of 1 WBE
Small roller bottles (105 × 210 mm, 2 l capacity)	15	225	675	300–325 to yield 250 of 1 WBE
Large roller bottles (105 × 420 mm, 4 l capacity)	25	550	1,250	600–650 to yield 500 of 1 WBE

Whole broth equivalent (WBE) is the extract concentration relative to the original fermentation volume

bottles or flasks, one for each roller bottle according to the proportions in Table 2.

8. Inoculation of roller bottles. Clean and turn on the biosafety cabinet. On a nearby table or laboratory cart, arrange each inoculum flask with its corresponding roller bottle and liquid media aliquot. Label each container with the strain number and medium type, using either preprinted adhesive labels or a lab marker. Labels should go on the bottles' inclined neck so they are not rubbed off by the roller machine cylinders.

   Inside the biosafety cabinet, aseptically pipette an aliquot of inoculum into its corresponding liquid medium. Close the liquid medium container and shake it vigorously to disperse the cells. Quickly pour (or pipette) the inoculated liquid into the corresponding vermiculite roller bottle and close it. Make sure the closures are securely fastened so they will not loosen on the roller machine (Fig. 2).

9. Incubation of roller bottles. The roller machine should be located in a humidified controlled temperature room or climatic chamber. Inoculated roller bottles are placed on the machine and the machine set at about 4–6 rpm. Excess speed is unnecessary and only wears out the drive mechanisms. As the bottles roll, observe that the vermiculite starts to coat the bottles. Some excess liquid may roll in the bottom of the bottle, but as each fungus grows, the liquid will be absorbed. Depending on the kind of fungus, the vermiculite is usually immobilized by a mycelial matrix in about 2–5 days (Figs. 3, 4, and 5). Once, the matrix is solidified, the bottles can be incubated on

a shelf, or the roller machine can be turned off and used as incubator shelf. The incubation time can vary depending on the kinds of fungi. For robustly growing ascomycetes, like *Aspergillus*, *Hypoxylon*, or *Botryosphaeria* species, 14 days may be adequate, but slower growing ascomycetes and basidiomycetes may require to 3–4 weeks or longer to fully colonize the bottles and exhaust the nutrients.

10. Extraction. In a chemical fume hood, add MEK (ca. 1.2–1.4 times the nominal fermentation volume) to the roller bottles. Close the bottles with the solvent-resistant caps and shake vigorously for 2–3 min and roll them on the roller apparatus for 3–4 h. Mycelium matted vermiculite clumps in the occasional recalcitrant sample are broken up with a long sturdy glass rod prior to rolling. After extraction, the extracts are filtered on a Buchner funnel, and the vermiculite mass is pressed down by hand with a plastic Erlenmeyer flask. The exhausted vermiculite is thoroughly dried in a fume hood and subsequently disposed in a biohazard container for incineration.

    The resulting extract, nominally at a 1× concentration, i.e. 1 ml broth equivalent (WBE) per ml, is sufficiently concentrated for screening, confirmation of hits, dereplication and at least preliminary chromatographic fractionation.

11. Preparation of screening samples. A sufficient aliquot is dried down under N_2 flow or vacuum, weighed, and redissolved in DMSO to give at least 6 ml of a 10 mg/ml solution, typically corresponding to a three- to sixfold concentrate over the original fermentation. This concentration is used for two main reasons: firstly, it yields fluid solutions for pipetting stations; secondly, assuming a 50- to 100-fold dilution into the assay medium because of limitations in DMSO concentration in most biochemical assays, it would result in crude extracts assayed at 100–200 μg/ml, and therefore, ultimately yielding low micromolar or more potent compounds, based on the presumption that the active metabolites represented at most a few percent of the crude extract. Approximately one-half of this DMSO solution is aliquoted into multiple 96-well conical-bottom assay plates, with 80 samples in the plate's center ten columns. Both 10 μl/well copies are prepared for single assays and 100 μl/well copies are prepared for laboratories running multiple screens over time. The remainder of the DMSO sample is retained as a backup in 2-ml mother plates. Assay and mother plates are heat-sealed and stored at −20°C until use. The bulk MEK extract is kept in reserve in glass bottles in a 4°C cold room.

12. Screening and follow-up of active extracts. Confirmation and prioritization of screening hits, which might include a dose–response, a counterscreen, and/or secondary assays, is carried

out on "fresh" DMSO aliquots recovered from the mother plates and a "new" sample is prepared from the bulk reserve MEK extract. Confirmed samples of interest are then analyzed by HPLC or LC/MS for purposes of dereplication and semi-quantitative analyses of known compounds.

Chromatographic fractionation of samples passing all tests then proceeds by using one-third of the remaining original MEK extract. This strategy is usually successful in either, full resolution of the active in ca. 20% of the cases; or in securing a chromatographic standard of the target compound to guide eventual refermentation, fermentation optimization, and scale-up in nearly all other cases. Enough MEK reserve remains for an occasional accident, or in anticipation of subsequent actives, for which one-half of the remainder would be used each time. In no instance should a sample be completely depleted; a few milliliter voucher of the original MEK should always be kept in permanent reserve for use in confirmation assays and for side-by-side with a replenishment fermentation and extraction.

Furthermore, by proceeding in this fashion, one not only has the benefits of duplicate samples but also samples with which to run comparative analyses in the rare cases of mismatch between DMSO and MEK reserves.

For chromatographic follow-up and identification of actives, the nature and composition of the vast majority of the FERMEX extracts is such that a first crude separation by gel filtration on Sephadex LH-20 (Pharmacia) in methanol followed by reverse-phase HPLC is a rapid and largely successful way of proceeding.

References

1. Alberts AW, Chen J, Kuron G, Hunt V, Huff J, Hoffman C, Rothrock J, Lopez M, Joshua H, Harris E, Patchett A, Monaghan R, Currie S, Stapley E, Albers-Schönberg G, Hensens O, Hirshfield J, Hoogsteen K, Liesch J, Springer J (1980) Mevinolin - a highly potent competitive inhibitor of hydroxymethylglutaryl-coenzyme-a reductase and a cholesterol-lowering agent. Proc Natl Acad Sci USA 77: 3957–3961
2. Chang RSL, Lotti VJ, Monaghan RL, Birnbaum J, Stapley EO, Goetz MA, Albers-Schönberg G, Patchett AA, Liesch JM, Hensens OD, Springer JP (1985) A potent nonpeptide cholecystokinin antagonist selective for peripheral tissues isolated from *Aspergillus alliaceus.* Science 230:177–179
3. Goetz MA, Monaghan RL, Chang RSL, Ondeyka J, Chen TB, Lotti VJ (1988) Asperlicin, a novel non-peptidal cholecystokinin antagonist from *Aspergillus alliaceus.* I. Fermentation, isolation and biological properties. J Antibiot 41:875–877
4. Hölker U, Höfer M, Lenz J (2004) Biotechnological advantages of laboratory-scale solid-state fermentation with fungi. Appl Microbiol Biotechnol 64:175–186
5. Mitchell DA, Krieger N, Berovìè M (eds) (2006) Solid-state fermentation bioreactors. Fundamentals of design and operation. Springer, Berlin
6. Ooijkaas LP, Weber FJ, Buitelaar RM, Tramper J, Rinzema A (2000) Defined media and inert supports: their potential as solid-state fermentation production systems. Trends Biotechnol 18:356–360
7. Robinson T, Singh D, Nigam P (2001) Solid-state fermentation: a promising microbial technology for secondary metabolite production. Appl Microbiol Biotechnol 55:284–289
8. Nielsen KF, Larsen TO, Frisvad JC (2004) Lightweight expanded clay aggregates (LECA),

a new up-scaleable matrix for production of microfungal metabolites. J Antibiot 57:29–36
9. Bigelis R, He H, Yang HY, Chang L-P, Greenstein M (2006) Production of antifungal antibiotics using polymeric solid supports in solid-state and liquid fermentation. J Ind Microbiol Biotechnol 33:815–826
10. Tamano K, Sano M, Yamane N, Terabayashi Y, Toda T, Sunagawa M, Koike H, Hatamoto O, Umitsuki G, Takahashi T, Koyama Y, Asai R, Abe K, Machida M (2008) Transcriptional regulation of genes on the non-syntenic blocks of *Aspergillus oryzae* and its functional relationship to solid-state cultivation. Fungal Genet Biol 45:139–151
11. Machida M, Yamada O, Gomi K (2008) Genomics of *Aspergillus oryzae*: Learning from the history of koji mold and exploration of Its future. DNA Res 15:173–183
12. Bills GF, Polishook JD (1991) Microfungi from *Carpinus caroliniana*. Can J Bot 69:1477–1482
13. Bills GF, Polishook JD (1992) Recovery of endophytic fungi from *Chamaecyparis thyoides*. Sydowia 44:1–12
14. Bills GF, Polishook JD (1994) Abundance and diversity of microfungi in leaf litter of a lowland rain forest in Costa Rica. Mycologia 86:187–198
15. Bills GF, Polishook JD (1995) Microfungi from decaying leaves of *Heliconia mariae* (Heliconiaceae). Brenesia 41–42:27–43
16. Bills GF, Polishook JD (1993) Selective isolation of fungi from dung of *Odocoileus hemionus* (mule deer). Nova Hedwigia 57:195–206
17. Bills GF, Christensen M, Powell M, Thorn G (2004) Saprobic soil fungi. In: Mueller G, Bills GF, Foster MS (eds) Biodiversity of fungi, inventory and monitoring methods. Elsevier Academic, Oxford, pp 271–302
18. Dombrowski AW, Bills GF, Sabnis G, Koupal LR, Meyer R, Giacobbe RA, Monaghan RL, Ondeyka JG, Lingham RB (1992) L-696,474, A novel cytochalasin inhibitor of HIV-1 protease: I. The producing organism and its fermentation. J Antibiot 45:671–678
19. Bills GF, Dombrowski AW, Horn WS, Jansson RK, Rattray M, Schmatz D, Schwartz RE (2002) *latin binomial* ATCC 74386 and fungus ATCC 74387 for producing sordarin compounds for fungi control. US Patent 6,436,395
20. Smith CJ, Morin NR, Bills GF, Dombrowski AW, Salituro GM, Smith SK, Zhao A, MacNeil DJ (2002) Novel sesquiterpenoids from the fermentation of *Xylaria persicaria* are selective ligands for the NPYY5 receptor. J Org Chem 67:5001–5004
21. Okuda T, Ando K, Bills G (2005) Fungal germplasm for drug discovery and industrial applications. In: An Z (ed) Handbook of industrial mycology, vol 22. Marcel Dekker, New York, pp 123–166
22. Singh SB, Cordingley MG, Ball RG, Smith JL, Dombrowski AW, Goetz MA (1991) Structure of stereochemistry of thysanone: a novel human rhinovirus 3 C-protease inhibitor from *Thysanophora penicilloides*. Tetrahedron Lett 32:5279–5282
23. Singh SB, Reamer RA, Zink D, Schmatz D, Dombrowski A, Goetz MA (1991) Variculanol: structure and absolute stereochemistry of a novel 5/12/5 tricyclic sesterterpenoid from *Aspergillus variecolor*. J Org Chem 56: 5618–5622
24. Singh SB, Smith JL, Sabnis GS, Dombrowski AW, Schaeffer JM, Goetz MA, Bills GF (1991) Structure and conformation of ophioblin K and 6-epiophiobolin K from *Aspergillus ustus* as a nematocidal agent. Tetrahedron 47:6931–6938
25. Lingham RB, Hsu A, Silverman KC, Bills GF, Dombrowski A, Goldman ME, Darke PL, Huang L, Koch G, Ondeyka JG, Goetz MA (1992) L-696,474, A novel cytochalasin as an inhibitor of the HIV-1 protease: III. Biological activity. J Antibiot 45:686–691
26. Ondeyka J, Hensens OD, Zink D, Ball R, Lingham RB, Bills G, Dombrowski A, Goetz M (1992) L-696,474, A novel cytochalasin as an inhibitor of HIV-1 protease. II. Isolation and structure. J Antibiot 45:679–685
27. Bills G, Hensens OD, Koupal L, Lingham RB, Ondeyka JG, Zink DL (1993) Antiviral agent. US Patent 5,183,826
28. Borris R, Lam Y-KT, Koupal L (1993) Antimigraine alkyl indole. US Patent 5,233,050
29. Lam YKT, Dai P, Borris R, Dombrowski A, Ransom R, Young G, Beer M, Middlemiss D, Smith J (1994) A new indole from *Penicillium daleae*. J Antibiot 47:724–726
30. Horn WS, Bierilo KK, Bills GF, Dombrowski AW, Helms GL, Jones ET, Linemeyer DL, Sesin DF, Schwartz RE, Wilson KE (1993) Characterization of the light- and base-mediated instability of calbistrin A. J Nat Prod 56: 1779–1785
31. Dombrowski AW, Koupal L (1993) Process for preparing cholesterol lowering compounds. US Patent 5,244,795
32. Singh SB, Goetz MA, Zink DL, Dombrowski AW, Polishook JD, Garcia ML, Schmalhofer W, McManus OB, Kaczorowski GJ (1994) Maxikdiol: a novel dihydroxyisoprimane as an

agonist of Maxi-K channels. J Chem Soc Perkin Trans 1:3349–3352

33. DarkinRattray SJ, Gurnett AM, Myers RW, Dulski PM, Crumley TM, Allocco JJ, Cannova C, Meinke PT, Colletti SL, Bednarek MA, Singh SB, Goetz MA, Dombrowski AW, Polishook JD, Schmatz DM (1996) Apicidin: A novel antiprotozoal agent that inhibits parasite histone deacetylase. Proc Natl Acad Sci USA 93:13143–13147
34. Singh SB, Zink DL, Polishook JD, Dombrowski AW, Darkin-Rattray SJ, Schmatz DM, Goetz MA (1996) Apicidins: Novel cyclic tetrapeptides as coccidiostats and antimalarial agents from *latin binomial.* Tetrahedron Lett 37:8077–8080
35. Singh SB, Zink DL, Liesch JM, Mosley RT, Dombrowski AW, Bills GF, Darkin-Rattray SJ, Schmatz DM, Goetz MA (2002) Structure and chemistry of apicidins, a class of novel cyclic tetrapeptides without a terminal alpha-keto epoxide as inhibitors of histone deacetylase with potent antiprotozoal activities. J Org Chem 67:815–825
36. Clapp WH, Harris GH, Bills GF, Curotto JE, Dombrowski AW, Dreikorn SJ, Kurtz MB, Mienz MS, Polishook JD, Strcicher SL, Thompson JR, Williams M, Zink DL (1997) Antifungal agents. US Patent 5,597,806
37. Tsipouras A, Goetz MA, Hensens OD, Liesch JM, Ostlind DA, Williamson JM, Dombrowski AW, Ball RG, Singh SB (1997) Sporandol: A novel antiparasitic binaphthalene from *Chrysosporium meridarium.* Bioorg Med Chem Lett 7:1279–1282
38. Tsipouras A, Ostlind DA, Hensens OD, Zink DL (1992) Novel fermentation product with antiparasitic activity. US Patent 5,089,530
39. Singh SB, Zink DL, Goetz MA, Dombrowski AW, Polishook JD, Hazuda DJ (1998) Equisetin and a novel opposite stereochemical homolog phomasetin, two fungal metabolites as inhibitors of HIV-1 integrase. Tetrahedron Lett 39:2243–2246
40. Dombrowski AW, Hastings JC, Hazuda DJ, Polishook JD, Singh SB (1998) In vitro HIV integrase inhibitors. US Patent 5,759,842
41. Jayasuriya H, Guan ZQ, Polishook JD, Dombrowski AW, Felock PJ, Hazuda DJ, Singh SB (2003) Isolation, structure, and HIV-1 integrase inhibitory activity of cytosporic acid, a fungal metabolite produced by a *Cytospora* sp. J Nat Prod 66:551–553
42. Hensens OD, Ondeyka JG, Dombrowski AW, Ostlind DA, Zink DL (1999) Isolation and structure of nodulisporic acid A(1) and A(2), novel insecticides from a *Nodulisporium* sp. Tetrahedron Lett 40:5455–5458
43. Ondeyka JG, Helms GL, Hensens OD, Goetz MA, Zink DL, Tsipouras A, Shoop WL, Slayton L, Dombrowski AW, Polishook JD, Ostlind DA, Tsou NN, Ball RG, Singh SB (1997) Nodulisporic acid A, a novel and potent insecticide from a *Nodulisporium* sp. isolation, structure determination, and chemical transformations. J Am Chem Soc 119:8809–8816
44. Ostlind DA, Felcetto T, Misura A, Ondeyka J, Smith S, Goetz M, Shoop W, Mickle W (1997) Discovery of a novel indole diterpene insecticide using first instars of *Lucilia sericata.* Med Vet Entomol 11:407–408
45. Bills G, Dombrowski A, Morris SA, Hensens O, Liesch JM, Zink DL, Onishi J, Meinz MS, Rosenbach M, Thompson JR, Schwartz RE (2000) Hyalodendrosides A and B, antifungal triterpenoid glycosides from a lignicolous hyphomycete, *Hyalodendron* species. J Nat Prod 63:90–94
46. Goetz MA, Zink DL, Dezeny G, Dombrowski A, Polishook JD, Felix JP, Slaughter RS, Singh SB (2001) Diterpenoid pyrones, novel blockers of the voltage-gated potassium channel Kv1,3 from fungal fermentations. Tetrahedron Lett 42:1255–1257
47. Singh SB, Zink DL, Dombrowski AW, Dezeny G, Bills GF, Felix JP, Slaughter RS, Goetz MA (2001) Candelalides A-C: Novel diterpenoid pyrones from fermentations of *Sesquicillium candelabrum* as blockers of the voltage-gated potassium channel Kv1.3. Org Lett 3:247–250
48. Sings HL, Harris GH, Dombrowski AW (2001) Dihydrocarbazole alkaloids from *Aspergillus tubingensis.* J Nat Prod 64:836–838
49. Singh SB, Herath K, Guan ZQ, Zink DL, Dombrowski AW, Polishook JD, Silverman KC, Lingham RB, Felock PJ, Hazuda DJ (2002) Integramides A and B, two novel nonribosomal linear peptides containing nine C-alpha-methyl amino acids produced by fungal fermentations that are inhibitors of HIV-1 integrase. Org Lett 4:1431–1434
50. Ondeyka JG, Dombrowski AW, Polishook JP, Felcetto T, Shoop WL, Guan ZQ, Singh SB (2006) Isolation and insecticidal/anthelmintic activity of xanthonol, a novel bis-xanthone, from a non-sporulating fungal species. J Antibiot 59:288–292
51. Jayasuriya H, Guan ZQ, Dombrowski AW, Bills GF, Polishook JD, Jenkins RG, Koch L, Crumley T, Tamas T, Dubois M, Misura A, Darkin-Rattray SJ, Gregory L, Singh SB (2007) Isolation, structure, and coccidiostat activity of coccidiostatin A. J Nat Prod 70: 1364–1367

52. Ondeyka JG, Dombrowski AW, Polishook JP, Felcetto T, Shoop WL, Guan Z, Singh SB (2003) Isolation and insecticidal activity of mellamide from *Aspergillus melleus.* J Ind Microbiol Biotechnol 30:220–224
53. Jones P, Shahab BA, Trinci APJ, Moore D (1988) Effect of polymeric additives, especially Junlon and Hostacerin, on growth of some basidiomycetes in submerged culture. Trans Br Mycolo Soc 90:577–583
54. Yang F-C, Yang M-J, Cheng SH (2009) A novel method to enhance the mycelia production of *Ganoderma lucidum* in submerged cultures by polymer additives and agitation strategies. J Taiwan Inst Chem Eng 40: 148–154

Chapter 6

The Chemical Identification and Analysis of *Aspergillus nidulans* Secondary Metabolites

James F. Sanchez and Clay C.C. Wang

Abstract

Filamentous fungi have long been recognized to be a rich source of secondary metabolites with potential medicinal applications. The recent genomic sequencing of several *Aspergillus* species has revealed that many secondary metabolite gene clusters are apparently silent under standard laboratory conditions. Several successful approaches have been utilized to upregulate these genes and unearth the corresponding natural products. A straightforward, reliable method to purify and characterize new metabolites therefore should be useful. Details are provided herein on the cultivation of *Aspergillus nidulans* and the LC/MS analysis of the metabolic profile. Following is an explanation of silica gel chromatography, HPLC, and preparative TLC. Finally, the NMR characterization of previously unknown *A. nidulans* metabolites is detailed.

Key words: Secondary metabolites, Natural products, Polyketides, Purification methods, *Aspergillus nidulans*

1. Introduction

Fungal genera such as *Aspergillus* are well known to generate a number of structurally complex secondary metabolites. Many of them have been observed to be significantly relevant to human health, including lovastatin, a well-known antihypercholesterolemic agent produced by *A. terreus*, and aflatoxin, a carcinogen produced by *A. flavus* (1, 2).

The recent sequencing of a number of fungal genomes has revealed, surprisingly, that fungi have many more secondary metabolism pathways than previously thought (3–6). As one example,

Nancy P. Keller and Geoffrey Turner (eds.), *Fungal Secondary Metabolism: Methods and Protocols*,
Methods in Molecular Biology, vol. 944, DOI 10.1007/978-1-62703-122-6_6,

Aspergillus nidulans contains 27 polyketide synthase (PKS) and 14 nonribosomal peptide synthetase (NRPS) genes (3), but fewer than ten secondary metabolites had been regularly observed. Since far fewer secondary metabolites have been identified from this heavily studied species, the products of many of these biosynthetic pathways are still unknown and are presumably silent under standard laboratory conditions. There is thus great opportunity to discover new metabolites that could potentially be new therapeutics or important mycotoxins.

A variety of methods have been developed to turn on silent gene clusters. Deletion of *A. nidulans* CclA, a Bre2 ortholog involved in histone 3 lysine 4 methylation, activated the expression of hidden secondary metabolite clusters. One novel cluster generated monodictyphenone, emodin, and emodin derivatives, while a second encoded two anti-osteoporosis polyketides, F9775A and F9775B (7). The latter polyketides were also revealed from a study of over 20 separate culture conditions, based on the understanding that different conditions can yield different products, an approach known as One Strain Many Compounds (OSMAC) (8).

Other targets besides histone proteins may be linked to the regulation of some secondary metabolites. In *A. nidulans* the single sumoylation gene *sumO* was demonstrated to influence the production of several metabolites (9). Deletion of this gene led to a decrease in austinol and dehydroaustinol production and also a substantial increase in the formation of the polyketide asperthecin. The significant amount of asperthecin allowed researchers to determine the genetic cluster and propose its biosynthesis.

As another approach, the native promoter of a transcriptional activator gene, *afoA*, was replaced by an inducible promoter, allowing the expression of the corresponding PKS and concurrent production of the polyketide asperfuranone (10).

With the generation of an increasing number of secondary metabolites, a reliable protocol is needed for their purification and identification. Here we describe the methods the Wang group carries out to isolate and characterize *A. nidulans* secondary metabolites, primarily polyketides, although the methods should be applicable to other fungal secondary metabolites.

2. Materials

2.1. Compositions of Trace Metal Solution and Common Media

1. Trace metal solution: $ZnSO_4 \cdot 7H_2O$, 2.2 g, H_3BO_3, 1.1 g, $MnCl_2 \cdot 4H_2O$, 0.5 g, $FeSO_4 \cdot 7H_2O$, 0.5 g, $CoCl_2 \cdot 5H_2O$, 0.16 g, $CuSO_4 \cdot 5H_2O$, 0.16 g, $(NH_4)_6Mo_7O_{24} \cdot 4H_2O$, 0.11 g, Na_4EDTA, 5.0 g. Add the solids in the listed order to 80 mL

of H_2O, dissolving each completely before adding the next. Heat the solution to boiling, cool to 60°C, and adjust the pH to 6.5 with saturated KOH. Cool to room temperature and adjust the volume to 100 mL with H_2O (see Note 1). Autoclave and store at room temperature.

2. 20× salts solution: $NaNO_3$, 120 g, KCl, 10.4 g, $MgSO_4{\cdot}7H_2O$, 10.4 g, KH_2PO_4, 30.4 g, H_2O, 1 L.
3. Glucose minimal media: Glucose (dextrose), 10 g, 20× salt solution, 50 mL, trace metal solution, 1 mL, H_2O, 1 L.
4. Czapek's minimal media: Sucrose, 30 g, 20× salt solution, 50 mL, trace metal solution, 1 mL, H_2O, 1 L.
5. Lactose minimal media: Lactose, 20 g, glucose (dextrose), 10 g, 20× salt solution, 50 mL, trace metal solution, 1 mL, H_2O, 1 L.

   For all liquid media, adjust the pH to 6.5 using concentrated KOH or HCl. Be sure to include any necessary supplements that the strain requires (e.g., pyridoxine). Autoclave and store at room temperature. The trace metal solution and supplements should also be autoclaved. The supplement uridine cannot be autoclaved and must instead be introduced to the autoclaved media through a sterile syringe filter.
6. Yeast Agar Glucose (YAG) Plates: Yeast extract, 5 g, glucose (dextrose), 20 g, agar, 15 g, trace metal solution, 2 mL, H_2O, 1 L. Be sure to include any necessary supplements.

   However, the yeast extract may already contain enough of the supplement in question. Our experience has shown that it is not necessary to add pyridoxine, but it is still necessary to add riboflavin or uridine and uracil. Autoclave and cool until it can be handled properly (but before it solidifies). Again, the supplement uridine cannot be autoclaved and must instead be introduced to the autoclaved media through a sterile syringe filter. Pour the liquid YAG media into Petri dishes (10 cm diameter), approximately halfway to the top. After they solidify, store them at 4°C.

2.2. Solutions for LC/MS Analysis and Purification

1. LC/MS "A" buffer: 950 mL H_2O, 50 mL acetonitrile (see Note 2), 500 μL formic acid.
2. LC/MS "B" buffer: 1 L acetonitrile, 500 μL formic acid.
3. HPLC "A" buffer: 950 mL H_2O, 50 mL acetonitrile (HPLC grade), 500 μL trifluoroacetic acid.
4. HPLC "B" buffer: 1 L acetonitrile, 500 μL trifluoroacetic acid.

3. Methods

3.1. Acquisition of Spore Suspension

1. Transfer a small amount of the strain of interest (it need not be quantified) from a "stock solution" (typically from glycerol or silica stock) onto a new YAG plate under sterile conditions.
2. Spread the strain around on the plate using a sterile bacterial cell spreader. Adding 1 mL of sterile water beforehand will make spreading easier.
3. This YAG plate then requires 2–3 days in a 37°C incubator for conidia to cover the top of a plate. Place 7 mL of sterile water on top of the plate, followed by gentle scraping with a spreader. The spore suspension should then be transferred to a sterile Falcon tube (see Note 3).
4. Measure the concentration of the suspension; this can be accomplished with a standard hemocytometer. A typical spore concentration for *A. nidulans* is approximately 1×10^6 spores/μL (see Note 4). Accordingly, a small amount of spore suspension will need to be diluted ~1,000-fold with water in order to count a reasonable number of spores.

3.2. Small-Scale Cultivation for Initial Analysis

1. Under sterile conditions, transfer 50 mL of the liquid media of choice into a sterile 125-mL flask. Enough spore suspension should then be transferred into the same flask such that the spore concentration is 5×10^6 spores/mL (or 2.5×10^8 spores in total).
2. Typically, place the flask in a shaker at ~100 rpm for 4 days at 37°C. By this time, the mycelia should have grown and appear as light-colored spheres.
3. If an inducible promoter is involved, the inducer should be added 18 h after initial cultivation. For *alcA* promoters, 133 μL of cyclopentanone is added to a 50 mL cultivation. The volume can be scaled accordingly.
4. After cultivation, 25 mL of filtered medium is typically collected in a 50 mL Falcon tube, using a piece of Miracloth for filtration. The media is then to be extracted twice with an equal amount of ethyl acetate (this can be done within the Falcon tube), and the ethyl acetate layer is transferred by pipette to tared vessels.
5. After evaporation, the mass of the residue can be determined (see Note 5). It can be dissolved in methanol at a concentration of 1 mg/mL to prepare it for LC/MS analysis (see Note 6).
6. As an alternative to liquid media, a solid agar plate can be used instead (see Note 7). The result may be a different metabolite profile. A total of 1×10^7 spores should be added to a plate of

10 cm diameter, spread onto the solid medium of choice. Place the plate in a 37°C incubator for 5 days.

7. Afterwards, the agar must be chopped into ~1 cm^2 pieces and collected into a flask. Add methanol until it covers all of the pieces. The flask is then sonicated for 1 h.
8. Collect the filtrate (again using Miracloth) and add a 1:1 mixture of methanol:dichloromethane to the chopped agar until the solvent reaches the top of the pieces. Sonicate again for an hour.
9. Filter again, this time using KimWipes instead of Miracloth. The combined solvent is then evaporated until water is the only solvent (see Note 8), which is then extracted as before.
10. When necessary (see Note 9) add concentrated HCl to the filtered media (whether from a liquid or solid source) until a pH of 2 is reached (pH paper can be used). Extract and evaporate as usual.
11. In all cases the solution should be syringe-filtered to remove particulates that can clog an LC/MS column. Place ~100 μL of solution into a disposable 1 mL syringe fitted with a 13 mm filter with a PTFE membrane. With the syringe plunger, push the solvent through the filter and into an LC/MS vial. The sample is then ready to be submitted to the LC/MS.
12. LC/MS typically uses two buffers, a polar "A" buffer and a nonpolar "B" buffer (see Subheading 2.2).
13. A gradient, an increasing amount of the "B" buffer over time, is used to separate and elute the molecules. A typical gradient in the Wang lab is 0% B from 0 to 5 min, 0–100% B from 5 to 35 min, 100% B from 35 to 40 min, 100–0% B from 40 to 45 min, and re-equilibration with 0% B from 45 to 50 min. The flow rate is 125 μL/min.

3.3. LC/MS Analysis

The most straightforward situation is when the metabolite is naturally secreted in abundant amounts by the organism under standard cultivation conditions (such as Glucose Minimal Media, 37°C, 4 days) and can be easily detected. This is true of sterigmatocystin and terrequinone A, both of which are produced by *A. nidulans* in standard cultivations and are UV-active, making them detectable by analytical chromatography.

Fortunately, a modern LC/MS machine (see Note 10) will allow one to detect non-UV-active compounds as well through a Total Ion Current (TIC) chromatogram. The *y*-axis of this chromatogram measures signal intensity as the molecules at any given point of the chromatogram are ionized for mass spectrometric analysis. As a result, compounds that escape detection under UV-based conditions, such as the emericellamides, reveal themselves as peaks in a TIC chromatogram. One important caveat to keep in mind is that this mode of molecule detection relies on its ability to be ionized, and this ability varies from molecule to molecule. However, a mass spectrometer can be programmed to form positively charged ions ("positive mode") or negatively charged ones ("negative mode"), and it is rare that a secondary metabolite will fail to be ionized in both cases. Some molecules, such as the emericellamides, are apparent in either positive or negative mode. Others are only detectable in one of the two modes. Therefore, if the object is to search for new metabolites whose properties are unknown, it is beneficial to examine three chromatograms: the UV chromatogram, the TIC chromatogram in positive mode, and the TIC chromatogram in negative mode.

If an unknown molecule emerges, it is unlikely that the structure can be solved at this point; however, important clues can be gleaned from the data provided. Aside from the mass, it can be determined if the compound is polar or nonpolar, depending on what time it elutes from the column. Also, the UV spectra can offer hints. For example, xanthones exhibit a maxima at ~380 nm, and this was helpful in locating xanthones from *A. nidulans* using the UV chromatogram.

3.4. Large-Scale Cultivation for Isolation of Metabolite

1. It will likely be necessary to grow the strain again in a larger scale to acquire enough material for NMR analysis. The amount of medium that is necessary will depend on the titer of the metabolite. A good starting point is to use about 2 L of medium.
2. For liquid culture, the researcher can simply use larger flasks and larger volumes of media. However, he or she may need to experiment with various parameters before the scale-up is successful. For instance, if a metabolite was acquired from 50 mL of media in a 125-mL flask, it does not automatically follow that the same result will come from using 1 L of media in a 2-L flask. The experimenter may instead need to use a number of smaller flasks.
3. For solid media it is normally not a problem to use larger plates (15 cm diameter) in place of smaller plates (10 cm diameter) for convenience. More sterile water (3 mL) should be added on top of each of these plates to facilitate spreading, and 2.25×10^7 μL of spore solution should be transferred.
4. The volumes of ethyl acetate involved will mean that separatory funnels must be used in the place of Falcon tubes, and the organic material should be collected in a large round-bottom flask. Extract at least twice; four times may be optimal.
5. A small amount (~0.5 mg) of dried material can be dissolved in methanol and submitted to the LC/MS to determine if the scaled-up cultivation was successful. Remember to syringe-filter the solution.

3.5. Silica Gel Column Chromatography for Semi-purification

The next step is to isolate the metabolite in question. Because a crude mixture is likely to contain a number of different products, multiple purification steps may be necessary, and careful planning is needed. The first round of purification is often standard organic chemistry silica gel chromatography. Although in some instances it is possible to isolate a metabolite at this step, the sheer number of chemicals in a crude mixture means that often we use the silica column step to obtain a semi-pure isolate that can be further purified in later stages using other methods.

1. Weigh out two portions of silica (particle size 0.063–0.200 mm). One portion will be 2–3 times the mass of the crude material. The other portion will be 30 times the crude mass.
2. Dissolve the crude material. If common organic solvents such as dichloromethane and methanol fail to dissolve the material, try adding water and then sonicate.
3. Put the small amount of silica into the dissolved material and evaporate the solvents completely. A substance resembling sand should be achieved. Remove any clumps with a spatula.
4. Put the large amount of silica in a large Erlenmeyer flask. Add in enough hexanes to form a slurry. ("Hexanes" is a common, commercially provided mixture of isomers of hexane.)
5. Pour the slurry into a glass chromatography column (see Note 11). Make sure there is a small amount of cotton taken from a cotton ball at the bottom of the column to prevent the silica from leaking out. Do not use too much cotton as it will significantly slow the flow of the solvent. The Wang group prefers not to use columns with frits at the bottom because an air bubble may form. It will be necessary to rinse the Erlenmeyer flask several times with hexanes to transfer all of the silica.
6. Allow solvent to drain out of the column. Using air pressure via an adapter with a regulator and release valve will greatly facilitate the running of the column, but always be careful that the valve at the bottom of the column is open before applying pressure. Stop when there is no longer any solvent that is above the silica.
7. Lightly tap the edge of the column with a blunt object such as the back of a hammer. This will cause the silica to settle. Some solvent will emerge above the level of the silica. Drain until again there is no solvent above the silica. Tap the edge of the column again to further pack the column. Drain.
8. Repeat tapping and draining until the silica settles no more (at which point no solvent will emerge above the silica). The column is now packed.
9. Add more hexanes to the top of the silica so that it is ~1 cm in height. Be careful not to disrupt the silica. It is a good idea to

use a pipette and slowly release the solvent onto the inner side of the column in order to preserve the integrity of the packed silica.

10. Pour in the small amount of silica that was mixed with the crude fungal extract. Make sure that all of this material is off the sides of the column. Extra hexanes can be used to completely transfer the material, if necessary. If so, drain the solvent until it is at the top of the material (see Note 12).
11. Add standard, commercially available sand so that it is ~1 cm in height. The sand serves as a physical barrier and will help prevent disruption of the packed silica when solvent is added.
12. Elute with the solvent systems that are necessary to drain the metabolite(s) of interest. For every 1.5 g of crude material, use 1 L of solvent mixture. A suitable solvent system is a mixture of ethyl acetate and hexanes. The most nonpolar metabolites will be eluted from a mixture of 10% ethyl acetate in hexanes. The more polar the metabolite, the more ethyl acetate will be needed, all the way up to 100% ethyl acetate (see Note 13).
13. For especially polar metabolites, it may be necessary to use methanol in ethyl acetate. Do not exceed 10% methanol and start gradually (see Note 14). If the experimenter is unsure which solvent composition is appropriate, it is advised to start with 10% ethyl acetate/hexanes, and then follow to 20% ethyl acetate/hexanes, and so forth. Each fraction should be collected separately and evaporated.
14. A small amount of material can be prepared for LC/MS analysis to keep track of the status of the purification.

3.6. Final, HPLC Purification of the Metabolite

1. Provided that the metabolite was eluted, the researcher can then proceed to HPLC purification. The evaporated fraction(s) from silica gel chromatography should be completely dissolved in methanol (see Note 15). Try to keep solvent volumes to a minimum. DMSO can also be added if the material fails to readily dissolve in the first solvent (see Note 16).
2. The solution is then filtered through a glass syringe into a vial for HPLC, the syringe fitted with the same filter used for LC/MS.
3. Refer to Subheading 2.2 for the composition of HPLC "A" and "B" buffers. A flow rate of 5.0 mL is used, with a UV detector at 250 nm.
4. The gradient to be chosen depends on the polarity on the compound to be purified (see Note 17).
5. Fractions containing the product (see Note 18) should be combined and evaporated, but first a sample should be submitted to the LC/MS to verify that it is indeed the product in question.

6. Frequently, the metabolite is now pure and, if a sufficient quantity was obtained, it is ready to be analyzed by NMR. However, sometimes the LC/MS and/or NMR reveal the presence of impurities, and further purification is necessary. Two strategies can be followed. One is to reuse the HPLC, this time employing a slower gradient, or even no gradient at all (isocratic). Another is to use preparative TLC (see next section).

3.7. Preparative TLC

1. Preparative TLC is appropriate for compounds that can be visualized under a UV light. The sample is dissolved in a small amount of methanol. A bit of cotton is taken from a cotton ball and packed down partly through the end of a glass pipette. This "paint brush" is used to apply the material as a line about 2 cm above the bottom of a pre-coated 20×20 cm TLC plate. No more than 10 mg of material should be applied to one plate.
2. If the appropriate solvent system is unknown, take a small amount of sample and run an analytical TLC. (Consult an organic chemistry laboratory textbook for instructions.) The right solvent composition will be similar to, but not necessarily identical to, the composition used to elute the product in column chromatography.
3. Place the plate in the developing chamber. Remove the plate when the solvent is about 2 cm from the top of the plate. This will take about an hour. Let dry for another hour.
4. If the product is colorless, it will be necessary to place the plate under a UV lamp (~250 nm) to visualize the location of the product. With a pencil, trace the outside of the product, then remove the plate from the UV lamp.
5. Using a razor blade, scrape off the silica that contains the product and collect in a flask. Pour in 10 mL of 95:5 dichloromethane:methanol. Add in a stir bar and stir for 1 h.
6. Collect the product by pouring the solvent over a filter atop a funnel and into a tared flask. Evaporate the solvent.

3.8. NMR Characterization of the Metabolite

Once pure product is obtained in sufficient quantity (as much as 5 mg may be needed), it is ready to be analyzed by NMR. The types of NMR experiments that should be performed depend on the nature of the product.

1. A proton NMR should be run first, and preferably also a carbon NMR. Be sure that the NMR tubes are clean and dry and that deuterated solvent is used. Consult an organic chemistry laboratory textbook if the operation of this machine is unclear.
2. If the researcher has an idea of the metabolite's identity (see Note 19), he or she can compare the acquired spectra with the metabolite's spectral data published in a journal article or a book (if such data exists).

3. If the metabolite's identity is still unknown, additional experiments, such as DEPT, HSQC, HMBC, COSY, and ROESY, may be needed to deduce the structure. The analysis of NMR data is beyond the scope of this chapter, and the reader is advised to consult a suitable text from a chemistry library.
4. It is often not necessary to fully solve the structure from these data. Rather, the researcher may solve only part of that structure, but that part is enough for him or her to enter into such programs as SciFinder Scholar and ascertain whether any published compounds also feature that part of the structure.
5. If the researcher is still unable to solve the structure, he or she may make use of a software program, such as ACD/Labs that can assist in this regard. Such programs can either compare the acquired NMR data with those in their databases or attempt to logically solve the structure. Success will rely on the amount of data that the experimenter is able to provide to the program.
6. Some computer programs also will predict the NMR spectra for a given chemical, so if the experimenter has an idea of the structure, he or she can see if the data match the computer prediction. A close match does not guarantee that the structure is correct, but it may help the researcher develop an understanding of the identity of the molecule.
7. Solving a structure through NMR can be a "back and forth" process. That is, the NMR data may provide some clues about the structure. The researcher can then consult journal articles, use software programs, or consult with peers, which may provide inspiration for the choice of further NMR experiments to advance in solving the structure (see Note 20).

4. Notes

1. The Wang group uses Barnstead's "DIamond RO" purification system whenever water is required, except for LC/MS, in which HPLC grade water is needed.
2. Be aware that many solvents are commercially sold in various "grades," which refers to the purity of the solvent. Solvent from different vendors with comparable "grades" should be interchangeable. Solvent prices are correlated with purity and because solvent is one of the major costs in these type of experiments, our choice of solvent grade is based on necessity. We reserve the use of ultra high purity solvents only when required. Acetonitrile should be HPLC grade. The other solvents we use are high purity (99.9%).

3. Fungal suspensions can be stored in a refrigerator for up to a month. After that, we tend to prepare a fresh suspension.
4. Some suspensions of mutant strains may be at least an order of magnitude less concentrated than wild-type suspensions.
5. Using a rotary evaporator is a convenient method to evaporate solvent. Its operation is straightforward, and instructions can be found in a number of organic chemistry laboratory texts. We use rotary evaporators from Buchi. For small scale evaporation we use the Turbovap system from Caliper Life Sciences that is able to dry down large number of samples in parallel unattended. One can either use nitrogen or compressed air to dry the samples. We found that the use of compressed air is sufficient and cost effective.
6. The final LC/MS concentration of 1 mg/mL need not be rigorously adhered to if the researcher finds it beneficial to increase or decrease the concentration somewhat.
7. Relative to the extract from liquid media, the extract from solid media is frequently larger in mass and more difficult to dissolve. The crude extract from solid media can be dissolved in DMSO at a concentration of 5 mg/mL, and then diluted fivefold with methanol.
8. It can be difficult to use a rotary evaporator on a mixture of water and organic solvents, because the mixture vigorously bubbles under vacuum and threatens to spill out of the flask. Evaporation will be easier if the flask is not more than a quarter full.
9. In many instances the secondary metabolite will be charged or otherwise be polar and cannot be extracted with an organic solvent. This problem can sometimes be circumvented by adjusting the pH of the aqueous solution so that previously charged species become neutral and thus more lipophilic. Lowering the pH to ~2 using concentrated HCl neutralizes negatively charged molecules. It might also be helpful to raise the pH of another portion of media to about 12, in case there are positively charged metabolites.
10. The Wang group maintains its own LC/MS. A Thermo Finnigan ion trap mass spectrometer with an electrospray source is used for initial screening. An Agilent time-of-flight mass spectrometer to obtain high resolution masses is also utilized. A sample can also be sent to a core facility for analysis. Often, the sample can be dry, and the facility will take care of sample preparation.
11. Columns come in a variety of sizes. Choose one that will result in the large portion of silica reaching a height of ~10 cm when it is poured in. ChemGlass is an excellent source of glassware.

12. There are two main ways to prepare a column, "dry loading" and "wet loading," and variations within these methods. A version of dry loading is described above. Although some people will insist on the superiority of a certain method, the "best" method is likely to be the one you are most comfortable with. We refer the interested reader to a seminal paper (11).
13. Often in the course of silica gel chromatography the product of interest will elute in more than one fraction. For instance, a product may elute in the 10% ethyl acetate fraction and the 20% ethyl acetate fraction. Since it is often desired to collect as much product as possible, it is a good idea to run multiple fractions of increasing polarity (such as 10, 20, and 30% ethyl acetate) to ensure that all of the product has eluted.
14. Small changes in the amount of methanol in an eluent will have a dramatic effect on the eluent's polarity, so measure carefully: An eluent with 5% methanol in ethyl acetate, for example, may have a dramatically different effect from one with 2% methanol.
15. When dissolving the material for HPLC, if there is a significant amount of insoluble material which the researcher suspects is not the product, it may be better to first centrifuge the solution and transfer the supernatant to the syringe. Otherwise, the insoluble particulates may quickly clog the filter.
16. When dissolving samples for the LC/MS or the HPLC, do not use solvents other than water, acetonitrile, methanol, and DMSO, as other solvents may damage the instruments.
17. Choosing the right gradient is a bit of an art and may require some trial and error. It may be useful to consider the LC/MS retention time and start with a gradient that was successful in purifying a compound with a similar retention time, based on your group's experience or the literature.
18. Depending on the size of your HPLC column, you may need to inject less than the entire volume of the solution and run the HPLC multiple times until all the solution has been purified, so that the column is not overloaded. It is nevertheless a good idea to first inject a small amount of solution first, to make sure that the gradient is appropriate and other parameters are acceptable. The column that the Wang lab uses is Phenomenex Luna 5 μm C18 (2), 250 × 21.2 mm.
19. It is advantageous to be aware of not only the secondary metabolites that are produced by the species in question but also many of those that are produced by other filamentous fungi. In our experience, a metabolite that is new to *A. nidulans* has much of the time been previously observed in another species. There is obviously a benefit, then, to be aware of many fungal metabolites and to think about whether there is any

possibility that one of them is the same as the metabolite that you have purified.

20. Infrared (IR) spectroscopy can also be used in conjunction with NMR to characterize the molecule (and some journals will require IR characterization). Although it is typically not as effective as NMR in illuminating the structure, it can be useful in determining the presence of various functional groups. The reader is advised to consult an organic chemistry textbook for more information. Many institutions will have their own IR spectrometers that are available to its members or to visitors upon request.

References

1. Kennedy J, Auclair K, Kendrew SG, Park C, Vederas JC, Hutchinson CR (1999) Modulation of polyketide synthase activity by accessory proteins during lovastatin biosynthesis. Science 284:1368–1372
2. Payne GA, Nystrom GJ, Bhatnagar D, Cleveland TE, Woloshuk CP (1993) Cloning of the afl-2 gene involved in aflatoxin biosynthesis from Aspgergillus flavus. Appl Environ Microbiol 59:156–162
3. Galagan JE, Calvo SE, Cuomo C, Ma LJ, Wortman JR, Batzoglou S et al (2005) Sequencing of Aspergillus nidulans and comparative analysis with A-fumigatus and A-oryzae. Nature 438:1105–1115
4. Machida M, Asai K, Sano M, Tanaka T, Kumagai T, Terai G et al (2005) Genome sequencing and analysis of Aspergillus oryzae. Nature 438:1157–1161
5. Nierman WC, Pain A, Anderson MJ, Wortman JR, Kim HS, Arroyo J et al (2005) Genomic sequence of the pathogenic and allergenic filamentous fungus Aspergillus fumigatus. Nature 438:1151–1156
6. Pel HJ, de Winde JH, Archer DB, Dyer PS, Hofmann G, Schaap PJ et al (2007) Genome sequencing and analysis of the versatile cell factory Aspergillus niger CBS 513.88. Nat Biotechnol 25:221–231
7. Bok JW, Chiang YM, Szewczyk E, Reyes-Dominguez Y, Davidson AD, Sanchez JF et al (2009) Chromatin-level regulation of biosynthetic gene clusters. Nat Chem Biol 5: 462–464
8. Sanchez JF, Chiang YM, Szewczyk E, Davidson AD, Ahuja M, Oakley CE et al (2010) Molecular genetic analysis of the orsellinic acid/F9775 gene cluster of Aspergillus nidulans. Mol Biosyst 6:587–593
9. Szewczyk E, Chiang YM, Oakley CE, Davidson AD, Wang CCC, Oakley BR (2008) Identification and characterization of the asperthecin gene cluster of Aspergillus nidulans. Appl Environ Microbiol 74:7607–7612
10. Chiang YM, Szewczyk E, Davidson AD, Keller N, Oakley BR, Wang CCC (2009) A gene cluster containing two fungal polyketide synthases encodes the biosynthetic pathway for a polyketide, asperfuranone, in Aspergillus nidulans. J Am Chem Soc 131:2965–2970
11. Still WC, Kahn M, Mitra A (1978) Rapid chromatographic technique for preparative separations with moderate resolution. J Org Chem 43:2923–2925

Chapter 7

Detection and Purification of Non-ribosomal Peptide Synthetase Products in *Neosartorya fischeri*

Beate Wollinsky and Shu-Ming Li

Abstract

Non-ribosomal peptide synthetases (NRPSs) are important enzymes of production machinery for natural products including clinically used antibiotics, antifungal, and anticancer agents. NRPS products are usually further modified by tailoring enzymes, resulting in the formation of diverse structures. We demonstrate here the production and isolation of metabolites produced by two bi-modular NRPSs, i.e., acetylaszonalenin and fumitremorgin-type alkaloids in *Neosartorya fischeri* NRRL181.

Key words: Acetylaszonalenin, Fumitremorgin-type alkaloids, HPLC, *Neosartorya fischeri*, Non-ribosomal peptide synthetase, Silica gel column chromatography

1. Introduction

Non-ribosomal peptide synthetases (NRPSs) are multienzyme complexes and play important roles in the formation of microbial secondary metabolites derived from amino acids (1–3). The peptides produced by the NRPS machinery are released from the enzymes by thioesterases (4) and often further modified by diverse tailoring enzymes such as hydroxylases, oxidoreductases as well as acetyl, methyl, prenyl, and glycosyl transferases (5). Like bacteria (6, 7), fungi are also important producers of secondary metabolites including NRPS products (8, 9). A number of fungal NRPS products or their derivatives are used clinically in modern medicine, e.g. β-lactam antibiotics (10) and ergot alkaloids (11). In contrast to bacteria, fungi produce more frequently cyclic dipeptides as NRPS products, e.g. acetylaszonalenin, echinulin, fumitremorgin-type alkaloids, brevianamides, notoamides, and roquefortine family

Nancy P. Keller and Geoffrey Turner (eds.), *Fungal Secondary Metabolism: Methods and Protocols*,
Methods in Molecular Biology, vol. 944, DOI 10.1007/978-1-62703-122-6_7, © Springer Science+Business Media, LLC 2012

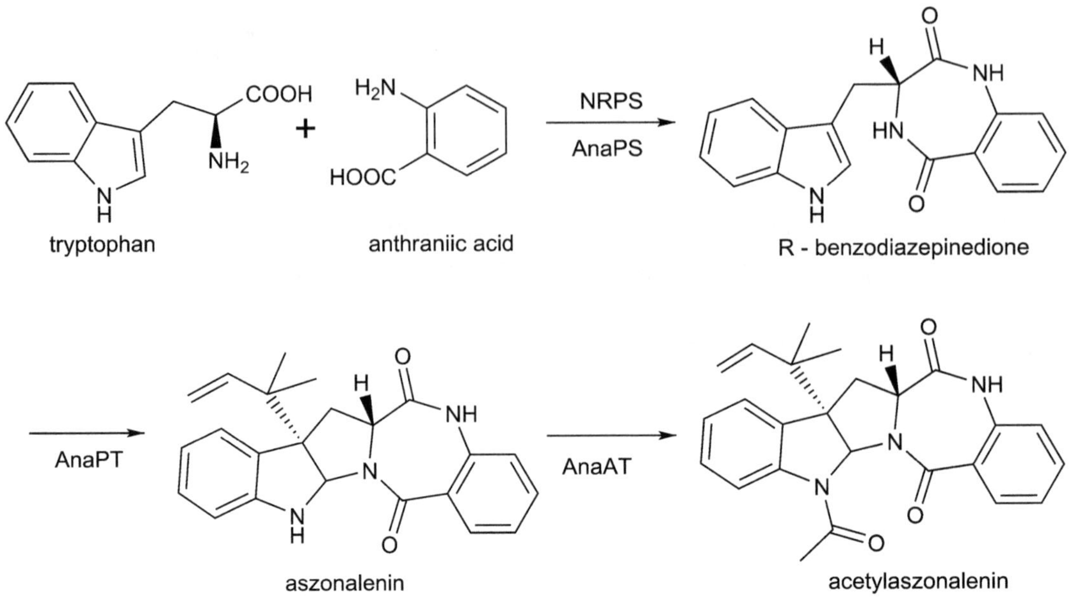

Fig. 1. Biosynthetic pathway of acetylaszonalenin in *Neosartorya fischeri*.

alkaloids (12). The responsible NRPSs for these compounds consist of two modules for two amino acids (13, 14).

Neosartorya fischeri (anamorph form: *Aspergillus fischerianus*) is a heat-resistant fungus from the family Trichocomaceae of the ascomycetes (15). The optimal temperature for its growth is found to be between 26 and 45°C (16). The teleomorph form (*N. fischeri*) predominates at 24°C and the anamorph form at 37°C (17). The organism was firstly isolated from canned strawberries in 1963 (18) and was reported to produce diverse mycotoxins including fumitremorgin-type alkaloids (19) and acetylaszonalenin (20).

Recently, the biosynthetic gene clusters of acetylaszonalenin and fumitremorgin-type alkaloids were identified by genome mining and characterized by biochemical and/or molecular biological approaches (12, 13, 21). It has been shown that the product of the NRPS AnaPS is modified subsequently by the prenyltransferase AnaPT and the acetyltransferase AnaAT, resulting in the formation of aszonalenin and acetylaszonalenin (Fig. 1) (22). In the biosynthesis of fumitremorgin-type alkaloids, brevianamide F (*cyclo*-L-Trp-L-Pro) is synthesized by the NRPS FtmPS (also called FtmA) and converted to diverse structures by prenylation, hydroxylation and cyclization (Fig. 2) (21). We describe here the production, detection, and purification of secondary metabolites from these two biosynthetic pathways, i.e. aszonalenin and acetylaszonalenin as well as 12,13-dihydroxyfumitremorgin C and fumitremorgin B. It was clearly demonstrated that different metabolites were detected as main products under different culture conditions.

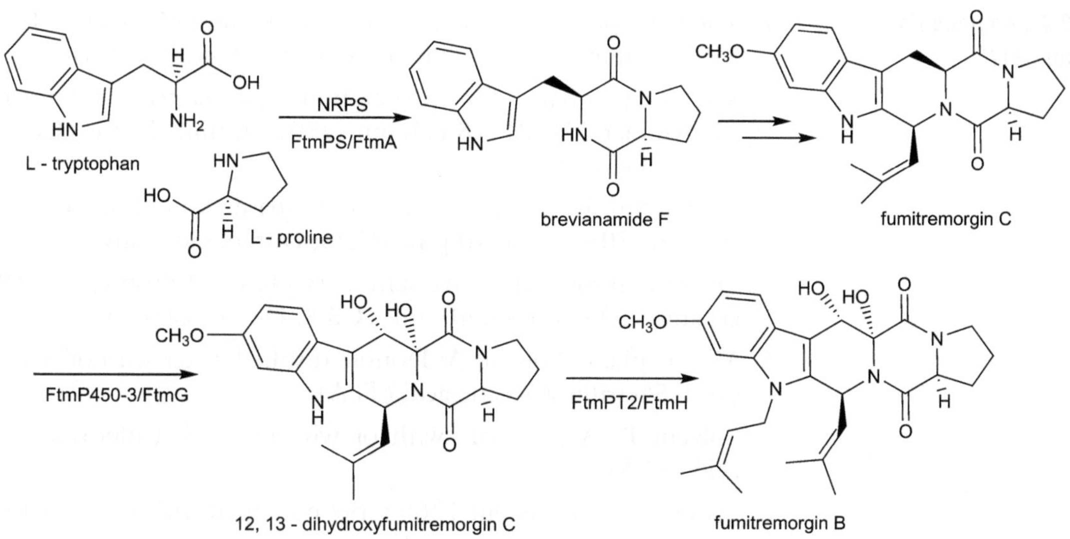

Fig. 2. Biosynthetic pathway of fumitremorgin-type alkaloids in *Neosartorya fischeri*.

2. Materials

2.1. Media for Fungal Cultivation

Prepare all media with double-distilled water and sterilize them for 20 min at 121°C. Adjust pH value with HCl (see Note 1). The final pH value should be about 4.8 for PD medium, 7.2 for CDY medium, and approximately 4.5 for MEP medium. Component amounts listed below are calculated for 1 l media. For solid media, add 15 g of agar to 1 l of liquid medium before autoclaving.

1. Potato dextrose medium (PD medium): Weigh 24.0 g potato dextrose broth.
2. Czapec dox medium with yeast extract (CDY medium): Weigh 35.0 g Czapec dox and 5.0 g yeast extract.
3. Malt extract peptone medium (MEP medium): Weigh 30.0 g malt extract and 3.0 g peptone.

2.2. Solutions for Preparing a Spore Suspension

Autoclave the solutions as for media (2.1).

1. 0.1% Tween: Weigh 0.3 g Tween 20 and add 300 ml double-distilled water.
2. 20% Glycerol: Mix 23.3 ml glycerol 86% and with 76.7 ml double-distilled water.

2.3. Silica Gel Column Chromatography

1. Column dimension: Ø = 3.5 × 45 cm.
2. Stationary phase: Silica gel 60, 40–63 µm (VWR, Darmstadt, Germany).
3. Mobile phase: Cyclohexane, ethyl acetate.

2.4. Components for HPLC

1. Stationary phase for analysis (analytical column): Multospher 120 RP-18 column (250×4 mm, 5 μm) (C&S, Düren, Germany).
2. Stationary phase for preparation (preparative column): Multospher 120 RP-18 column (250×10 mm, 5 μm) (C&S, Düren, Germany).
3. Precolumn for analytical stationary phase: Multoprep RP-18 column (30×4 mm, 40 μm) (C&S, Düren, Germany).
4. Precolumn for preparative stationary phase: Multoprep RP-18 column (50×10 mm, 40 μm) (C&S, Düren, Germany).
5. Mobile phase: Solvent A: Double-distilled water with or without 0.5% trifluoroacetic acid (TFA).

   Solvent B: Acetonitrile with or without 0.5% trifluoroacetic acid (TFA).
6. HPLC system: Agilent 1200 series equipped with an autosampler and photodiode array detector.

3. Methods

3.1. Cultivation of *N. fischeri* NRRL 181

3.1.1. Preparation of Stock Solution of Spores

1. Prepare five plates with solid CDY medium.
2. Spread 1,000–10,000 spores to each plate.
3. Cultivate for approximately 2 weeks at 25°C in darkness.
4. Wash each plate thoroughly (spores and mycelia) three times with 4 ml of 0.1% Tween (see Notes 2 and 3).
5. Collect and transfer the suspensions to 50 ml falcon tubes with some glass beads and vortex them exhaustively (see Note 4).
6. Filter the suspension over sterile cotton wool.
7. Centrifuge for 5 min at 3,000 × *g* and resuspend the pellet in 500 μl sterile 20% glycerol (see Note 5).
8. Use fresh spores for inoculation or store them at −80°C (see Note 6).

3.1.2. Fungal Cultivation

1. For screening of secondary metabolite production in submerged liquid cultures, choose 250 ml flasks and add 100 ml medium described above.
2. To get more crude extract for silica gel column chromatography, use large scale, e.g. 1 l medium in 2 l flask.
3. After autoclaving and cooling to room temperature, inoculate with approximately 1×10^5 spores from the spore suspension for 100 ml medium.
4. Cultivate the fungus for 2 weeks at 25 or 37°C in darkness (see Subheading 3.5).

3.2. Extraction of the Secondary Metabolites

1. Filter the culture broth to get culture filtrate and mycelia (see Note 7).
2. Extract the culture filtrate twice with equal volume of ethyl acetate (see Notes 8 and 9).
3. Evaporate the organic phase to dryness with a rotary evaporator (see Note 10).
4. Dissolve the residue in methanol for HPLC analysis.
5. Extract the mycelia after step 1 with methanol: acetone (1:1) in a ratio of 1 g wet fresh cell/ml.
6. Evaporate the filtrate to dryness with a rotary evaporator (see Note 10).
7. Dissolve the residue in methanol for HPLC analysis (see Note 11).

3.3. Silica Gel Column Chromatography

Four compounds, aszonalenin, acetylaszonalenin, 12,13-dihydroxyfumitremorgin C, and fumitremorgin B, were isolated from different cultures by silica gel chromatography in a similar way as follows.

1. The metabolites must be extracted from culture broth with ethyl acetate and from mycelia with methanol and acetone (see Subheading 3.2).
2. Use 20–30 g of silica gel for 1 g of the extract.
3. Add cyclohexane to the silica gel and degas it.
4. Fill the column with the silica gel suspension.
5. Dissolve the extract in 3–5 ml of methanol and add to 2 g silica gel.
6. Evaporate the solvent on an evaporator and apply the residue (silica gel with bound extract) onto the column (see Note 12).
7. Start the column chromatography with cyclohexane using twice the volume of the stationary phase. For example, if your column has a diameter of 3.5 cm and is filled with silica gel to 30 cm in length, you should use 600 ml solvent for each fraction.
8. Increase the polarity of the eluent by the addition of ethyl acetate to cyclohexane (Table 1).
9. The four compounds are found in different fractions (Table 1).
10. Evaporate each fraction with a rotary evaporator to dryness and weigh them (see Note 13).
11. To detect the substance in the fractions, dissolve the residues in methanol (concentration about 20 μg/μl) (see Note 14) and analyze on HPLC.
12. If necessary, purify the fractions on HPLC as given under Subheading 3.5.

Table 1
Silica gel column chromatography

Eluent (v/v)	Substances
Cyclohexane	–
Cyclohexane:ethyl acetate (3:1)	Aszonalenin
Cyclohexane:ethyl acetate (1:1)	Acetylaszonalenin, fumitremorgin B
Cyclohexane:ethyl acetate (1:9)	12,13-Dihydroxyfumitremorgin C

3.4. High Performance Liquid Chromatography

1. Dissolve your sample in methanol.
2. Centrifuge the sample for 10 min at 10,000×*g* or filter the sample (see Note 15).
3. Transfer the sample for HPLC into an autosampler tube or a new reaction tube for manual injection.
4. Use an analytical column for analysis of extracts and fractions at a flow rate of 1 ml/min. To isolate compounds, you can also use the analytical column under the same condition. To save time, it is convenient to use the preparative column. In this case, you can usually use the same gradient, but with a flow rate of 2.0–2.5 ml.
5. Equilibrate the HPLC column with the mobile phase (see Note 16). Use the same ratio of solvent A and B with which you will start your analysis. The equilibration time depends on the system set up, usually 5–10 min.
6. Use a linear gradient from 0–100% acetonitrile in 20 min. After each run wash the column with acetonitrile for 5 min (see Note 17).
7. Detection at 254 nm.

3.5. Different Metabolites Were Produced Under Different Culture Conditions

1. For screening in submerged liquid cultures, different media and temperature (25 or 37°C) are used.
2. The fungus is cultivated with shaking (120 rpm) or without shaking (surface cultures) for about 2 weeks.
3. The secondary metabolites in culture broth and mycelia are extracted and analyzed on HPLC.
4. Different main products are detected with different conditions (Fig. 3).

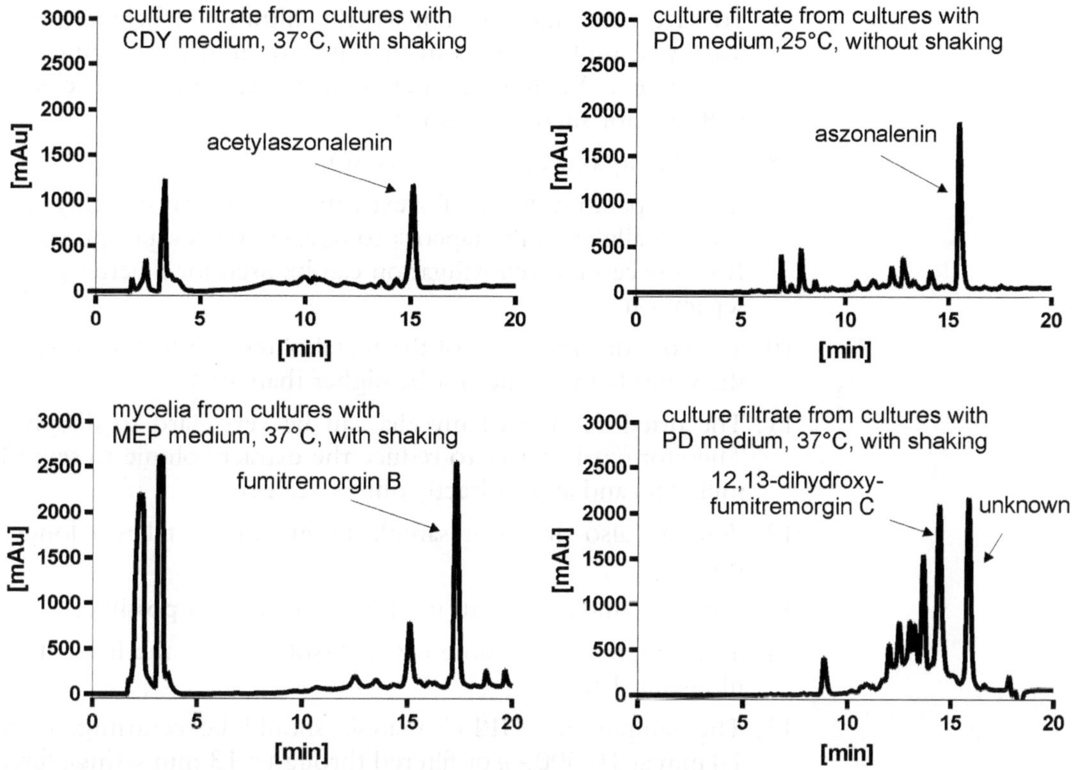

Fig. 3. HPLC chromatograms of fungal extracts from cultures obtained under different conditions. The substances are detected with a photodiode array detector and illustrated for absorption at 254 nm.

4. Notes

1. For adjusting the pH value, dissolve the compounds in 80% of the end volume, adjust the pH value and add water to the end volume.
2. To prevent the spore aerosols, add the 0.1% Tween solution slowly to the plates.
3. Use an inoculation loop to scrape the spores from the plates.
4. Use glass beads with a diameter of 2.85–3.3 mm.
5. The volume of 20% glycerol to be used depends on the amount of the pellet, usually 100 μl for spores from one plate.
6. For a better comparison of fungal growth or secondary metabolite production, it is advisable to use the same amount of spores for inoculation. Therefore, the titer of the spore stock solution should be determined by use different dilutions. Plating on solid CDY medium for a viable colony count.

7. To save time and to improve the filtration efficiency, it is usually helpful to carry out the filtration under vacuum (suck off). Use a Büchner funnel with a filter paper (pore size 5–8 μm) for suction filtration.
8. Dry the organic phase with Na_2SO_4.
9. Use a separating funnel for extraction. Shake thoroughly and occasionally open the tapcock to release excess vapor pressure. If it is necessary, centrifugation can be used for a better phase separation.
10. To avoid decomposing of the metabolites, the temperature of the water bath should not be higher than 40°C.
11. The dried extracts of mycelia can be very hard to dissolve. Therefore, it is better to reduce the extract volume to several milliliters and apply directly onto silica gel.
12. You can also dry your sample by air, but it takes a longer time.
13. For low sample amount, use bottles as small as possible.
14. The sample can be more easily dissolved with the help of an ultrasonic bath.
15. The samples for HPLC analysis should be centrifuged for 10 min at 10,000 × *g* or filtered through a 13 mm syringe filter (0.2 μm PTFE). Insoluble components are thereby removed, which could plug the HPLC equipment.
16. Use always double-distilled water and solvents of HPLC Gradient Grade.
17. The metabolites can be usually better separated in the presence of 0.5% TFA. However, keep in mind that TFA can influence of stability of some compounds.

Acknowledgment

This work in the author's laboratory was financially supported by a grant from the Deutsche Forschungsgemeinschaft (Li 844/1-3).

References

1. Walsh CT, Fischbach MA (2010) Natural products version 2.0: connecting genes to molecules. J Am Chem Soc 132:2469–2493
2. Koglin A, Walsh CT (2009) Structural insights into nonribosomal peptide enzymatic assembly lines. Nat Prod Rep 26:987–1000
3. Strieker M, Tanovic A, Marahiel MA (2010) Nonribosomal peptide synthetases: structures and dynamics. Curr Opin Struct Biol 20: 234–240
4. Du L, Lou L (2010) PKS and NRPS release mechanisms. Nat Prod Rep 27:255–278
5. Walsh CT, Chen H, Keating TA, Hubbard BK, Losey HC, Luo L, Marshall CG, Miller DA, Patel HM (2001) Tailoring enzymes that modify nonribosomal peptides during and after

chain elongation on NRPS assembly lines. Curr Opin Chem Biol 5:525–534

6. Challis GL, Naismith JH (2004) Structural aspects of non-ribosomal peptide biosynthesis. Curr Opin Struct Biol 14:748–756
7. Finking R, Marahiel MA (2004) Biosynthesis of nonribosomal peptides. Annu Rev Microbiol 58:453–488
8. Keller NP, Turner G, Bennett JW (2005) Fungal secondary metabolism - from biochemistry to genomics. Nat Rev Microbiol 3: 937–947
9. von Döhren H (2004) Biochemistry and general genetics of nonribosomal peptide synthetases in fungi. Adv Biochem Eng Biotechnol 88:217–264
10. Brakhage AA, Sprote P, Al Abdallah Q, Gehrke A, Plattner H, Tuncher A (2004) Regulation of penicillin biosynthesis in filamentous fungi. Adv Biochem Eng Biotechnol 88:45–90
11. Wallwey C, Li S-M (2011) Ergot alkaloids: structure diversity, biosynthetic gene clusters and functional proof of biosynthetic genes. Nat Prod Rep 28:496–510
12. Li S-M (2010) Prenylated indole derivatives from fungi: structure diversity, biological activities, biosynthesis and chemoenzymatic synthesis. Nat Prod Rep 27:57–78
13. Ames BD, Walsh CT (2010) Anthranilate-activating modules from fungal nonribosomal peptide assembly lines. Biochemistry 49: 3351–3365
14. Maiya S, Grundmann A, Li S-M, Turner G (2006) The fumitremorgin gene cluster of *Aspergillus fumigatus*: identification of a gene encoding brevianamide F synthetase. Chembiochem 7:1062–1069
15. Hocking AD, Pitt JI (1984) Food spoilage fungi. II Heat resistent fungi. CSIRO Food Res Q 44:73–82
16. Nielsen PV, Beuchat LR, Frisvad JC (1989) Influence of atmospheric oxygen content on growth and fumitremorgin production by a heat-resistant mold *Neosartorya fischeri*. J Food Sci 54:679–682, 685
17. Beuchat LR (1986) Extraordinary heat resistance of *Talaromyces flavus* and *Neosartorya fischeri* ascopores in fruit products. J Food Sci 51:1506–1510
18. Kavanagh J, Larchet N, Stuart M (1963) Occurence of a heat-resistant species of *Aspergillus* in canned strawberries. Nature 198:1322
19. Nielsen PV, Beuchat LR, Frisvad JC (1988) Growth of and fumitremorgin production by *Neosartorya fischeri* as affected by temperature, light, and water activity. Appl Environ Microbiol 54:1504–1510
20. Wakana D, Hosoe T, Itabashi T, Nozawa K, Okada K, Takaki GMdC, Yaguchi T, Fukushima K, Kawai KI (2006) Isolation of isoterrein from *Neosartorya fischeri*. Mycotoxins 56:3–6
21. Li S-M (2011) Genome mining and biosynthesis of fumitremorgin-type alkaloids in ascomycetes. J Antibiot 64:45–49
22. Yin W-B, Grundmann A, Cheng J, Li S-M (2009) Acetylaszonalenin biosynthesis in *Neosartorya fischeri*. Identification of the biosynthetic gene cluster by genomic mining and functional proof of the genes by biochemical investigation. J Biol Chem 284:100–109

Chapter 8

Production, Detection, and Purification of Clavine-Type Ergot Alkaloids

Christiane Wallwey and Shu-Ming Li

Abstract

Ergot alkaloids are indole derivatives with diverse structures and biological activities. This chapter describes the procedure from fungal cultivation to purified ergot alkaloids, as exemplified by fumigaclavine A in *Penicillium commune*. Furthermore, useful notes for working with purified ergot alkaloids are given.

Key words: Ergot alkaloids, Fumigaclavine, *Claviceps purpurea*, *Aspergillus fumigatus*

1. Introduction

Ergot alkaloids are secondary metabolites which are produced by fungi of the families Clavicipitaceae (e.g., *Claviceps* and *Neotyphodium*) and Trichocomaceae (including *Aspergillus* and *Penicillium*) (1, 2). It was believed that some plants from the families Convolvulaceae, Poaceae, and Polygalaceae produce also ergot alkaloids as results of horizontal gene transfer from fungi to higher plants during the evolutionary process (2). Recent investigations have however revealed that plant-associated clavicipitaceous fungi are responsible for the ergot alkaloid production in these plants (3–6).

Ergot alkaloids are chemically a complex family of indole derivatives with diverse structures, which show different biological and pharmacological activities (7). They are toxins and caused severe epidemics in the Middle Ages (8–10). On the other hand, they are in widespread use in modern medicine (11). The characteristic structural feature of all ergot alkaloids is the tetracyclic ergoline ring. Based on their structures, ergot alkaloids can be divided into

Nancy P. Keller and Geoffrey Turner (eds.), *Fungal Secondary Metabolism: Methods and Protocols*, Methods in Molecular Biology, vol. 944, DOI 10.1007/978-1-62703-122-6_8, © Springer Science+Business Media, LLC 2012

three classes: the clavine-type alkaloids, the ergoamides, and the ergopeptines (7). The clavine-type alkaloids consisting merely of the ergoline ring, e.g. fumigaclavines, are detected as end pathway products in the fungi of the family Trichocomaceae (7). The members of the family Clavicipitaceae produce D-lysergic acid amides (ergoamides) and/or peptides (ergopeptines) as end products. Functions for most of the structural genes have been assigned to reaction steps in the biosynthesis of ergot alkaloids by gene inactivation experiments or biochemical characterization of the overproduced proteins (Fig. 1) (7). The biosynthetic pathways of ergot alkaloids in Clavicipitaceae und Trichocomaceae begin with the same precursors and share common reaction steps to chanoclavine-I aldehyde, which is converted to festuclavine in *Aspergillus fumigatus* (12), to pyroclavine in *Penicillium commune* (13), and to agroclavine in *Claviceps purpurea* (14).

We describe, as an example, the production of fumigaclavine A in *P. commune* strain NRRL 2033 on wheat grains. By using this medium, fumigaclavine A was detected as the dominant metabolite. Well-established methods for ergot alkaloid analysis such as thin layer chromatography (TLC) and high performance liquid chromatography (HPLC) are used for detection.

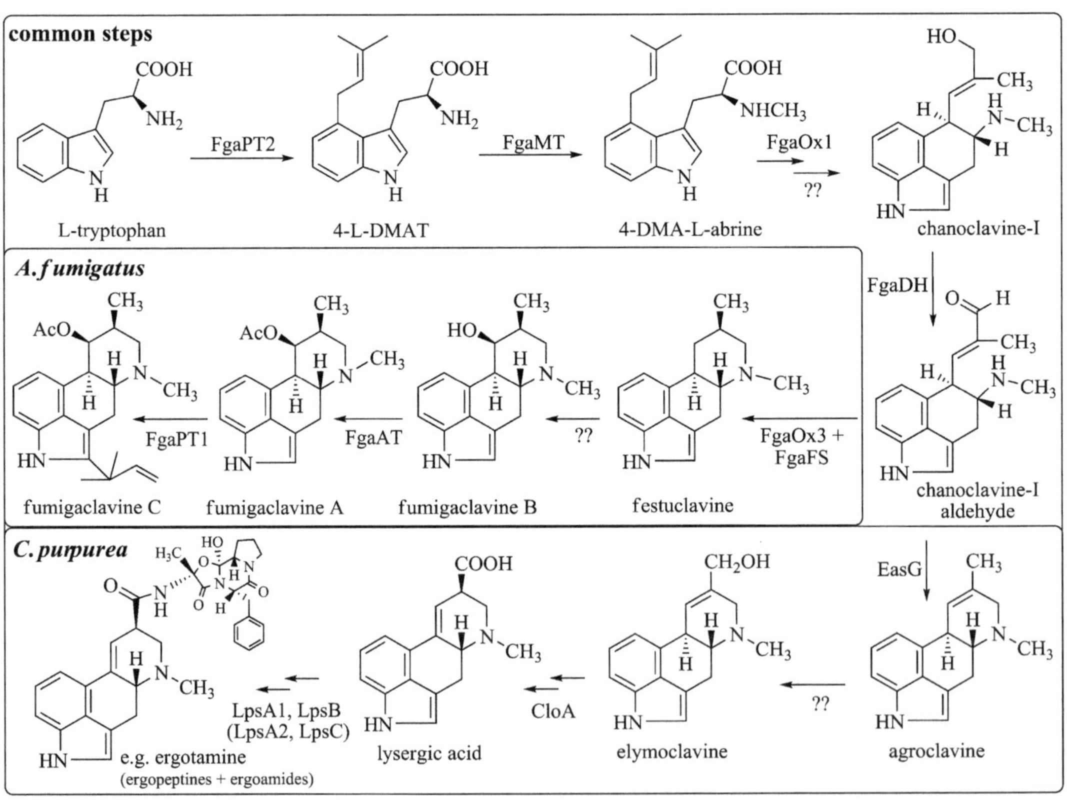

Fig. 1. Putative biosynthetic pathways of ergot alkaloids in *Aspergillus fumigatus* and *Claviceps purpurea*.

2. Materials

2.1. Solutions for Preparing a Spore Suspension

1. Czapec-Dox medium with yeast extract (CDY medium): Weigh 35.0 g Czapec-Dox and 5.0 g yeast extract for 1 l double-distilled water (see Note 1). Add 15 g of agar to 1 l of liquid medium before autoclaving for 20 min at 121°C.
2. 0.1% Tween: Weigh 0.3 g Tween 20 and add 300 ml double-distilled water.
3. 20% Glycerol: Mix 23.3 ml glycerol 86% with 76.7 ml double-distilled water.

2.2. Media for Cultivation of Fungi

1. Glucose Peptone with Soy flour (GPS medium): Weigh 50.0 g glucose, 10.0 g peptone, 5.0 g soy flour, 2.0 g potassium nitrate (KNO_3), 0.5 g magnesium sulfate heptahydrate ($MgSO_4 \times 7H_2O$), and 0.1 g calcium chloride ($CaCl_2$) and dissolve in 1 l double-distilled water. Adjust the pH value to 6.2 with ammonium hydroxide (see Note 2 and 3).
2. Wheat grains: Add 100 g of wheat grains into a 1 l Erlenmeyer flask. Add 100 ml double-distilled water before autoclaving.

   All media have to be autoclaved for 20 min at 121°C.

2.3. Extraction of Ergot Alkaloids from Cultures

1. Ammonium hydroxide, 30–33%, extra pure.
2. Methylene chloride, ≥99%.
3. Ethyl acetate, ≥99%.
4. Sodium sulfate, anhydrous.
5. 0.01 M Hydrochloric acid (HCl): Dilute 37% hydrochloric acid (equates to 12 M) 1–1,200. It may be easier to prepare first a 1 M stock solution and then dilute this to 0.01 M.

2.4. Thin Layer Chromatography and Detection

1. Analytical stationary phase: Merck TLC plate silica gel 60 F_{254}, aluminum sheets, 20 × 20 cm (Place for approximately 15 samples. If necessary cut the plate to the size you need).
2. Preparative stationary phase: TLC ready-to-use layers, SIL G-200 UV254, glass plate, 2.0 mm, Fluorescent indicator (Carl Roth, Karlsruhe).
3. Mobile phase for separation of fumigaclavine A and B: Prepare acetone, methyl alcohol, and ethyl acetate in a ratio of 1:1:1.
4. Mobile phase for separation of fumigaclavine precursors: Prepare methyl alcohol, chloroform, and ammonium hydroxide in a ratio of 80:20:0.2 (see Notes 4 and 5).
5. Van Urk reagent: Weigh 125 mg *para*-dimethylaminobenzaldehyde and 10 mg iron(III) chloride ($FeCl_3$) and dissolve in 100 ml 65% (v/v) sulfuric acid.

6. 1% (w/v) Sodium nitrite ($NaNO_2$) solution: Weigh 0.5 g of sodium nitrite and dissolve in 50 ml double-distilled water.

2.5. High Performance Liquid Chromatography

1. Stationary phase for analysis: Multospher 120 RP-18-column (250×4 mm, 5 μm) (C&S, Düren).
2. Stationary phase for preparation: Multospher 120 RP-18-column (250×10 mm, 5 μm) (C&S, Düren).
3. Precolumn for analytical stationary phase: Multoprep RP-18 column (30×4 mm, 40 μm) (C&S, Düren).
4. Precolumn for preparative stationary phase: Multoprep RP-18 column (50×10 mm, 40 μm) (C&S, Düren).
5. Mobile phase: Solvent A is double-distilled water with 0.5% trifluoroacetic acid (TFA) and solvent B is acetonitrile with 0.5% TFA (see Notes 6 and 7).
6. HPLC system: Agilent 1200 series equipped with an autosampler and photodiode array detector.

3. Methods

3.1. Preparation of Stock Solution of Spores

1. Prepare five plates with solid CDY medium.
2. Spread 10^3 to 10^4 spores to each plate.
3. Cultivate for 7–14 days at 25°C in the dark.
4. Wash each plate thoroughly (spores and mycelia) three times with 4 ml of 0.1% Tween.
5. Collect and transfer the suspensions to 50 ml falcon tubes with some glass beads with a diameter of 2.85–3.3 mm and vortex them exhaustively.
6. Filter the suspension over sterile cotton wool.
7. Centrifuge for 5 min at 3,000×*g* and resuspend the pellet in 500 μl sterile 20% glycerol.
8. Use fresh spores for inoculation or store them at −80°C.

3.2. Cultivation of Fungi for Production of Ergot Alkaloids

1. Prepare the necessary amount of media for cultivation, e.g. 100 ml medium into a 250 ml Erlenmeyer flask for screening of secondary metabolites. For production of fumigaclavine A both GPS medium and wheat grains are suitable.
2. Inoculate 100 ml medium with 1×10^5 spores of *P. commune* for surface cultures.
3. Cultivate at 25°C for 14 days in the dark without shaking.

3.3. Extraction of Ergot Alkaloids

3.3.1. Liquid Cultures

1. Separate the mycelia from media by filtration, e.g., sucking off with a Büchner funnel through a filter paper with a pore size of 5–8 μm.
2. Adjust the pH value of the filtrate to 8–9 with ammonium hydroxide.
3. Extract the filtrate twice with the same volume of methylene chloride.
4. Combine the extracts and evaporate to dryness.

3.3.2. Culture on Wheat Grains

1. Overlay the wheat grains with 300 ml ethyl acetate and shake gently for 4 h at 25°C.
2. Afterwards separate the organic phase by filtration.
3. Dry the organic phase over sodium sulfate.
4. Evaporate to dryness.

3.4. Analysis of Ergot Alkaloids in the Extracts

3.4.1. Removal of Non-alkaloid Substances

1. Dissolve the residue of the extraction in 10 ml ethyl acetate.
2. Extract it twice with 1 volume of 0.01 M hydrochloric acid.
3. Adjust the aqueous phase to pH 9 with ammonium hydroxide.
4. Extract the aqueous phase twice with 1 volume of methylene chloride.
5. Combine the extracts and evaporate to dryness.

3.4.2. Thin Layer Chromatography

1. Draw a starting line with a soft pencil on the bottom of your TLC plate (see Note 8). Be careful that you do not scratch the silica layer.
2. Then mark the required areas for sample application on the starting line. It makes sense to use an area of 1 cm for each sample and 0.5 cm space between two samples.
3. Pour the mobile phase into the chromatography chamber as soon as possible so that there is enough time for saturation of the chamber (see Note 9). Saturation depends on the chamber size, e.g., 30 min for a chamber with the following dimensions: $L \times H \times W$ $22 \times 22 \times 10$ cm (Table 1).
4. Dissolve your sample in a volatile solvent. For clavine-type alkaloids methanol is a convenient solvent.
5. Use a glass capillary to apply your sample as a band with a width of 1 cm as illustrated in Fig. 2 (see Notes 10 and 11).
6. Wait until the solvent is volatilized.
7. Place the prepared TLC plate straight into the TLC chamber.
8. Close the chamber as fast as possible.
9. The time of running depends on the mobile phase and the respective question, but always stop the run before the solvent front achieves the top of the plate.

Table 1
Examples of mobile phases for TLC

Separation of fumigaclavine A and B	Separation of fumigaclavine precursors
Acetone:methyl alcohol:ethyl acetate (1:1:1)	Methyl alcohol:chloroform:ammonium hydroxide (80:20:0.2)

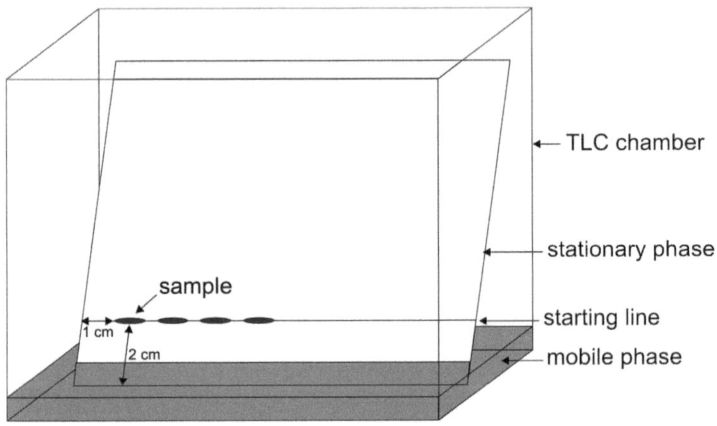

Fig. 2. Schematic illustration of the TLC assembly and the sample application.

10. Take out the plate and mark the solvent front directly with a pencil.
11. Dry the plate completely before detection.
12. For detection of the ergot alkaloids, spray the TLC plate first with van Urk reagent (see Note 12).
13. To intensify the spots, spray the TLC plate with 1% sodium nitrite solution (see Note 13).

3.5. Purification of Ergot Alkaloids

Purification of ergot alkaloids from fungal extract is possible with preparative TLC (see Subheading 3.4.1 and Note 14) or high performance liquid chromatography (HPLC).

3.5.1. High Performance Liquid Chromatography

1. Dissolve your sample in methanol.
2. Centrifuge the sample for 10 min at 10,000 × g (see Note 15).
3. Transfer the sample for HPLC into an autosampler tube or a new reaction tube for manual injection.
4. Equilibrate the HPLC column with the mobile phase (see Note 16). Use the same ratio of solvent A and B with which

Table 2
Examples of HPLC elution profiles

	Method for analysis of fungal extracts	Method for separation of chanoclavine-I and chanoclavine-I aldehyde (15)
Isocratic step	Not used	25% Solvent B for 8 min
Gradient step	30–100% Solvent B in 25 min	25–35% Solvent B in 25 min
Wash step	100% Solvent B for 5 min	100% Solvent B for 5 min
Equilibration step	30% Solvent B for 5 min	25% Solvent B for 5 min

you start the analysis of your sample. The equilibration time depends on the system set up.

5. Sample injection and analysis (Table 2).
6. Detection at 282 nm.

3.6. Purification of Clavine-Type Alkaloids from Enzyme Assays

If you carry out enzyme assays with ergot alkaloids (see Note 17 and 18), you can remove enzymes and cofactors after incubation by extraction and thereby concentrate the sample before analysis.

1. Adjust the pH value of the assay to 8–9 with 1 M sodium hydroxide.
2. Extract twice with 2 volumes of ethyl acetate (see Note 19).
3. Combine the ethyl acetate phases and evaporate to dryness (see Note 20 and 21).
4. Dissolve the sample in a volume of solvent, which is appropriate for your application.
5. Analyze the samples by TLC (see Subheading 3.4.1) or HPLC (Fig. 3, see also Subheading 3.4.1).

3.7. Sample Preparation for Nuclear Magnetic Resonance Spectroscopy

Clavine-type alkaloids and their precursors can exist in protonated or deprotonated form. For example, they are protonated after isolation by HPLC due to the presence of TFA in the mobile phase. After isolation by TLC, they exist as free bases (see Note 22). For nuclear magnetic resonance (NMR) measurement, the free base is usually dissolved in deuterated chloroform ($CDCl_3$). The protonated form is hardly soluble in $CDCl_3$ and is therefore dissolved in deuterated methyl alcohol (CD_3OD) (see Note 23) (12).

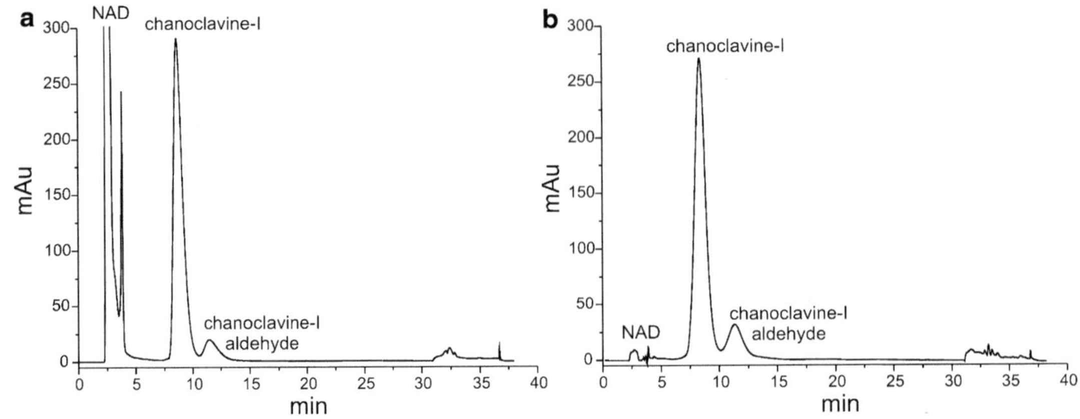

Fig. 3. HPLC chromatograms of enzyme assays for conversion of chanoclavine-I to chanoclavine-I aldehyde, (**a**) directly after incubation (**b**) after removal of NAD by extraction with ethyl acetate.

4. Notes

1. When you prepare media, first put water and a magnetic stirrer into your flask and stir it before adding the solid components. Otherwise the components can agglomerate and need more time to dissolve.
2. If you have to adjust the pH value of a solution, first make up to 80–90% of the final volume, then adjust the pH value and make up to the end volume.
3. If the pH value of the solution differs significantly from the required pH, you can start with concentrated acids or bases so that you do not need a large volume. When the pH value of the solution is close to the required pH, you should change to diluted acids and bases, otherwise it is very likely that you will overshoot the desired pH value.
4. Many mobile phases need to be used fresh, especially mobile phases with components which volatilize easily, since this leads to changes in the composition of the mobile phase.
5. Keep volume contraction in mind if you have mobile phases with water and alcohol.
6. TFA in the mobile phase for reverse-phase HPLC sharpens the peaks and improves resolution. Furthermore some components cannot be eluted completely from the column without TFA.
7. Use cheaper solvents with a lower purity for extraction of the fungal media to save money. Solvents with a higher purity are necessary for HPLC and preparation of samples for NMR measurement to avoid unnecessary impurities in the spectra.

8. You have to pay attention that the distance between the sample and the bottom of the TLC plate is sufficient. Furthermore, be sure that the sample is above the level of the mobile phase, so that the sample cannot diffuse into the solvent.
9. For reproducible *Rf* values the chromatography chamber must be saturated. The *Rf* value depends on the degree of saturation of the chamber. The eluent front runs faster in a saturated chamber than in an unsaturated one.
10. Make sure that there is enough space between the sample and the edges of the plate, otherwise the so called "edge-effects" can appear.
11. Check under UV light to ensure that there is enough sample on your TLC plate before starting the TLC. If not, you can add more sample.
12. After development, check the plate again under UV light at 254 nm and mark the spots which are visible by fluorescence quenching. The van Urk reaction allows specific detection of indole derivatives. After spraying the TLC plate with the van Urk reagent, indole derivatives appear as purple spots under visible light.
13. If you store van Urk reagent and sodium nitrite in the dark and in closed bottles, the reagents can be used for longer. The oxygen in the air oxidizes sodium nitrite to sodium nitrate.
14. Sometimes the resolution is better with analytical stationary phase. Then you can also use this plate for isolation of compounds. But keep in mind that the sample capacity is much lower.
15. Before transferring the samples for the HPLC into the autosampler tubes, centrifuge them for 10 min at 10,000 × *g*. Thereby insoluble components are removed, which can plug the HPLC equipment.
16. For the same reason, use always double-distilled water and solvents with a purity of at least HPLC Gradient Grade.
17. Take note of the glycerol, DMSO, or cofactor concentration in enzyme assays. A high concentration of these compounds in assays can inhibit the enzyme reaction. Maybe you have to determine the best concentrations of these compounds for your enzyme.
18. Furthermore, check the pH value of your assay. Some cofactors, e.g., NAD, NADP, FMN, or FAD, can change the pH value significantly. Sometimes it is necessary to adjust the pH of the cofactors before adding it to the assay.
19. Because of the faster separation of the phases, ergot alkaloids in the enzyme assays can be better extracted with ethyl acetate than with methylene chloride.

20. Use a "Speed-Vac" to evaporate small volumes. First start to rotate the samples then start the vacuum pump. Otherwise the samples can be sucked out of the reaction tube.
21. For the same reason, it is important to release firstly the vacuum and then to stop the rotation after sample drying with "Speed-Vac."
22. Residues of silica gel particles influence the quality of NMR spectra. Therefore if it is possible, isolate the compounds for NMR measurement by HPLC.
23. If you have only low amounts of sample for NMR measurement, you can use smaller NMR tubes so that you need only 300 μl instead of 750 μl solvent for dissolving the sample.

Acknowledgment

The work in the author's laboratory was financially supported by a grand from the Deutsche Forschungsgemeinschaft (Li844/3-1).

References

1. Schardl CL, Panaccione DG, Tudzynski P (2006) Ergot alkaloids—biology and molecular biology. Alkaloids Chem Biol 63:45–86
2. Gröger D, Floss HG (1998) Biochemistry of ergot alkaloids—achievements and challenges. Alkaloids Chem Biol 50:171–218
3. Ahimsa-Müller MA, Markert A, Hellwig S, Knoop V, Steiner U, Drewke C, Leistner E (2007) Clavicipitaceous fungi associated with ergoline alkaloid-containing convolvulaceae. J Nat Prod 70:1955–1960
4. Leistner E, Steiner U (2009) Fungal origin of ergoline alkaloids present in dicotyledonous plants (Convolvulaceae). In: Anke T, Weber D (eds) The mycota XV: physiology and genetics, 1st edn. Springer, Berlin, Heidelberg, pp 197–208
5. Markert A, Steffan N, Ploss K, Hellwig S, Steiner U, Drewke C, Li S-M, Boland W, Leistner E (2008) Biosynthesis and accumulation of ergoline alkaloids in a mutualistic association between *Ipomoea asarifolia* (Convolvulaceae) and a Clavicipitalean fungus. Plant Physiol 147:296–305
6. Kucht S, Gross J, Hussein Y, Grothe T, Keller U, Basar S, Konig WA, Steiner U, Leistner E (2004) Elimination of ergoline alkaloids following treatment of *Ipomoea asarifolia* (Convolvulaceae) with fungicides. Planta 219:619–625
7. Wallwey C, Li S-M (2011) Ergot alkaloids: structure diversity, biosynthetic gene clusters and functional proof of biosynthetic genes. Nat Prod Rep 28:496–510
8. Lee MR (2009) The history of ergot of rye (*Claviceps purpurea*) II: 1900–1940. J R Coll Physicians Edinb 39:365–369
9. Lee MR (2010) The history of ergot of rye (*Claviceps purpurea*) III: 1940–80. J R Coll Physicians Edinb 40:77–80
10. Lee MR (2009) The history of ergot of rye (*Claviceps purpurea*) I: from antiquity to 1900. J R Coll Physicians Edinb 39:179–184
11. Haarmann T, Rolke Y, Giesbert S, Tudzynski P (2009) Ergot: from witchcraft to biotechnology. Mol Plant Pathol 10:563–577
12. Wallwey C, Matuschek M, Xie X-L, Li S-M (2010) Ergot alkaloid biosynthesis in *Aspergillus fumigatus*: conversion of chanoclavine-I aldehyde to festuclavine by the festuclavine synthase FgaFS in the presence of the old yellow enzyme FgaOx3. Org Biomol Chem 8:3500–3508
13. Unsöld IA, Li S-M (2006) Reverse prenyltransferase in the biosynthesis of fumigaclavine C in *Aspergillus fumigatus*: gene expression, purification and characterization of fumigaclavine C synthase FgaPT1. Chembiochem 7:158–164

14. Matuschek M, Wallwey C, Xie X, Li SM (2011) New insights into ergot alkaloid biosynthesis in *Claviceps purpurea*: an agroclavine synthase EasG catalyses, via a non-enzymatic adduct with reduced glutathione, the conversion of chanoclavine-I aldehyde to agroclavine. Org Biomol Chem 9:4328–4335

15. Wallwey C, Matuschek M, Li S-M (2010) Ergot alkaloid biosynthesis in *Aspergillus fumigatus*: conversion of chanoclavine-I to chanoclavine-I aldehyde catalyzed by a short-chain alcohol dehydrogenase FgaDH. Arch Microbiol 192:127–134

Chapter 9

Analysis of Volatile Compounds Emitted by Filamentous Fungi Using Solid-Phase Microextraction-Gas Chromatography/Mass Spectrometry

Ludmila V. Roze, Randolph M. Beaudry, and John E. Linz

Abstract

Here, we describe a solid-phase microextraction-gas chromatography/mass spectrometry (SPME-GC/MS) analytical approach that identifies and analyzes volatile compounds in the headspace above a live fungal culture. This approach is a sensitive, solvent-free, robust technique; most importantly from a practical standpoint, this approach is noninvasive and requires minimal sample handling. Aliquots of liquid fungal cultures are placed into vials equipped with inert septa and equilibrated at a constant temperature, and headspace gases are sampled using an SPME fiber inserted through the septum into the headspace above the fungal culture for a standardized period of time. The outer polymer coating of a fused silica fiber absorbs volatiles from the headspace; the volatiles are then desorbed in the hot GC inlet and chromatographed in the usual manner. The separated compounds are subsequently identified by mass spectrometry. All steps in volatile profiling of a single sample from volatile sorption on a fiber to obtaining a list of volatiles can take as little as 15 min or can be extended to several hours if longer sorption is required for compounds present at very low levels and/or have low rates of diffusion.

Key words: Filamentous fungi, Aspergillus, Volatiles, Solid-phase microextraction, Gas chromatography/mass spectrometry

1. Introduction

Volatile cell metabolites are an important group of compounds with a high vapor pressure at normal temperatures. These compounds are partitioned between gas and liquid phase under growth conditions optimal for many microorganisms including filamentous fungi. This large group of diverse organic compounds includes acids, alcohols, aldehydes, esters, short-chain fatty acids, lipid oxides, terpenes, and phenolics. Volatile compounds in their liquid

Nancy P. Keller and Geoffrey Turner (eds.), *Fungal Secondary Metabolism: Methods and Protocols*,
Methods in Molecular Biology, vol. 944, DOI 10.1007/978-1-62703-122-6_9, © Springer Science+Business Media, LLC 2012

phase and gas phase have been shown to play an important role in fungal development, defense, protection against stress, communication, and pathogenicity (1–14).

Volatile organic compounds represent a significant portion of the cell metabolome although the number of cell volatiles and their ratio to the entire number of cell metabolites is not clear. Bacteria, including *Staphylococcus*, *Bacillus*, *Pseudomonas*, and *Xanthomonas*, produce up to 60 volatile compounds per strain (15). A total of 346 volatiles were reported to be emitted by different bacteria (16). Interestingly, based on genome-scale reconstruction of the metabolic network in *Escherichia coli* and *Saccharomyces cerevisiae*, these organisms are in general able to produce 436 and 584 metabolites (the proportion of volatile compounds was not determined), respectively (17, 18); however currently available metabolic networks do not account for specialized (secondary) metabolism that also produces volatile compounds.

Volatile profiling is a powerful tool to discover new metabolites. These might include, for instance, novel antimicrobial compounds, biofuels promising chemicals, and new biochemical pathways (19–27). Importantly, species-specific volatiles could be used as biomarkers for species identification (28–32).

A number of studies demonstrate that volatile profiling analysis could also be used for gaining rapid access to changes in the metabolism of living organisms including plants, yeast, and bacteria and characterizing the metabolic state of the cell or the organ (9, 19, 33–35).

Gas chromatography/mass spectrometry is a premier technique for separation and analysis of volatile compounds. Here, we provide a protocol for a solid-phase microextraction-gas chromatography/mass spectrometry (SPME-GC/MS) analytical approach to analyze volatile compounds produced by the filamentous fungus *Aspergillus parasiticus* grown in a liquid medium. The protocol was developed in an attempt to characterize the association between secondary metabolism and other metabolic and cellular processes (9). We used SPME-GC/MS to identify 24 volatile compounds in the headspace of *A. parasiticus* cultures and via comparative volatile profiling we demonstrated that a genetic block in aflatoxin biosynthesis (a secondary metabolic pathway) or disruption of the global regulator *veA* (a global regulator of secondary metabolism and development) redirects intracellular carbon flow. However, this protocol can also be utilized to gain rapid access (nearly real time) to information on intracellular metabolism in any fungus grown under similar conditions.

To conduct this procedure, a fused-silica fiber coated with polydimethylsiloxane/divinylbenzene (PDMS/DVB, the polymer stationary phase) was inserted into the headspace above the fungal culture, the fiber absorbs volatiles from the growth environment, and the volatiles are then desorbed in the hot GC inlet and gas chromatography is carried out in the usual manner. The separated compounds are subsequently identified by mass spectrometry. The most advantageous practical feature of this technique is that the procedure does not require cell disruption that includes homogenization and extraction steps. In addition, the procedure is sensitive, rapid, and in many cases does not require quantification of the compound with the use of a standard, because the relative amounts of volatiles in the headspace above the culture can be assessed based on instrument response (36) (see Note 1). Experience in handling and operating a gas chromatograph/mass spectrometer and associated computer software is required to successfully complete the procedure.

2. Materials

2.1. Strain and Culture Media

1. *A. parasiticus* strain SU-1 (ATCC 56775): a wild-type aflatoxin producer.
2. YES liquid medium: 2% yeast extract and 6% sucrose, pH 5.8.
3. Potato dextrose agar: 0.4% potato starch, 2% dextrose, 1.5% agar (Becton, Dickinson and Company, Sparks, MD, USA).
4. Sterile Tween-20 solution: 0.05%, sterilized by autoclaving.
5. Sterile glycerol solution: 20% glycerol, sterilized by autoclaving.

2.2. Equipment

1. Graduated 22 ml clear screw cap vials (Supelco, Bellefonte, PA, USA).
2. Mininert® valves (Supelco).
3. SPME fiber assembly 65 μm PDMS-DVB (Supelco).
4. SPME holder (Supelco).
5. Gas chromatograph HP-6890 (Hewlett-Packard Co., Wilmington, DE, USA).
6. Mass spectrometer FCD-650 (LECO Corp., St. Joseph, MI, USA).
7. Shaking water bath.

3. Methods

The procedure described below utilizes one strain of aspergilli, *A. parasiticus* SU-1. However, we anticipate that any fungal species grown in liquid culture may be used for volatile profiling instead of SU-1.

Before the work on analysis of the volatiles begins, whether the goal of the study is to compare volatiles produced by the fungus at different growth stages, by different mutants, or by different fungal species, attention should be given to characterization of growth rate and growth stages of the fungal culture. Knowing the growth characteristics of the culture will provide with an accurate and appropriate sampling regimen and allow valid comparison in volatile profiling.

3.1. Preparation of Spore Stock and Fungal Culture

For preparation of the spore stock, *A. parasiticus* conidiospores (or spores) were spread onto potato dextrose agar solidified in a 150 cm^2 cell culture flask with canted neck (Corning Inc., Corning, NY, USA) using a sterile loop, and grown for 10–14 days at 30°C in the dark. Spores were harvested in sterile 0.05% Tween-20, filtered through sterile glass wool, centrifuged at 3,000 × *g* for 10 min at RT, re-suspended in 20% sterile glycerol at concentration of 10^6–10^8 spores per milliliter, and stored at −80°C.

A. parasiticus SU-1 conidiospores at 10^4 spores/milliliter were inoculated into a 250-ml flask containing 100 ml of YES liquid medium as an aflatoxin-inducing growth medium. The flask was capped with a foam plug that was then covered with a piece of foil; each flask contained five 6 mm glass beads (Sigma-Aldrich, St Louis, MO) (prewashed in 1N HCl as recommended by the manufacturer's manual) to produce more uniform growth in duplicate flasks (37), and the culture was grown at 30°C in the dark at 150 rpm for 72 h.

3.2. Sample Preparation and Sampling of Volatiles

Twelve milliliter of fungal cultures were harvested at 72 h of growth and were dispensed into 22 ml clear screw cap vials equipped with Mininert® valves (Fig. 1). Vials with cultures were pre-equilibrated at 30°C (the same temperature was used for fungal growth) in a water bath for at least 30 min, with shaking at 50 rpm, before headspace gases were sampled (see Note 2).

A 65 μm silica-fused fiber coated with PDMS/DVB (the polymer stationary phase) was conditioned at 250°C for 30 min according to the manufacturer's instructions before use. Sampling was performed by placing the conditioned fiber, attached to the fiber holder, through the Mininert Valve into the headspace above the fungal culture; it is important that the fiber does not touch the culture. The vial containing the culture and the inserted fiber with the holder was incubated for 3 min in a 30°C water bath (see Note 3). Vials were continuously swirled at 100 rpm during incubation and SPME fiber exposure.

3.3. Gas Chromatography and Mass Spectrometry Parameters

The SPME fiber was removed from the vial, and volatiles were desorbed in the injection port of a gas chromatograph for 3 min; absorption and desorption times were optimized as described previously (9, 36). Volatiles were separated on a 29-m-long, 250 μm (i.d.) capillary column (HP-5MS, Hewlett-Packard Co.,

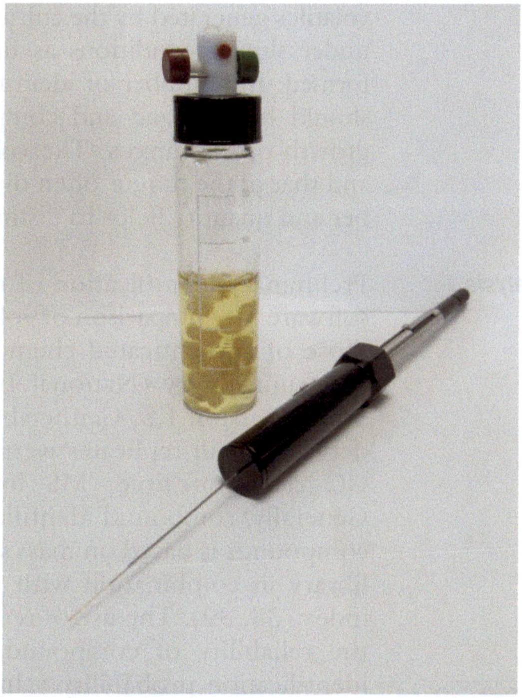

Fig. 1. A vial with a Mininert valve, containing *A. parasiticus* SU-1 culture, and SPME assembly with fiber exposed.

Wilmington, DE, USA) having a film thickness of 0.25 μm. The first 20 cm of the column was cooled with liquid nitrogen during the desorption process to cryofocus the volatiles (see Note 4). Ultrapure helium (99.999%) was used as a carrier gas at a flow rate of 1.5 ml/min. The initial temperature of the column (40°C) was increased upon removal of liquid nitrogen at 60°C/min to obtain a final temperature of 250°C, and this was maintained for 1 min. The volatile metabolites exiting the GC column were fragmented with an electron ionization source (70 eV), and ion masses were detected by time-of-flight mass spectrometry (FCD-650, LECO Corp., St. Joseph, MI, USA). A total of two to five biological replicates were performed for each sample. The headspace samples above the cultures were withdrawn at regular intervals during growth and volatiles were analyzed within 1–2 h following equilibration period; this sampling method resulted in stable and reproducible measurements.

3.4. Controls

Appropriate controls must be performed for the SPME fiber, glass vial, and screw cap and valve; these potentially confounding volatiles are not generated by the fungus and should be excluded from volatiles detected in the fungal culture headspace. A control for volatiles generated by the culture medium alone during incubation under similar conditions as fungal cultures should also be performed; the number of identified compounds and their quantity should be analyzed and compared to volatiles detected during growth of the fungus. The volatile profiles of the medium alone and that of the fungus often overlap, but the analysis of their number and quantity helps to distinguish those specific to the fungus.

3.5. Data Analysis

Preliminary identification of metabolites was achieved by using software for comparison of volatile compounds mass spectra with those of authenticated chemical standards contained in a mass spectrum library (National Institute for Standard Technology, Search Version 1.5, Gaithersburg, MD, USA). Only compounds detected in all replicates were identified by comparison of their GC retention time, MS ion spectra, and retention index. Generally, compound identification without the use of standard compounds is based on mass spectra similarity search in the mass library in combination with the retention time, and retention index (38, 39). The use of retention index significantly increases the reliability of compound identification. Compounds with identification probability values below 70% were rejected from further analysis. Probability of correct identification is estimated based on the use of match factors which are assessed from the test results; the factors provide a measure for similarity of the spectra of the unknown compounds and the spectra present in the mass spectral library (40).

The compounds hexamethyl-cyclotrisiloxane; methyl 10-[(bromomethyl)dimethylsilyloxy]-11-methoxy-3,7,11-trimethyl-2,6-dodecadienoate; desmethyldoxepin; trisiloxane, 1,1,3,3,5,5-hexamethyl-; tetramethyl-silane; and trimethyl-2,6-octadienyl-silane were detected in all control and culture samples; these compounds were considered as contaminants. Figure 2 shows volatile compounds identified in headspace of *A. parasiticus* SU-1 grown in YES liquid medium at 30°C for 72 h.

4. Notes

1. SPME-GC/MS technique is ideal for qualitative identification of volatile compounds; the technique can also be used to compare relative levels of volatile compounds based on instrument response (36).

   However, Bartelt (41) in his report demonstrated that SPME fiber device could be used to measure absolute concentration of organic volatiles in headspace using GC peak areas. In this work, the calibration factor K was determined

GC/MS data			NIST library matching result		Peak area, x 10^5 µV*s
Compound #	R.T., min	Selective ion, m/z	Library hit	Proba - bility, %	
Alcohols					
1	1.46	31	1-Propanol	82.86	4.20
2	1.66	31	1-Butanol	79.68	1.50
Leucine-derived volatiles					
3	1.64	58	Butanal, 3-methyl-	93.07	0.53
4	2.28	88	Butanoic acid, 3-methyl-, ethyl ester	97.03	0.21
Isoleucine-derived volatiles					
5	1.66	58	Butanal, 2-methyl-	80.1	0.72
6	2.27	102	Butanoic acid, 2-methyl-, ethyl ester	75.23	1.22
Valine-derived volatiles					
7	1.44	72	Propanal, 2-methyl-	93.73	0.53
8	1.58	33	1-Propanol, 2-methyl-	79.04	3.08
9	1.94	116	Propanoic acid, 2-methyl-, ethyl ester	82.4	2.67
10	1.96	18	Propanoic acid, 2-methyl-	80.23	0.16
Lipid-derived volatiles					
11	1.53	82	Furan, 2-methyl-	91.25	0.20
Esters					
12	1.54	43	Ethyl Acetate	84.3	2.30
13	1.78	57	Propanoic acid, ethyl ester	97.86	1.15
14	2.08	71	Butanoic acid, ethyl ester	95.45	0.32
15	2.87	88	Hexanoic acid, ethyl ester	88.81	0.14
Organic acids					
16	1.61	60	Acetic acid	83.29	0.50
Aldehydes and ketones					
17	1.26	30	Formaldehyde	88.33	0.42
25 unknown compounds					

Fig. 2. SPME-GS/MS headspace analysis of volatile compounds produced by *A. parasiticus* SU-1 grown at 30 °C for 72 h in YES liquid medium in the dark (26). The experiment was conducted in triplicate; for each experiment the culture was grown in two individual flasks. The results are presented as an average of six measurements of relative area R.T., retention time.

for 71 volatile compounds ranging in size from 1 to 16 carbons and commonly associated with biological systems; the constant K provides relationship between the absolute amounts of gas-phase analyte absorbed by a fiber with poly (dimethylsiloxane) coating and the headspace concentration under equilibrium. A regression model was developed to predict calibration factors with the use of the Kovats retention indices and to quantitate headspace concentration for different compounds including esters, ketones, aldehydes, alcohols, acids, and hydrocarbons.

2. Reaching equilibrium for volatiles contained in the fungus, medium, and headspace above the culture in the vial is a rate-limiting step; it takes approximately 30 min for fungal cultures and for fruit samples (36). The sorption of many volatiles into the fiber achieves equilibrium in 3 min. Reaching equilibrium for other volatiles may take longer (up to several hours). Whether you are at equilibrium or still in accumulation phase, as long as the sorption time and temperature is controlled and constant, relative quantification is correct as long as the mass spectrometer is not saturated.
3. A short extraction time was used for four reasons: (a) to reduce matrix effects described by Song et al. (36), (b) to reduce the total quantity of the major volatile constituents (while still detecting minor constituents), (c) to keep the volatiles analyzed in the linear range of the mass spectrometer, and (d) to increase sample throughput.
4. A few seconds before a sample was injected into the GC injection port, the first 20 cm of the column was simply suspended 1 cm over a pool of liquid N_2 (approximately 15 ml) held in a shallow styrofoam cup (with a diameter 8 cm and height 3 cm); the cryofocusing was continued during all time of desorption (3 min). To terminate the cryofocusing and initiate chromatography, the cup with liquid N_2 was removed from under the column, and then the oven door was closed as the oven temperature program and collection of mass spectra were initiated manually. The peak width at half height was approximately 0.4 s and the variation in retention time for analytes was no greater than 0.5 s.

Acknowledgements

This work was supported by grant from the NIH (CA52003-18) to John E. Linz and Ludmila V. Roze and by funds from the Michigan Agricultural Experiment Station.

References

1. Abanda-Nkpwatt D, Krimm U, Coiner HA, Schreiber L, Schwab W (2006) Plant volatiles can minimize the growth suppression of epiphytic bacteria by the phytopathogenic fungus *Botrytis cinerea* in co-culture experiments. Environ Exp Bot 56:108–119
2. Bahn YS, Xue C, Idnurm A, Rutherford JC, Heitman J, Cardenas ME (2007) Sensing the environment: lessons from fungi. Nat Rev Microbiol 5:57–69
3. Fan LH, Song J, Beaudry RM, Hildebrand PD (2006) Effect of hexanal vapor on spore viability of *Penicillium expansum*, lesion development on whole apples and fruit volatile biosynthesis. J Food Sci 71:M105–M109
4. Huang JQ, Jiang HF, Zhou YQ, Lei Y, Wang SY, Liao BS (2009) Ethylene inhibited aflatoxin biosynthesis is due to oxidative stress alleviation and related to glutathione redox state changes in *Aspergillus flavus*. Int J Food Microbiol 130:17–21
5. Mendgen K, Wirsel RSG, Jux A, Hoffmann J, Boland W (2006) Volatiles modulate the development of plant pathogenic rust fungi. Planta 224:1353–1361
6. Norrman J (1968) Morphogenetic effects of some volatile organic compounds on *Pestalotia rhododendri*. Arch Mikrobiol 61: 128–142
7. Roze LV, Beaudry RM, Arthur AE, Calvo AM, Linz JE (2007) *Aspergillus* volatiles regulate aflatoxin synthesis and asexual sporulation in *Aspergillus parasiticus*. Appl Environ Microbiol 73:7268–7276
8. Roze LV, Beaudry RM, Keller NP, Linz JE (2004) Regulation of aflatoxin synthesis by FadA/cAMP/protein kinase A signaling in *Aspergillus parasiticus*. Mycopathologia 158:219–232
9. Roze LV, Chanda A, Laivenieks M, Beaudry RM, Artymovich KA, Koptina AV, Awad DW, Valeeva D, Jones AD, Linz JE (2010) Volatile profiling reveals intracellular metabolic changes in *Aspergillus parasiticus*: veA regulates branched chain amino acid and ethanol metabolism. BMC Biochem 11:33
10. Roze LV, Koptina AV, Laivenieks M, Beaudry RM, Jones DA, Kanarsky AV, Linz JE (2011) Willow volatiles influence growth, development, and secondary metabolism in *Aspergillus parasiticus*. Appl Microbiol Biotechnol 92(2):359–370
11. Wright MS, Greene-McDowelle DM, Zeringue HJ, Bhatnagar D, Cleveland TE (2000) Effects of volatile aldehydes from *Aspergillus*-resistant varieties of corn on *Aspergillus parasiticus* growth and aflatoxin biosynthesis. Toxicon 38:1215–1223
12. Zeringue HJ, Brown RL, Neucere JN, Cleveland TE (1996) Relationships between C-6-C-12 alkanal and alkenal volatile contents and resistance of maize genotypes to *Aspergillus flavus* and aflatoxin production. J Agric Food Chem 44:403–407
13. Zeringue HJ Jr, McCormick SP (1990) Aflatoxin production in cultures of *Aspergillus flavus* incubated in atmospheres containing selected cotton leaf-derived volatiles. Toxicon 28:445–448
14. Zonneveld BJM (1988) Effect of carbon-dioxide on fruiting in *Aspergillus nidulans*. Trans Br Mycol Soc 91:625–629
15. Kai M, Haustein M, Molina F, Petri A, Scholz B, Piechulla B (2009) Bacterial volatiles and their action potential. Appl Microbiol Biotechnol 81:1001–1012
16. Schulz S, Dickschat JS (2007) Bacterial volatiles: the smell of small organisms. Nat Prod Rep 24:814–842
17. Feist AM, Henry CS, Reed JL, Krummenacker M, Joyce AR, Karp PD, Broadbelt LJ, Hatzimanikatis V, Palsson BO (2007) A genome-scale metabolic reconstruction for *Escherichia coli* K-12 MG1655 that accounts for 1260 ORFs and thermodynamic information. Mol Syst Biol 3:121
18. Forster J, Famili I, Fu P, Palsson BO, Nielsen J (2003) Genome-scale reconstruction of the *Saccharomyces cerevisiae* metabolic network. Genome Res 13:244–253
19. Kang JH, Liu G, Shi F, Jones AD, Beaudry RM, Howe GA (2010) The tomato odorless-2 mutant is defective in trichome-based production of diverse specialized metabolites and broad-spectrum resistance to insect herbivores. Plant Physiol 154:262–272
20. Mercier J, Jimenez JI (2007) Potential of the volatile-producing fungus *Muscodor albus* for control of building molds. Can J Microbiol 53:404–410
21. Mitchell AM, Strobel GA, Moore E, Robison R, Sears J (2010) Volatile antimicrobials from *Muscodor crispans*, a novel endophytic fungus. Microbiology 156:270–277
22. Ozaki K, Ohta A, Iwata C, Horikawa A, Tsuji K, Ito E, Ikai Y, Harada K (2008) Lysis of cyanobacteria with volatile organic compounds. Chemosphere 71:1531–1538
23. Singh SK, Strobel GA, Knighton B, Geary B, Sears J, Ezra D (2011) An endophytic *Phomopsis*

sp. possessing bioactivity and fuel potential with its volatile organic compounds. Microb Ecol 61:729–739

24. Strobel G (2006) Harnessing endophytes for industrial microbiology. Curr Opin Microbiol 9:240–244
25. Last RL, Jones AD, Shachar-Hill Y (2007) Towards the plant metabolome and beyond. Nat Rev Mol Cell Biol 8:167–174
26. Zhang CL, Wang GP, Mao LJ, Komon-Zelazowska M, Yuan ZL, Lin FC, Druzhinina IS, Kubicek CP (2010) *Muscodor fengyangensis* sp. nov. from southeast China: morphology, physiology and production of volatile compounds. Fungal Biol 114:797–808
27. Athukorala SN, Fernando WG, Rashid KY (2009) Identification of antifungal antibiotics of *Bacillus* species isolated from different microhabitats using polymerase chain reaction and MALDI-TOF mass spectrometry. Can J Microbiol 55:1021–1032
28. Demyttenaere JC, Morina RM, De Kimpe N, Sandra P (2004) Use of headspace solid-phase microextraction and headspace sorptive extraction for the detection of the volatile metabolites produced by toxigenic *Fusarium* species. J Chromatogr A 1027:147–154
29. Demyttenaere JC, Morina RM, Sandra P (2003) Monitoring and fast detection of mycotoxin-producing fungi based on headspace solid-phase microextraction and headspace sorptive extraction of the volatile metabolites. J Chromatogr A 985:127–135
30. Laothawornkitkul J, Taylor JE, Paul ND, Hewitt CN (2009) Biogenic volatile organic compounds in the Earth system. New Phytol 183:27–51
31. Thorn RM, Reynolds DM, Greenman J (2011) Multivariate analysis of bacterial volatile compound profiles for discrimination between selected species and strains *in vitro*. J Microbiol Methods 84:258–264
32. Turner AP, Magan N (2004) Electronic noses and disease diagnostics. Nat Rev Microbiol 2:161–166
33. Dudareva N, Pichersky E (2008) Metabolic engineering of plant volatiles. Curr Opin Biotechnol 19:181–189
34. Rossouw D, Naes T, Bauer FF (2008) Linking gene regulation and the exo-metabolome: a comparative transcriptomics approach to identify genes that impact on the production of volatile aroma compounds in yeast. BMC Genomics 9:530
35. Tikunov Y, Lommen A, de Vos CH, Verhoeven HA, Bino RJ, Hall RD, Bovy AG (2005) A novel approach for nontargeted data analysis for metabolomics. Large-scale profiling of tomato fruit volatiles. Plant Physiol 139:1125–1137
36. Song J, Fan LH, Beaudry RM (1998) Application of solid phase microextraction and gas chromatography time-of-flight mass spectrometry for rapid analysis of flavor volatiles in tomato and strawberry fruits. J Agric Food Chem 46:3721–3726
37. Skory CD, Chang PK, Linz JE (1993) Regulated expression of the *nor-1* and *ver-1* genes associated with aflatoxin biosynthesis. Appl Environ Microbiol 59:1642–1646
38. Retention index guide http://massfinder.com/wiki/Retention_index_guide.
39. Kovats E (1958) Gas-chromatographische charakterisierung organischer verbindungen. 1. Retentionsindices aliphatischer halogenide, alkohole, aldehyde und ketone. Helv Chim Acta 41:1915–1932
40. Stein SE (1994) Estimating probabilities of correct identification from results of mass-spectral library searches. J Am Soc Mass Spectrom 5:316–323
41. Bartelt RJ (1997) Calibration of a commercial solid-phase microextraction device for measuring headspace concentrations of organic volatiles. Anal Chem 69:364–372

Chapter 10

Tools for Manipulation of Secondary Metabolism Pathways: Rapid Promoter Replacements and Gene Deletions in *Aspergillus nidulans*

C. Elizabeth Oakley, Heather Edgerton-Morgan, and Berl R. Oakley

Abstract

Targeted gene deletions and promoter replacements are proving to be a valuable tool for awakening and analyzing silent secondary metabolism gene clusters in *Aspergillus nidulans* and, as molecular genetic methods for manipulating the genomes of other fungi are developed, they will likely be as valuable in those organisms. Here we describe procedures for constructing DNA fragments by PCR that can be used to replace genes or promoters quickly and on a large scale. We also describe transformation procedures for *A. nidulans* that allow these fragments to be introduced into target strains efficiently such that many genes or promoters can be replaced in a single experiment.

Key words: Secondary metabolism, Polymerase chain reaction, PCR, Gene deletion, Promoter replacement, Transformation, *Aspergillus*, Gene expression, Polyketide synthase, Non-ribosomal peptide synthetase

1. Introduction

The sequencing of fungal genomes has revealed that the genes involved in the production of a particular secondary metabolite are clustered in the genome and that, for each fungus, there are often many more gene clusters than known secondary metabolites (1–4). Many gene clusters are, thus, cryptic, not producing detectable products under normal lab growth conditions. Since secondary metabolites are a rich source of medically important compounds, it is important to obtain expression of these compounds. For compounds that are expressed normally, it is important to identify the genes involved and to decipher the biosynthetic pathways that produce them. Historically, these have been daunting tasks, but for

Nancy P. Keller and Geoffrey Turner (eds.), *Fungal Secondary Metabolism: Methods and Protocols*,
Methods in Molecular Biology, vol. 944, DOI 10.1007/978-1-62703-122-6_10, © Springer Science+Business Media, LLC 2012

Aspergillus nidulans they have been greatly facilitated by advances in gene targeting procedures (5, 6).

With respect to activating cryptic clusters, some are regulated at the chromatin level and they can be activated by deletion of certain genes involved in chromatin modification (7). Some gene clusters carry transcription factors within them that, when overexpressed, can activate expression of all the genes in the cluster resulting in the production of the product of the cluster (6, 8–10). Once expression of a pathway has been obtained, sets of targeted deletions can be used to determine which genes are part of the pathway (7, 10–15). Targeted deletions of polyketide synthase (PKS) genes or non-ribosomal peptide synthetase (NRPS) genes allow one to identify the gene cluster responsible for the synthesis of the compound. Targeted deletions of genes surrounding the responsible PKS or NRPS allow the genes required for the synthesis of the compound to be identified. In addition, many of the deletions will block the biosynthetic pathway and cause the accumulation of intermediate compounds in the pathway, and this allows the steps of the biosynthetic pathway to be determined. In addition, the intermediate compounds may be valuable themselves.

For these approaches to be maximally useful, it is important to be able to delete genes efficiently and to replace their promoters so that their levels of expression can be controlled. We have developed procedures that allow linear molecules to be created by fusion PCR (5, 16–18) that can be used to delete genes or replace their promoters (Fig. 1). This approach eliminates the need to clone genes in the conventional sense. Appropriate sequences can be amplified from genomic DNA and fused to genes or cassettes to construct transforming fragments. This approach is scalable in that transforming fragments to delete many genes (or replace the promoters of many genes) can be generated at the same time using a single PCR machine. It is also scalable in that one set of protoplasts can be generated for a strain and then aliquoted and transformed with the fusion PCR fragments. For example, if one has 10 candidate PKS genes that might be involved in the production of a particular secondary metabolite, one can create deletion constructs for all 10 of them by fusion PCR and, in a day, transform the constructs into an *A. nidulans* recipient strain, deleting each of the candidate PKS genes.

Our strategy for deleting genes in *A. nidulans* is to replace the target gene with a selectable marker as shown in Fig. 1. Sequences flanking the gene that one wishes to delete are amplified from genomic DNA. The primers are based on *A. nidulans* genomic sequences available at AspGD (http://www.aspergillusgenome.org/), CADRE (http://www.cadre-genomes.org.uk/index.html), or the Broad Institute (http://www.broadinstitute.org/annotation/genome/aspergillus_group/MultiHome.html). Primers P3 and P4 are designed with "tails" that are not part of the target

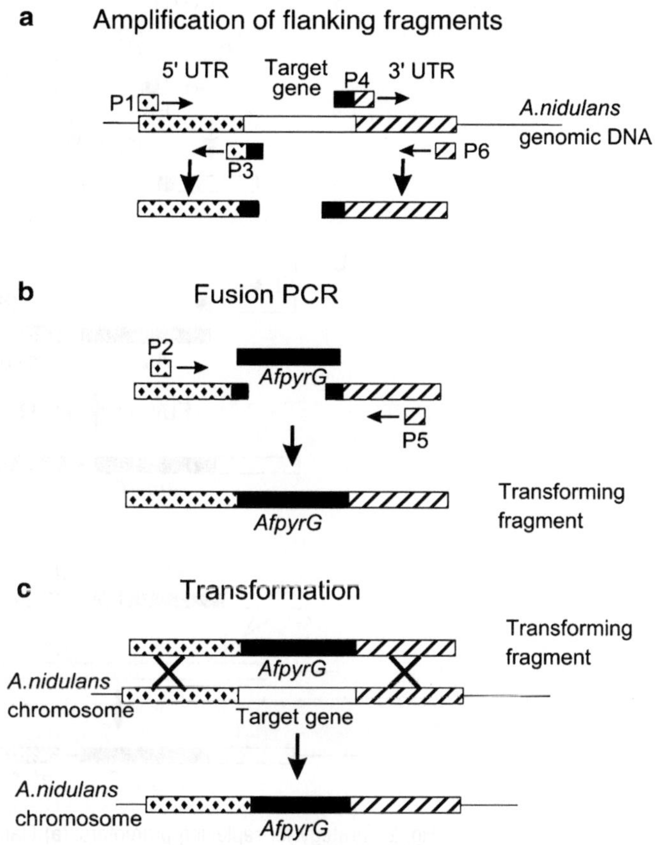

Fig. 1. Strategy for deleting genes. (**a**) Genomic sequences flanking the target gene are amplified from genomic DNA. Primers are designed based on the genomic sequences in the databases. Primers P3 and P4 are designed to have "tails" (*black region*) that are identical to sequences at the end of the fragment to which they will be fused, in this case the *A. fumigatus pyrG* gene (*AfpyrG*) which is used as a selectable marker. (**b**) Fusion PCR is used to amplify a transforming fragment. (**c**) Homologous recombination during transformation results in replacement of the target gene by the selectable marker.

genomic sequence, but, rather, are identical to the ends of the sequence that one wishes to fuse with the flanking sequences. In the case of gene replacements, the flanks are fused to a selectable marker. Selectable markers can be genes that complement *A. nidulans* mutations that inhibit growth or they can confer resistance to compounds toxic to *A. nidulans*. It is best to use a marker that is not an *A. nidulans* gene because *A. nidulans* genes can recombine by homologous recombination with the mutant allele in the genome and complement the mutation without integrating at the target gene. We routinely use the *Aspergillus fumigatus pyrG* gene (*AfpyrG*) (19), *A. fumigatus riboB* (*AfriboB*), and *A. fumigatus pyroA* (*AfpyroA*) genes as selectable markers (5) complementing *A. nidulans pyrG*89, *riboB*2, and *pyroA*4 mutations.

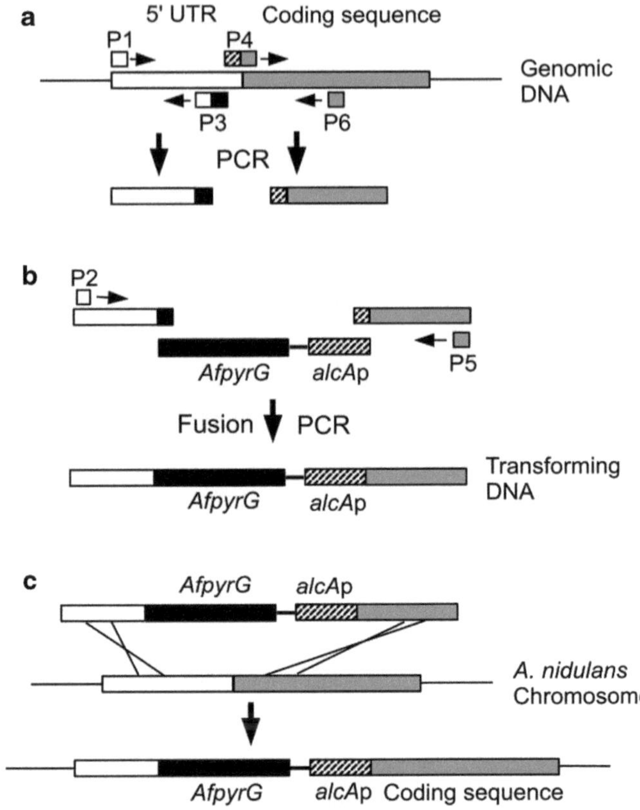

Fig. 2. Strategy for replacing promoters. (**a**) Flanking sequences for the target gene are amplified from genomic DNA. Primers are designed based on genomic sequences in the databases. Primers P1 and P3 amplify a region of the 5′ untranslated region (5′ UTR). The full extent of the endogenous promoter is usually unknown. Any portion of the promoter that is not deleted will be moved far away from the start codon in any case. P3 does not have to be immediately upstream of the start codon because the region between P3 and the start codon will be deleted upon integration (**c**). P4 amplifies beginning with the start codon of the coding sequence. P3 and P4 are designed with tails to allow them to be fused to a cassette containing a selectable marker and promoter as shown in **b**. In this case the selectable marker is *AfpyrG* and the promoter is from the *A. nidulans alcA* gene. (**c**) Transformation and homologous integration of the fusion PCR product result in the coding sequence of the gene being put under control of the *alcA* promoter.

Promoters can be replaced using a similar strategy (Fig. 2). A flanking fragment is amplified from the 5′ untranslated region. As drawn in Fig. 2, the flanking fragment does not include promoter sequences, but there is no harm in including promoter sequences. In the final construct, they will be separated from the coding sequence and should not affect transcription. A second fragment is amplified that begins with the start codon and extends approximately 1,000 bp. The fragment can end within the coding sequence or downstream if the coding sequence is less than 1,000 bp. The two amplified fragments are fused by PCR to a cassette containing a selectable marker and the promoter of choice. In Fig. 2, *AfpyrG* (19) is the selectable marker and the promoter is from the *A. nidulans alcA* gene (20, 21), but the procedure is flexible with respect to choice of promoter and selectable marker. The cassette can, itself, be constructed by fusion PCR and the construct can be subcloned into a plasmid for long-term maintenance if desired.

2. Materials

2.1. Polymerase Chain Reaction

1. PCR enzymes: AccuPrime Pfx DNA polymerase and AccuPrime *Taq* High Fidelity (Invitrogen), KOD Hot Start DNA Polymerase (EMD Chemicals, Inc.), Phusion Hot Start II High Fidelity DNA polymerase (Thermo Scientific). Some enzyme kits include nucleotides; in other cases they must be purchased separately.
2. PCR cleanup kit: QIAquick PCR purification kit (Qiagen).
3. Thermocycler with programmable ramp rates (see Note 1).
4. PCR tubes: 200 μl thin-wall PCR tubes appropriate for thermocycler.
5. PCR primers (design of primers discussed under Subheading 3).
6. *A. nidulans* genomic DNA (can be prepared by a number of procedures but must be clean enough to allow long-term storage and reliable PCR). We generally prepare *A. nidulans* DNA in large amounts, purifying it over two CsCl density gradients (22). It can be stored for years at 4°C in TE buffer (10 mM Tris pH 8.0, 1 mM EDTA).
7. Microcentrifuge for PCR cleanup.

2.2. Growth and Transformation (See Note 2)

1. 1.2 M KOH solution: Boil 200 ml of deionized distilled water. Allow it to cool to room temperature. Add 50 ml to a clean beaker. Add 6.73 g of KOH and allow it to dissolve. Transfer it to a clean graduated cylinder and add boiled deionized distilled water to a final volume of 100 ml. This solution will absorb CO_2 and should, consequently, be made fresh before use.
2. KCl/citrate solution (~1.2 M to be diluted to make protoplasting solution): Dissolve 8.2 g of KCl and 2.1 g of citric acid monohydrate in approximately 50 ml of deionized distilled water. Adjust to pH 5.8 with 1.2 M KOH solution. Transfer to a clean graduated cylinder and add deionized distilled water to a final volume of 100 ml. Mix thoroughly. This solution can be autoclaved, making sure that there is no volume change which will alter the molarity of the solution, or it can be stored at 4°C for up to 2 weeks in a tightly closed container.

3. 0.6 M KCl solution: Dissolve 4.47 g of KCl in 100 ml of deionized distilled water. This solution can be autoclaved, making sure that there is no volume change, and can be stored at room temperature in a tightly closed container (see Note 3).
4. 0.6 M KCl, 50 mM $CaCl_2$ solution: Dissolve 4.47 g of KCl and 0.74 g of $CaCl_2$ dihydrate in 100 ml of deionized distilled water. Sterilization and storage are the same as for 0.6 M KCl.
5. Polyethylene glycol (PEG) solution: Dissolve 4.47 g of KCl and 0.74 g of $CaCl_2$ dihydrate in 20 ml of deionized distilled water in a clean, rinsed 100 ml graduated cylinder. Add 0.804 ml of 1.0 M Tris–HCl solution and 0.196 ml of 1.0 M Tris base solution. Mix thoroughly. Add 25 g of PEG approximate molecular weight 3,350 (Sigma). Add deionized distilled water to approximately 90 ml. Seal the top with parafilm and mix by inversion. The PEG will dissolve slowly and the suspension can be mixed by inversion at 10-min intervals until the PEG dissolves. Gentle warming speeds the dissolving of the PEG. After the PEG goes into solution, add deionized distilled water to 100 ml. Seal the top with parafilm and mix thoroughly by inversion. Aliquot the solution into clean, rinsed bottles or vials (~20 ml with large tops to allow easy access to the solution) and autoclave, making sure that the volume does not change. After the vials have cooled, seal their tops with Parafilm M and store at room temperature. Do not store at 4°C or below as the PEG will precipitate.
6. 1.2 M Sucrose solution: Dissolve 41.1 g of sucrose in 50 ml of deionized distilled water. When dissolved make up to a final volume of 100 ml. Aliquot into conveniently sized bottles. Sterilize by autoclaving making sure that there is no volume change. Store in containers tightly sealed with Parafilm M. Storage can be at room temperature or 4°C, but it will be used at 4°C.
7. Protoplasting enzyme: VinoTaste Pro (Novo Nordisk). In the United States VinoTaste Pro is available from Gusmer Enterprises (www.gusmerenterprises.com) (see Note 4).
8. Miracloth (Calbiochem): We normally cut miracloth into 4 in. squares, place them into a glass Petri dish, and autoclave.
9. Strains: Correct gene targeting is greatly facilitated by deletion of the *nkuA* gene (5). *nkuA* deletion strains are available from the Fungal Genetics Stock Center (http://www.fgsc.net/).
10. Millex-GV 0.22 μm pore size PVDF (Durapore) filter units (Millipore) (33 mm syringe-driven disposable) or Stericup-GV or Stericup-GP 0.22 μm pore size vacuum-driven disposable filter units (150 ml) (Millipore). All these units have low protein binding filter membranes (see Note 5).

11. YG medium (made with distilled H_2O): 5 g/l yeast extract, 20 g/l D-glucose, 400 μl/l of trace element solution (23).
12. Uridine stock solution (100×): 24.42 g uridine dissolved in 100 ml final volume of deionized distilled water. Filter sterilize with a 0.22 μm pore size filter unit. Can be stored at room temperature.
13. Riboflavin stock solution (100×): 25 mg/100 ml of deionized distilled water. Filter sterilize with a 0.22 μm pore size Stericup-GV or a Stericup-GP filter unit. Can be stored at room temperature, but riboflavin is light sensitive and must be stored in the dark.
14. Pyridoxine stock solution (1,000×): Dissolve 50 mg of pyridoxine in 100 ml of deionized distilled water. This solution may be autoclaved and stored at room temperature.
15. Gyratory water bath shaker (e.g., New Brunswick Innova 3100 or C76).
16. Sterile transfer pipets (1 ml volume).
17. Variable speed refrigerated centrifuge with a swinging bucket rotor capable of centrifuging 15 and 50 ml disposable centrifuge tubes.
18. Variable speed microcentrifuge (preferably with swing-out head).
19. 50 ml sterile disposable centrifuge tubes or equivalent sterilized, reusable, capped centrifuge tubes.
20. 15 ml sterile disposable centrifuge tubes.
21. Phase contrast microscope.
22. Vortex Genie vortex mixer.
23. Selection plates for transformation: Selection plates allow for the selection of transformants. To prevent protoplast lysis, they contain 0.6 M KCl. To select for complementation of *pyrG89*, we generally use YAG (5 g/l yeast extract, 20 g/l D-glucose, 400 μl/l of trace element solution (23), 15 g/l agar) plus 0.6 M KCl although minimal medium with appropriate supplements can also be used. For other selections we use minimal medium (below) with appropriate supplements plus 0.6 M KCl (see Note 6).
24. Minimal medium (24).

Add 10 g D-glucose, 2 ml of a 26% w/v solution of $MgSO_4$ $7H_2O$, 15 g Agar, and 400 μl of trace metal solution (23) to 900 ml distilled water. Autoclave. Cool to 60°C. Add 100 ml of separately autoclaved 10× stock salts. Mix thoroughly before pouring plates. 10× stock salts: 60 g/l $NaNO_3$, 5.2 g/l KCl, 15.2 g/l KH_2PO_4, adjust pH to 6.0–6.5 with saturated KOH solution.

3. Methods

3.1. Design PCR Primers (See Note 7)

Primers P1, P2, P5, and P6 (Figs. 1 and 2) are designed based on the genomic sequences available in the databases. We have used three different databases, AspGD (http://www.aspgd.org/), CADRE (http://www.cadre-genomes.org.uk/index.html), and the Broad Institute (http://www.broadinstitute.org/annotation/genome/aspergillus_group/MultiHome.html). The genome sequences are available on each database, but there are differences in the way they are presented and retrieved that individual researchers may prefer (see Note 8).

1. To design primers to replace a gene (Fig. 1) first download, from a database, a sequence extending from 1,100 bp 5′ to (upstream of) the start codon of the target gene to 1,100 bp 3′ to (downstream of) the stop codon of the gene.
2. Choose approximate positions of primers in the downloaded sequence. Primer P1 is from a region between 1,000 and 1,100 bp upstream of the start codon. P2 is a sequence approximately 1,000 bp upstream of the start codon. A portion of P3 corresponds to a sequence near to, and slightly upstream of, the start codon and the remainder corresponds to the cassette that will be fused to the flanking fragment. A portion of P4 corresponds to a sequence near to but slightly downstream of the stop codon and another portion corresponds to the cassette. P5 corresponds to a sequence approximately 1,000 bp downstream of the stop codon and P6 corresponds to a sequence between 1,000 and 1,100 bp downstream of the stop codon.
3. Design primers with appropriate melting temperatures. Use the following formula to calculate melting temperature (T_m): $T_m = 4 \times$ (total number of C's and G's) $+ 2 \times$ (total number of A's + T's) (see Note 9). For primers P1, P2, P5, and P6 use a melting temperature of approximately 60°C. If the G + C content is 50% the primer would be a 20mer. Primers 3 and 4 each have two regions. One region anneals to genomic DNA and this portion of the primer should be designed to have a melting temperature of approximately 60°C. The other region of primers 3 and 4 corresponds to a cassette (either a selectable marker or a selectable marker plus the promoter that will be introduced). For the gene replacement shown in Fig. 1, the black region of P3 corresponds to one end of the cassette containing *AfpyrG* and the black region of P4 corresponds to the other end of *AfpyrG*. These primer regions should also have a melting temperature of approximately 60°C. Each region of P3 and P4 should, thus, have a melting temperature of 60°C,

but the melting temperatures of the entire primers will, of course, be much higher (see Note 10).

4. For a promoter replacement (Fig. 2), download a sequence extending from 1,200 bp upstream of the start codon of the target gene to 1,100 bp downstream of the start codon. Precise choice of primer position depends on how much of the native promoter one wishes to remove. Since the procedure moves any remnant of the native promoter far away from the coding sequence of the target gene, the amount of the promoter that will be removed is generally not important. Be careful, however, not to remove control sequences for adjacent genes. In our example, we will assume that approximately 50 bp of the native promoter will be removed. A portion of P3 will thus correspond to a sequence approximately 50 bp upstream of the start codon and the remainder will correspond to the end of the selectable marker in the cassette. P1 will be from a region between 1,050 and 1,150 bp upstream of the start codon and P2 will correspond to a sequence approximately 1,050 bp upstream of the start codon. A portion of P4 will correspond to the end of the promoter in the cassette and the other portion will correspond to a portion of the coding sequence of the target gene beginning with the start codon. P5 will be approximately 1,000 bp downstream of the start codon and P6 will be 1,000–1,100 downstream of the start codon.

5. Design primers to have the same T_m as for gene replacements. For promoter replacements the sequence of primer 4 is constrained by the fact that the start codon has to be included. This means that in some cases T_m has to be 58°C or 62°C.

3.2. Amplify Flanking Sequences

This amplification uses *A. nidulans* genomic DNA. We use AccuPrime *Pfx* DNA polymerase (Invitrogen) (see Note 11). For a given target gene, primers P1 and P3 are used to amplify one flanking fragment and P4 and P6 are used to amplify the other fragment. Flanking fragments for several genes can be amplified at the same time in the same thermocycler.

1. Set up a 50 μl reaction in a thin-walled 200 μl PCR tube according to the enzyme manufacturer's instructions with 100 ng *A. nidulans* genomic DNA, 300 nM final concentration of each primer, and 1.25 Units of AccuPrime *Pfx*.
2. Use the following PCR amplification profile (thermocycler lid heated to 105°C).
   (a) 94°C for 2 min to activate the enzyme.
   (b) 30 cycles with the following repeating steps:
   - 94°C for 20 s.
   - Cool to 70°C at the maximum ramp rate allowed by the thermocycler. Remain at 70°C for 1 s.

- Ramp down from 70°C to the annealing temperature at a rate of 0.1°C/s (see Note 12). The annealing temperature is 5°C below the T_m for the primers. With respect to primers P3 and P4 use the T_m that corresponds to the region of annealing of the primer to the genomic DNA. If the T_m for the two primers is different, use an annealing temperature that is 5°C below the lower T_m.
- Anneal for 30 s.
- Ramp from the annealing temperature to an extension temperature of 68°C at a rate of 0.2°C/s.
- Extend at 68°C for 1 min/kb of anticipated product.
- Ramp from 68 to 94°C at maximum rate.

(c) Final extension at 68°C for 4 min.

3. Purify PCR products. It is necessary to remove excess primers, etc. before using the amplified molecules for fusion PCR. Residual primers can cause extra bands during the fusion PCR amplification. We use a QIAquick PCR purification kit (Qiagen) and follow the manufacturer's instructions except that we elute with TE buffer instead of the buffer EB supplied in the kit. Examine fragments by agarose gel electrophoresis to evaluate purity and concentrations of fragments. Note, however, that successful fusion PCR can be carried out with DNA preparations that have some spurious amplified bands.

3.3. Create Transforming Fragments by Fusion PCR

1. Choose an appropriate thermostable DNA polymerase. We have used three DNA polymerases successfully for fusion PCR: AccuPrime *Taq* High Fidelity, KOD Hot Start DNA Polymerase, and Phusion Hot Start II High Fidelity DNA polymerase (see Note 13). We have published a detailed protocol for fusion PCR with AccuPrime *Taq* High Fidelity DNA polymerase (18) and will not repeat it here. Phusion and KOD are unusually heat-resistant, rapid, and processive enzymes. This allows relatively short extension times even when one is amplifying long fragments. In our hands, KOD and Phusion give essentially identical results, and we use similar PCR profiles for them.
2. Set up 50 μl reactions in 200 μl thin-walled PCR tubes. The reactions are set up according to the manufacturers' instructions. We use 0.5 μl of each DNA fragment (both flanks and the cassette). Use nested primers (P2 and P5 in Figs. 1 and 2). The use of nested primers greatly reduces the amplification of extraneous bands.
3. Use the following fusion PCR amplification profile (thermocycler lid heated to 105°C). The profile is for KOD. Values in parentheses are for Phusion.

(a) 94°C for 2 min to activate the enzyme (98°C for 30 s).

(b) 25 cycles repeating the following steps:

- 94°C for 10 s (98°C for 10 s).
- Ramp down to 75°C at the maximum ramp rate allowed by the thermocycler. Remain at 75°C for 1 s.
- Ramp down from 75°C to the annealing temperature at 0.1°C/s. The annealing temperature is the lowest primer $T_m + 2°C$ ($T_m + 3°C$).
- Anneal for 10 s.
- Ramp up to an extension temperature of 68°C (72°C) at 0.2°C/s.
- Extend for 30 s/kb of anticipated final product.
- Ramp up at maximum ramp rate to 94°C (98°C).

(c) Final extension at 68°C for 4 min (72°C for 5 min).

4. Purify PCR products as discussed in Subheading 3.2, step 3. Evaluate fusion PCR products by agarose gel electrophoresis. Note: Accurate gene targeting can be achieved even if the fusion PCR product has multiple bands (5). Products can be stored frozen.

3.4. Prepare Protoplasts for Transformation

One of the advantages of using fusion PCR to generate transforming fragments is that many can be generated in parallel using a single thermocycler. If several fragments are to be transformed into the same strain, it is convenient to make enough protoplasts to allow parallel transformations with each fragment. The procedure below easily allows ten parallel transformations.

1. Inoculate 20 ml of nonselective media in a 50 ml Erlenmeyer flask with 1×10^8 conidia. We normally use YG medium. For strains carrying the *pyrG89* allele, we supplement YG with 10 mM uridine and 9 mM uracil (final concentrations) (see Note 14).
2. Incubate for 16 h at 30°C on a gyratory shaker shaking at 120 rpm. This can be an air shaker or a water bath shaker (e.g., New Brunswick Innova 3100 or C76).
3. Cool variable speed centrifuge to 4°C. Make sure that 1.2 M sucrose is at 0–4°C.
4. Prepare 2× protoplasting solution. Approximately 60 min before hyphae are to be harvested, prepare protoplasting solution. Dissolve 2 g of VinoTaste Pro in 10 ml of 1.2 M KCl/citrate solution in a clean centrifuge tube that holds at least 15 ml. VinoTaste Pro often contains lipid. To remove it centrifuge the solution at $5{,}000 \times g$ for 15 min at room temperature and remove the lipid layer from the top of the solution. The solid enzyme can be stored for long periods at 4°C prior to

preparation of the protoplasting solution, but after long storage a higher concentration of enzyme may be required.

5. Sterilize the solution by filtering it through a sterile, disposable, syringe-driven 0.22 μm low protein binding filter unit (Millex GV PVDF filter material). It is useful to run a volume of 1.2 M KCl/citrate solution through the filter to wash away any contaminants that might be associated with the filter before filtering the protoplasting solution. To filter 10 ml of protoplasting solution, it may be necessary to use more than one filter unit. An alternative (particularly for larger volumes of protoplasting solution) is to use a 150 ml sterile, vacuum-driven unit (Stericup-GV or Stericup-GP).
6. Harvest the hyphae by filtering the culture through sterilized Miracloth. Wash with at least 3× the culture volume of sterile deionized distilled water or growth medium. This helps to remove any residual detergent. Use a flamed spatula to press residual liquid from the hyphal mass.
7. Resuspend the hyphal mass in 8 ml of nonselective medium (YG with any necessary supplements). Add 8 ml of sterile 2× protoplasting solution. Mix thoroughly.
8. Incubate at 30°C shaking at 100 rpm on a gyratory water bath shaker.
9. Mix thoroughly at 15-min intervals to break up clumps. It is important to break up hyphal clumps to maximize protoplast production. We do this by squirting the protoplasting solution and hyphae in and out of a 1 ml transfer pipet at 15-min intervals. We squirt once for each ml of protoplasting solution. Vigorous squirting is unlikely to damage protoplasts, but do not squirt so vigorously as to create a foam because this will result in denaturing the protoplasting enzyme.
10. Monitor protoplast formation by phase contrast microscopy (Fig. 3). Monitoring protoplast formation is important in that it enables one to determine when sufficient protoplasting has occurred and in rare instances in which there are problems with protoplasting, one can determine the nature of the problem (e.g., hyphae are not digested or protoplasts lyse immediately when they form). Protoplasting should be optimal after 1–1.5 h. Longer protoplasting will usually not result in more protoplasts and regrowth of cell wall may occur.
11. Place flask with protoplasts on ice for 5 min with gentle agitation to cool the suspension.
12. Pipet 15 ml of 1.2 M sucrose into a sterile 50 ml centrifuge tube.
13. Place a centrifuge tube containing 1.2 M sucrose in an ice bucket and use a sterile 10 ml pipet to layer the protoplast suspension onto the sucrose cushion (Fig. 4).

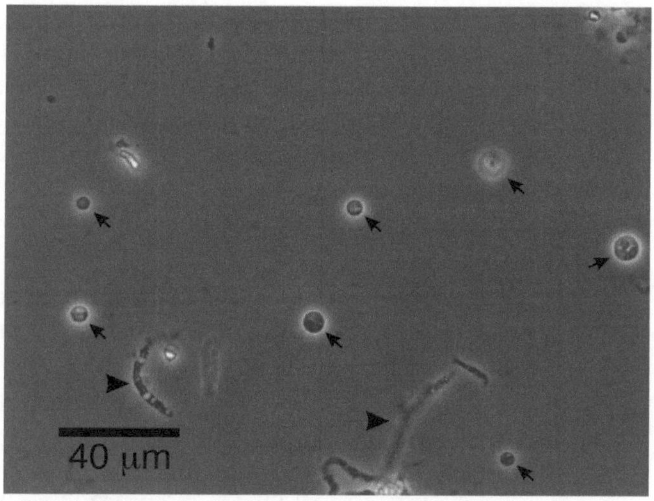

Fig. 3. Phase contrast image of protoplasts. The image was taken slightly more than 1 h after protoplasting was begun. Protoplasts are indicated by *arrows* and are of various sizes as is typical. Remnants of digested hyphae are present, some of which are indicated with *arrowheads*.

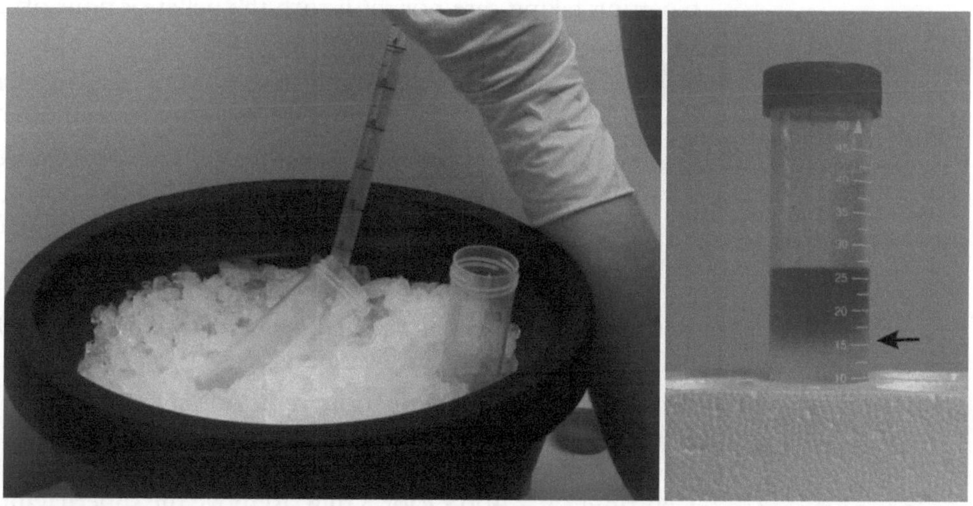

Fig. 4. Layering of a protoplast suspension onto a 1.2 M sucrose cushion. *Left*: The protoplast suspension is allowed to flow slowly from the pipet onto the sucrose cushion in a centrifuge tube in an ice bucket. The suspension, cushion, and interface (*arrow*) are visible at the *right*.

14. Centrifuge at 2,800 × *g* (RMax) at 4°C for 10 min. Accelerate and decelerate gently to prevent disturbing the interface between the sucrose and the protoplast suspension. After centrifugation set the temperature of the centrifuge to 25°C.
15. Harvest the protoplasts. During centrifugation most of the hyphal debris will sediment to the bottom of the sucrose cushion. The protoplasts will form a layer at the top of the sucrose

cushion. The thickness of the layer will depend on how much the interface is disturbed during handling. The protoplasts can be harvested by gently removing the protoplast solution down to the protoplast layer and then removing the protoplast layer. Alternatively, a pipet can be extended through the protoplasting solution down to the bottom of the protoplast layer and the layer gently pulled into the pipet by suction with a pipet bulb.

16. Transfer protoplasts to a 15 ml or a 50 ml sterile centrifuge tube. Add a volume of 0.6 M room-temperature KCl that is equal to or greater than the volume of the protoplasts.
17. Centrifuge for 10 min at 2,300 × *g* (RMax). The centrifuge was set to 25°C (step 14 above) and gradually warms during the centrifugation. During this centrifugation, shake the PEG solution (the PEG stratifies during storage and some may come out of solution) and filter it using a 33 mm Millex-GV disposable filter (syringe mounted, 0.22 μm pore size). Filtered PEG solution should not be used for subsequent transformations without re-filtering.
18. Decant supernatant. Either remove the supernatant gently by pipetting taking care not to disturb the pellet or pour off gently making sure not to dislodge the pellet.
19. Resuspend the pellet in 2 ml of 0.6 M KCl. Transfer to 2 × 1.5 ml sterile microcentrifuge tubes (1 ml in each tube).
20. Wash three times at room temperature with 0.6 M KCl. Using a variable speed microcentrifuge, preferably with a swing-out head, centrifuge at 2,400 × *g* for 3 min. After centrifugation remove the supernatant carefully with a 1,000 μl micropipettor. Add 1.0 ml of fresh, sterile 0.6 M KCl with a micropipettor. Resuspend each pellet by squirting it in and out of the 1,000 μl pipet tip six to ten times (see Note 15).
21. After the third centrifugation remove the supernatant and resuspend each pellet in 500 μl of 0.6 M KCl, 50 mM $CaCl_2$ solution. Combine the two suspensions into one tube. Centrifuge at 2,400 × *g* for 3 min. Remove the supernatant and resuspend the pellet in a volume appropriate for the number of transformations to be carried out. We normally use 40 μl of protoplasts per transformation. Thus for ten transformations resuspension of the pellet in 450 μl would be appropriate. This gives enough extra volume to compensate for losses during pipetting into individual tubes.
22. Transfer 40 μl of protoplast solution into individual 1.5 ml sterile microcentrifuge tubes (one tube per transformation).

3.5. Transform

1. To each tube add DNA in no more than 6 μl of TE buffer. Mix thoroughly by vortexing (6–7 × 1-s bursts with a Vortex Genie). The protoplasts are relatively resistant to physical shearing and

it is important to mix the transforming DNA thoroughly with the protoplasts.

2. Add 20 μl of PEG solution (at room temperature). Mix by vortexing as in step 1.
3. Place tubes in an ice-water bath for 25 min.
4. Add 400 μl of room-temperature PEG solution. Mix by squirting in and out of a micropipet tip approximately ten times. It is important to mix thoroughly.
5. Incubate at room temperature for 30 min.
6. Spread transformation mixture onto selection plates (100 μl/plate).
7. Incubate. Transformation rates are higher if one incubates at 30°C overnight before shifting plates to 37°C. However, if plates are initially incubated at 37°C, more than adequate number of transformants are usually obtained and the colonies grow more quickly.

3.6. Determine if Transformants Have Correct Gene Replacements

This can be done with Southern hybridizations using standard methods or by diagnostic PCR. A diagnostic PCR procedure is as follows.

1. After streaking transformants into single colonies, prepare DNA by the method of Lee and Taylor (25). This procedure is somewhat lengthy because it requires lyophilization of hyphae, but it is reliable.
2. Choose appropriate primers to verify correct gene targeting (Fig. 5). In many cases, primers P1 and P6 can be used. Because nested primers (P2 and P5) are used for fusion PCR amplification, P1 and P6 are not in the transforming fragment but they are in the genome just outside of the targeted region. The size of the band amplified by P1 and P6 can be easily predicted for the untransformed gene and for the correctly targeted gene. Correct targeting can be further verified by designing a primer in the DNA flanking the target gene, but outside of the transforming sequence, and a second primer within the selectable marker of the transforming sequence (PA and PB in Fig. 5c). This primer pair will amplify a band of a predictable size in correct transformants but will not give specific amplification in wild-type controls. A similar amplification can be carried out with another primer in the selectable marker and a primer outside of the transforming sequence in the coding sequence or 3′UTR (Fig. 5d). Use the primer design principles stated in Subheading 3.1.
3. Diagnostic PCR: The fidelity of the thermostable polymerase is not important for diagnostic PCR. This allows a variety of enzymes to be used. In general, the manufacturer's instructions should be followed. We normally use the PCR profile in Subheading 3.2, step 2 above.

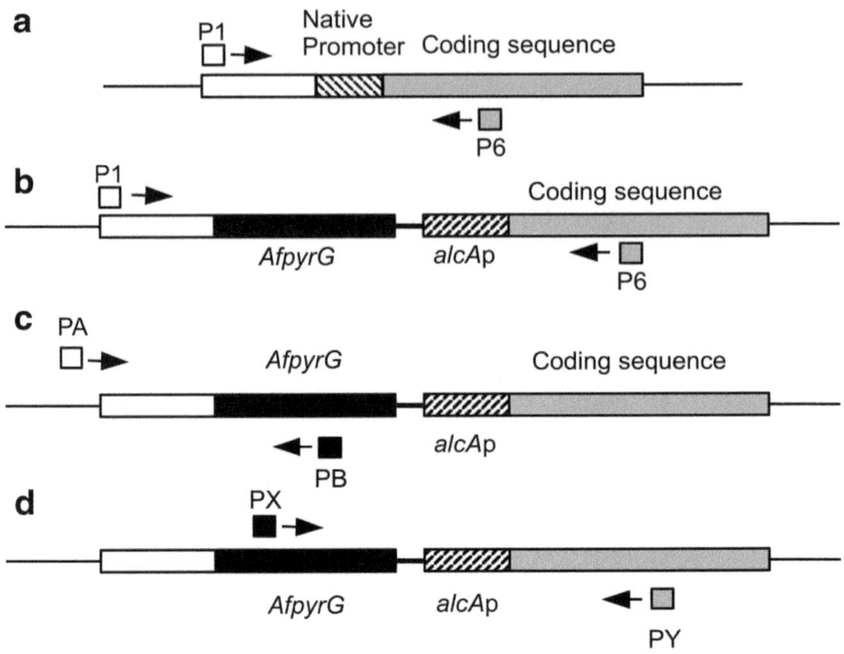

Fig. 5. Diagnostic PCR for a promoter replacement. Similar strategies (not shown) can be used for gene replacements. In **a** and **b**, P1 and P6 are the same as the P1 and P6 primers in Figs. 1 and 2. (**a**) shows the wild-type locus and (**b**) shows a correct promoter replacement. If a correct promoter replacement has occurred the band amplified by P1 and P6 will be longer than in the wild type and of a size that can be calculated precisely from sequence information of the locus and inserted sequences (*AfpyrG* and *alcA*p). (**c**) For correct promoter replacements, a primer that binds within the *AfpyrG* gene (PB) and a primer outside of the locus (PA) will amplify a band, the size of which can be calculated precisely. If the locus is untransformed, no amplification will be obtained from these primers. The same logic applies for (**d**), but in the opposite direction.

4. Notes

1. Thermocycler: In our experience the brand of the thermocycler is not important. It is, however, important that ramp rates be programmable. This allows the same cycling conditions to be used on different machines and this allows consistent results to be obtained.
2. *A. nidulans* protoplasts are *extremely* sensitive to detergents. All glassware must be clean and thoroughly rinsed to remove all traces of detergent.
3. We generally seal the containers with Parafilm M to eliminate the possibility of evaporation which would change the molarity of the solution. It is often convenient to aliquot the solution into clean, rinsed small bottles or vials before autoclaving. One or two vials can, thus, be used for each experiment. If larger containers are used for multiple experiments, the chances of contamination increase.

4. In the past we have used Vinoflow FCE (Novo Nordisk), an enzyme used in the wine industry (5, 18). This enzyme is no longer commercially available, but we have found that VinoTaste Pro (Novo Nordisk) works as well (although we generally need to use a higher concentration of VinoTaste Pro than Vinoflow FCE).
5. Low protein binding filters are important for filtering the protoplast solution to prevent removal of the protoplasting enzyme.
6. Other osmotic balancers may be used instead (e.g., 1.0 M sucrose).
7. We use primers of standard purity. We do not find it necessary to use HPLC or gel-purified primers. A 10 nM scale synthesis provides a more than adequate amount of primers.
8. We generally design our flanking DNA fragments to be approximately 1,000 bp in length. These are the fragments that flank the cassettes and are essentially amplified regions of genomic DNA except for short tails that allow them to be fused to cassettes. A length of 1,000 bp allows efficient transformation and accurate gene targeting (5), but it is short enough to allow linear transforming DNA molecules to be constructed easily by fusion PCR (i.e., the final product will not be too long to amplify efficiently).
9. A number of formulas are used to calculate T_m. In the absence of information about ionic conditions, calculation of precise T_m is impossible, but the formula given works reliably for our purposes.
10. With primers of the T_m we specify, one does not need to worry about secondary structure at the annealing temperature and concerns about repeats are minimal (i.e., they will be short enough to minimize spurious annealing at the temperatures used for PCR). It is important not to design primers that are too long. Their melting temperatures may be high enough to allow incorrect annealing and this may result in extra amplified bands.
11. The AccuPrime buffer system greatly reduces nonspecific priming and *Pfx* is a high-fidelity polymerase which minimizes the possibility of the introduction of mutations during PCR amplification.
12. Ramp rates are important. The specified ramp rates allow correct primers to anneal and extend before incorrect primers. The use of maximum ramp rates for all steps can result in amplification of spurious bands.
13. AccuPrime *Taq* High Fidelity is actually a mixture of two thermostable DNA polymerases, *Taq* DNA polymerase as well as

an enzyme from *Pyrococcus* sp. that has 3′ to 5′ exonuclease activity that allows proofreading. The advantage of AccuPrime *Taq* High Fidelity DNA polymerase over the other two enzymes is that the AccuPrime buffer system minimizes amplification of bands in addition to the desired fusion PCR product. AccuPrime *Taq* High Fidelity DNA polymerase has two disadvantages relative to the other polymerases. First, its fidelity is considerably lower than the other two enzymes. This gives an increased chance for PCR-induced mutations. Second, the AccuPrime *Taq* High Fidelity DNA polymerases are less thermostable and have lower rates of polymerization than KOD or Phusion. Extension times have to be longer with AccuPrime *Taq* High Fidelity DNA polymerases than the other two enzymes, and they have to be extended even more in later rounds of polymerization because of the denaturing of the enzymes.

14. Uridine is maintained as a stock solution and is added after autoclaving. Uracil is added directly to the medium before autoclaving. For strains carrying *riboB2* we supplement with riboflavin (added after autoclaving from a sterile stock solution) to a final concentration of 7 μM. Pyridoxine is added to a final concentration of 2.43 μM and can be added before or after autoclaving. The conidia should be free of detergent if possible. If they are harvested in a tween + saline solution, as is often the case, the conidia should be washed in sterile deionized distilled water by centrifugation and resuspension three times before inoculation.
15. During protoplast washing steps centrifugation of protoplasts at higher g forces or longer times than specified can lead to protoplast lysis.

Acknowledgement

This work was supported by grants R01GM31837 and PO1GM084077 from the NIH and by the Irving S. Johnson fund of the Kansas University Endowment.

References

1. Keller NP, Turner G, Bennett JW (2005) Fungal secondary metabolism—from biochemistry to genomics. Nat Rev Microbiol 3:937–947
2. Galagan JE, Calvo SE, Cuomo C et al (2005) Sequencing of *Aspergillus nidulans* and comparative analysis with *A. fumigatus* and *A. oryzae*. Nature 438:1105–1115
3. Machida M, Asai K, Sano M et al (2005) Genome sequencing and analysis of *Aspergillus oryzae*. Nature 438:1157–1161
4. Nierman WC, Pain A, Anderson MJ et al (2005) Genomic sequence of the pathogenic and allergenic filamentous fungus *Aspergillus fumigatus*. Nature 438:1151–1156

5. Nayak T, Szewczyk E, Oakley CE et al (2006) A versatile and efficient gene-targeting system for *Aspergillus nidulans*. Genetics 172:1557–1566
6. Chiang YM, Chang SL, Oakley BR et al (2010) Recent advances in awakening silent biosynthetic gene clusters and linking orphan clusters to natural products in microorganisms. Curr Opin Chem Biol 15:1–7
7. Bok JW, Chiang YM, Szewczyk E et al (2009) Chromatin-level regulation of biosynthetic gene clusters. Nat Chem Biol 5:462–464
8. Yu JH, Butchko RA, Fernandes M et al (1996) Conservation of structure and function of the aflatoxin regulatory gene *aflR* from *Aspergillus nidulans* and *A. flavus*. Curr Genet 29: 549–555
9. Bergmann S, Schumann J, Scherlach K et al (2007) Genomics-driven discovery of PKS-NRPS hybrid metabolites from *Aspergillus nidulans*. Nat Chem Biol 3:213–217
10. Chiang YM, Szewczyk E, Davidson AD et al (2009) A gene cluster containing two fungal polyketide synthases encodes the biosynthetic pathway for a polyketide, asperfuranone, in *Aspergillus nidulans*. J Am Chem Soc 131: 2965–2970
11. Szewczyk E, Chiang YM, Oakley CE et al (2008) Identification and characterization of the asperthecin gene cluster of *Aspergillus nidulans*. Appl Environ Microbiol 74:7607–7612
12. Chiang YM, Szewczyk E, Nayak T et al (2008) Molecular genetic mining of the *Aspergillus* secondary metabolome: discovery of the emericellamide biosynthetic pathway. Chem Biol 15:527–532
13. Chiang YM, Szewczyk E, Davidson AD et al (2010) Characterization of the *Aspergillus nidulans* monodictyphenone gene cluster. Appl Environ Microbiol 76:2067–2074
14. Sanchez JF, Chiang YM, Szewczyk E et al (2010) Molecular genetic analysis of the orsellinic acid/F9775 gene cluster of *Aspergillus nidulans*. Mol Biosyst 6:587–593
15. Sanchez JF, Entwistle R, Hung JH et al (2011) Genome-based deletion analysis reveals the prenyl xanthone biosynthesis pathway in *Aspergillus nidulans*. J Am Chem Soc 133: 4010–4017
16. Kuwayama H, Obara S, Morio T et al (2002) PCR-mediated generation of a gene disruption construct without the use of DNA ligase and plasmid vectors. Nucleic Acids Res 30:E2
17. Yang L, Ukil L, Osmani A et al (2004) Rapid production of gene replacement constructs and generation of a green fluorescent protein-tagged centromeric marker in *Aspergillus nidulans*. Eukaryot Cell 3:1359–1362
18. Szewczyk E, Nayak T, Oakley CE et al (2006) Fusion PCR and gene targeting in *Aspergillus nidulans*. Nat Protoc 1:3111–3120
19. Weidner G, d'Enfert C, Koch A (1998) Development of a homologous transformation system for the human pathogenic fungus *Aspergillus fumigatus* based on the *pyrG* gene encoding orotidine 5′-monophosphate decarboxylase. Curr Genet 33:378–385
20. Doy CH, Pateman JA, Olsen JE et al (1985) Genomic clones of *Aspergillus nidulans* containing *alcA*, the structural gene for alcohol dehydrogenase and *alcR*, a regulatory gene for ethanol metabolism. DNA 4:105–114
21. Waring RB, May GS, Morris NR (1989) Characterization of an inducible expression system in *Aspergillus nidulans* using *alcA* and tubulin-coding genes. Gene 79:119–130
22. Oakley BR, Rinehart JE, Mitchell BL et al (1987) Cloning, mapping and molecular analysis of the *pyrG* (orotidine-5′-phosphate decarboxylase) gene of *Aspergillus nidulans*. Gene 61:385–399
23. Vishniac W, Santer M (1957) The thiobacilli. Bacteriol Rev 21:195–213
24. Pontecorvo G, Roper JA, Hemmons DW et al (1953) The genetics of *Aspergillus nidulans*. Adv Genet 5:141–238
25. Lee SB, Taylor JW (1990) Isolation of DNA from fungal mycelia and single spores. In: Innis MA, Gelfand DH, Sninsky JJ, White TJ (eds) PCR protocols: a guide to methods and applications. Academic Press, Inc., San Diego; New York, pp 282–287

Chapter 11

Fast and Easy Method for Construction of Plasmid Vectors Using Modified Quick-Change Mutagenesis

Jin Woo Bok and Nancy P. Keller

Abstract

Plasmid vector construction is an essential step for molecular microbiology yet often a time-consuming process. Manipulation of the fungal genome to express genes to activate secondary metabolite production often requires creation of plasmid constructs in a reiterative fashion. Here we introduce a modified Quick-change site-directed mutagenesis method that allows for rapid and accurate construction of fungal transformation vectors.

Key words: Plasmid vector, Restriction enzyme, Polymerase chain reaction, Quick-change site-directed mutagenesis

1. Introduction

Plasmids are self-replicable extrachromosomal circular DNAs containing genes of resistance to microbes, degradation of complex organic compounds, and production of toxins and enzymes (1). Plasmids were quickly recognized as a convenient means to package foreign DNA for molecular cloning strategies allowing for mass production of certain genes and/or encoding proteins in a desired host (2–4). PCR technology has resulted in improved techniques in plasmid creation (5–8); however there are limits in terms of speed due to a multitude of protocol steps, desired DNA sequence changes, and/or restriction enzyme digestibility (9, 10). In vitro site-directed mutagenesis has been broadly used to selectively change DNA sequences for vector modification in lieu of PCR methods (11–15) but these approaches such as use of single-stranded DNA (ssDNA) as the template (16–18) can be time consuming and/or technically difficult.

Nancy P. Keller and Geoffrey Turner (eds.), *Fungal Secondary Metabolism: Methods and Protocols*,
Methods in Molecular Biology, vol. 944, DOI 10.1007/978-1-62703-122-6_11,

An advance in plasmid creation has been the newly developed Quick-change site-directed mutagenesis method by Agilent Technologies which is specifically optimized for site-specific mutation in virtually any double-stranded plasmid (19). Here we provide a modified Quick-change site-directed mutagenesis, which is applicable for the insertion or replacement of up to several kb DNA into a plasmid vector by a single PCR amplification (20, 21). This method does not need ssDNA template, subcloning, unique restriction enzyme sites, multiple transformations, or in vitro vector methylation pretreatments. We also present a fungal genomic DNA extraction method, a modified high-yield gel extraction method, and a simple plasmid mini-preparation method which when combined with our Quick-change modifications lead to fast and reproducible vector constructions for studies in fungal molecular biology.

2. Materials

2.1. *Aspergillus nidulans* Genomic DNA Isolation

1. 20× salts solution: $NaNO_3$ 120 g, KCl 10.4 g, $MgSO_4{\cdot}7H_2O$ 10.4 g, KH_2PO_4 30.4 g, and then add ddH_2O to 1 L, autoclaved and stored at room temperature.
2. Trace elements: $ZnSO_4{\cdot}7H_2O$ 2.2 g, H_3BO_3 1.1 g, $MnCl_2{\cdot}4H_2O$ 0.5 g, $FeSO_4{\cdot}7H_2O$ 0.5 g, $CoCl_2{\cdot}5H_2O$ 0.16 g, $CuSO_4.5H_2O$ 0.16 g, $(NH_4)6Mo_7O_{24}4H_2O$ 0.11 g, Na_4EDTA 5.0 g, added in order to 80 mL of H_2O, dissolving each completely before adding the next. Heat the solution to boiling, cool to 60°C, and adjust the pH to 6.5–6.8 with KOH pellets. Cool to room temperature and adjust volume to 100 mL with ddH_2O.
3. *Aspergillus* glucose minimal medium (GMM): 20× salt solution (50 mL/L), trace elements (1 mL/L), D-glucose (10 g/L), agar (for solid media) (15 g/L) in 1 L ddH_2O, adjust pH to 6.5 and then autoclave for 15 min, at 121°C on liquid cycle.
4. *Aspergillus* wild-type strain (or strain of choice).
5. Phenol:$CHCl_3$:isoamyl alcohol = 25:24:1 (Ambion, Austin, TX, cat. no. 9732), abbreviated as PCI.
6. RNase (5′, Gaithersburg, MD, cat. no. 2500130).

2.2. Plasmid Preparation

1. LB medium: Tryptone 10 g, yeast extract 5 g, NaCl 10 g in 1 L ddH_2O, adjust pH to 7 and then autoclave for 15 min, at 121°C on liquid cycle.
2. Resuspension buffer: 50 mM Tris–HCl, pH 8.0, 10 mM EDTA, 20 μg RNase A.
3. Lysis buffer: 200 mM NaOH, 1% SDS.

4. Neutralization buffer: 3 M potassium acetate, pH 5.5.
5. Isopropanol.
6. 70% ethanol.
7. 10 mM Tris–HCl, pH 8.

2.3. Gel Extraction

1. Qiagen gel extraction kit (Valencia, CA, cat. no. 28706).

2.4. Quick Change

1. dNTP mixture (10 mM of each dATP, dCTP, dTTP, and dGTP, Promega, Fitchburg, WI, cat. no. U151B).
2. PfuUltra II Fusion HS DNA polymerase (Agilent Technologies, Santa Clara, California, cat. no. 6000672-51).
3. pME3188 YFP split vector (22).
4. *A. nidulans* genomic DNA.
5. G50 spin column (GE Healthcare, Buckinghamshire, UK, cat. no. 27-5340-03).

3. Methods

The key step in creating plasmid vectors for fungal transformations is a reliable mutagenesis method. In addition, it is important to have good quality of template DNA for PCR steps and good recovery of PCR products from agarose gel for vector construction. The basic procedure utilizes a supercoiled double-stranded plasmid DNA vector and a PCR product containing the desired DNA fragment with flanking sequences (see Fig. 1). The flanking sequences of PCR product recovered by modified gel extraction, each complementary to opposite strands of the vector, are extended during temperature cycling by PfuUltra II Fusion HS DNA polymerase (23). Extension of the PCR template generates nicked circular DNA molecules. Following temperature cycling, the reaction mixture is treated with *Dpn* I endonuclease (target sequence: 5′-Gm6ATC-3′) used to digest the methylated parental DNA template and to select for target gene containing nonmethylated synthesized DNA (24, 25). The nicked plasmid vector DNA incorporating the desired mutations is then transformed into competent cells for nick repair. The mutated plasmid vector will be extracted from transformed bacteria by a modified simple mini-plasmid preparation. The mutated region is confirmed by PCR amplification followed by sequencing of the amplicon.

3.1. *A. nidulans* Genomic DNA Isolation

1. Add 20 mL GMM with appropriate supplements and yeast extract at 5 g/L into a petri dish.
2. Inoculate with several loopful of *Aspergillus* conidia from solid media.

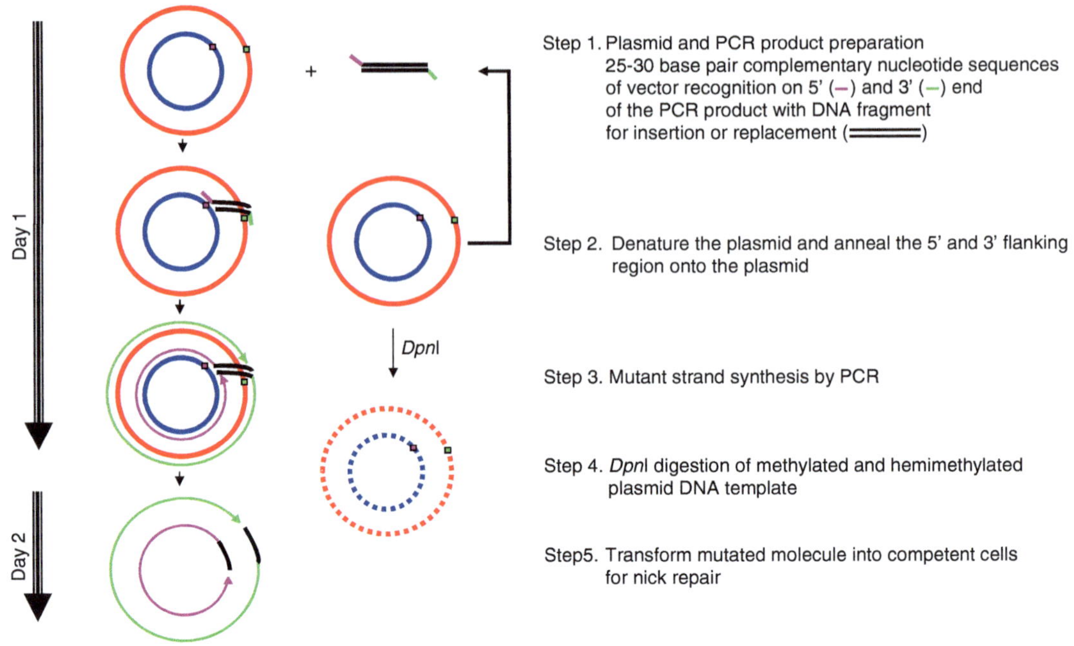

Fig. 1. Overview of plasmid construction by the modified Quick-change method.

3. Incubate at 37°C for 16–24 h (see Note 1).
4. Pour off media and scrape hyphal material together. Put material in the lid of petri dish and remove as much liquid as possible.
5. Take cell material and place between paper towels and squeeze out the remaining liquid. Put cell material in Eppendorf tube and freeze in −80°C for at least 30 min or in liquid nitrogen for several minutes.
6. Lyophilize overnight (see Note 2).
7. Break up lyophilized hyphae into a fine powder using toothpicks.
8. Add 700 μL of LETS buffer containing 20 mM EDTA (pH 8), 0.5% SDS, 10 mM Tris–HCl (pH 8), and 0.1 M LiCl. Mix by using toothpick as well as inverting the tubes several times and leave the samples on bench for 5 min.
9. Add 700 μL of PCI. Mix by shaking 10–15 times. Leave samples on bench for 5 min.
10. Spin for 10 min at 4°C at maximum speed of microcentrifuge.

    (Optional: Transfer supernatant (SN) to a new tube, and add equal volume of PCI and repeat steps 9 and 10).
11. Transfer SN to a new tube and add 1 mL 95% ethanol. Mix well and spin for 5 min to pellet DNA.

12. Decant SN and add 70% ethanol to wash pellet.
13. Spin for 1 min at maximum speed to pellet DNA.
14. Discard SN and dry pellet on bench for 5 min (see Note 3).
15. Resuspend in 50–70 μL 10 mM Tris–HCl (pH 8) and add 2 μL RNase (10 mg/mL stock) (see Note 4).
16. Inactivate DNase and digest RNA at 65°C for 30 min.

3.2. Plasmid Preparation (See Note 5)

1. Inoculate 3 mL LB medium containing proper antibiotic with bacterial clone, incubate overnight at 37°C and 200 rpm, and harvest bacteria by spinning at maximum speed for 30 s.
2. Remove supernatant as much as possible (see Note 6) and resuspend the pellet completely in 100 μL resuspension buffer.
3. Add 100 μL lysis buffer and mix gently 3–4 times by inversion.
4. Add 125 μL neutralization buffer and mix gently for 3 min.
5. Spin down to remove bacterial debris by centrifugation at maximum speed for 2 min and transfer supernatant in 1.5 mL new centrifuge tube.
6. Add 200 μL isopropanol and mix well to pellet plasmid DNA. Leave the sample for a couple of minutes on bench.
7. Collect DNA pellet by spin down at maximum speed for 2 min and decant supernatant.
8. Add 500 μL 70% ethanol to wash pellet.
9. Collect the DNA pellet by spinning at maximum speed for 30 s and remove SN (see Note 7).
10. Air-dry the pellet for 3 min on bench and add 50 μL 10 mM Tris–HCl buffer, pH 8.0 (see Note 8).

3.3. Gel Extraction

1. Excise the DNA band from the agarose gel with a clean, sharp scalpel and put into a 1.5 mL microfuge tube (see Note 9).
2. Weigh the gel slice in a colorless tube. Add 3 volumes of Buffer QG to 1 volume of gel for DNA fragments (see Note 10).
3. Incubate at 50°C for 10 min (or until the gel slice has completely dissolved). To help dissolve gel, mix by vortexing every 2–3 min during the incubation (see Note 11).
4. Check that the color of the mixture is yellow.

   If the color of the mixture is orange or purple, add 10 μL 3 M sodium acetate, pH 5.0, and mix. The color should turn to yellow. The adsorption of DNA to QIAquick membrane is only efficient at $pH \leq 7.5$. Buffer QG now contains a pH indicator which is yellow at pH 7.5, and orange or violet at higher pH, allowing easy determination of the optimal pH for DNA binding.

5. Add 1 gel volume of isopropanol to the sample and mix.
   Do not centrifuge the sample at this stage (see Note 12).
6. To bind DNA, apply the sample to the QIAquick column with a 2 mL collection tube and centrifuge for 1 min (see Note 13).
7. Discard the flow-through and place QIAquick column back in the same collection tube.
8. Add 0.5 mL of buffer QG to QIAquick column and spin for 1 min.
   This step will remove all traces of agarose.
9. To wash, add 750 μL of Buffer PE to QIAquick column and centrifuge for 1 min.
10. Discard the flow-through and centrifuge QIAquick column for an additional 1 min at maximum speed.
11. Place QIAquick column into a clean 1.5 mL microfuge tube (see Note 14).
12. To elute DNA, add 50 μL of 10 mM Tris–Cl, pH 8.5 or H_2O to the center of the QIAquick membrane and centrifuge the column for 1 min at maximum speed. Alternatively, for increased DNA concentration, add 30 μL of elution buffer to the column, let the column stand for 1 min, and then centrifuge for 1 min. Elution efficiency is dependent on pH. The maximum elution efficiency is achieved between pH 7.0 and 8.5. When using water for elution, make sure that the pH is within this range, and store DNA at −20°C as DNA may degrade in the absence of a buffering agent. The purified DNA can also be eluted in TE buffer (10 mM Tris–HCl, 1 mM EDTA, pH 8.0) but the EDTA may inhibit subsequent enzymatic reactions.
13. Eluted samples are loaded on gel to check quality (Fig. 2) (see Note 15).

3.4. Quick Change

3.4.1. Primer Design

IDT provides good-quality primers; of course, you can order primers through your vender.

Insertion or replacement of DNA fragments in plasmid vector needs two primers for PCR amplification. Keep primer length between 45 and 60 bp. You should not try to exceed 60 bp because above that size IDT forces you to order 100 nmol scale with the corresponding price increase. Ideally, you would want 25–30 bp on either side of the mismatch, and approximately equal melting temperature (TM) for both “arms” of the primer (see Note 16). The overall primer TM (*not counting the mismatched pairs*) should be in the range of 60–75°C.

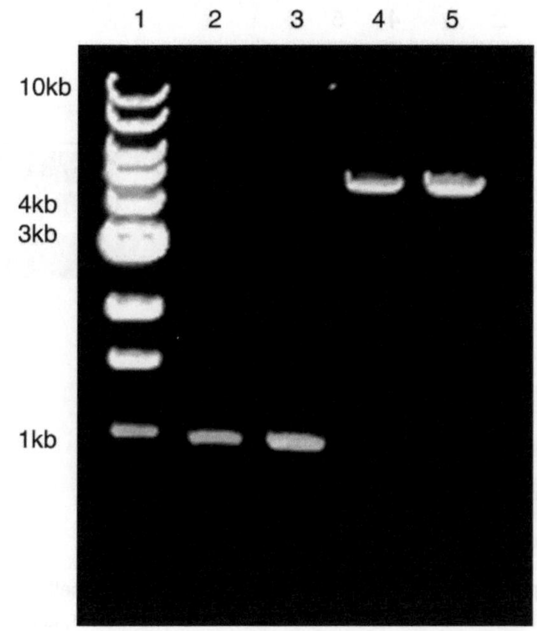

Fig. 2. Gel extraction of PCR products by Qiagen protocol and modified Qiagen protocol; lane1: 1 kb size marker, *lanes 2* and *4*: Qiagen protocol, *lanes 3* and *5*: modified Qiagen protocol.

3.4.2. PCR Condition

Contained in PCR tubes are the following:

1. 25–50 ng template (see Notes 17 and 18).
2. 250 ng PCR product (see Note 19).
3. 0.75 μL dNTP mix.
4. 2.5 μL 10× PfuUltra II Fusion ("Pfu Fusion" for brevity) buffer (see Note 20).
5. 1.0 μL DMSO (optional) (see Note 21).
6. 0.5 μL of Pfu polymerase.
7. Add sterile water to 25 μL and one drop of mineral oil (see Note 22).

Heat the reaction to 95°C for 5 min to denature. Perform 36 cycles of PCR [95°C for 1 min (denaturation), 58°C for 1 min (annealing), and 68°C for 1.5 min/kbp template (elongation)]. Followed by final elongation at 65°C for 10 min hold at 4°C (see Notes 23–26).

3.4.3. Dpn I Treatment and Transformation (See Note 27)

After removing mineral oil from PCR sample by rolling on parafilm, pass through the sample by G50 column to eliminate salts from previous reaction. Load 3 μL of sample in agarose gel to see success

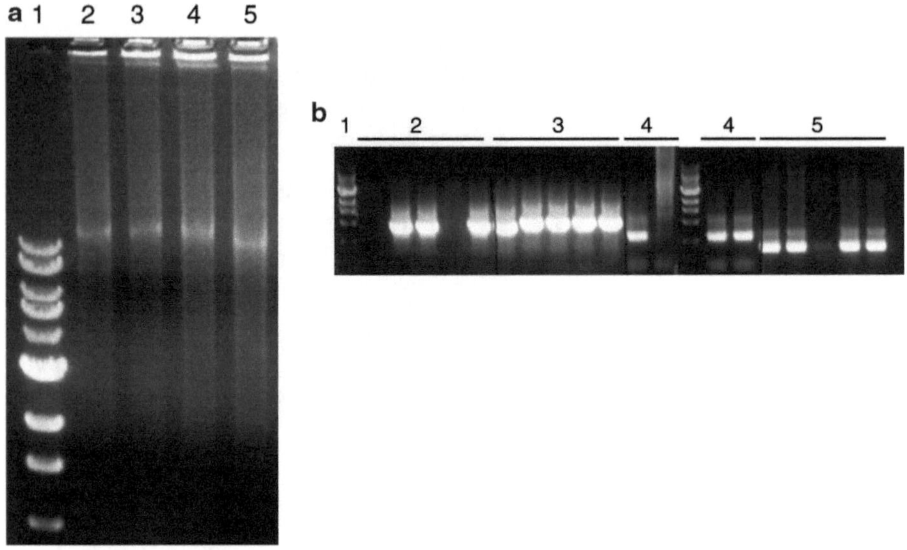

Fig. 3. PCR amplified plasmid vectors by modified Quick-change method and their confirmation by PCR. (**a**) 3 μL load of PCR product after modified Quick-change; *lane 1*: 1 kb size marker, *lane 2*: *veA* replaced with AN2012 in pME3188, *lane 3*: *veA* replaced with AN10356 in pME3188, *lane 4*: *veA* replaced with AN4252 in pME3188, *lane 5*: *veA* replaced with AN4500 in pME3188. (**b**) PCR confirmation of each vector obtained by modified plasmid DNA preparation; *lanes 2, 3, 4*, and *5* are matched to the same gene in Fig. 3a, respectively.

of the reaction (Fig. 3). Set up *Dpn*I digestion for 3 h as follows (see Note 28):

Sample	13 μL
NEB buffer #4	3 μL
*Dpn*I (NEB)	1.5 μL
Sterile distilled water	13 μL

Use 15 μL for transformation with chemical competent cell or use 3–5 μL for transformation with electroporation competent cell after passing G50 column (see Note 29), and then immediately add 0.5 mL SOC, let cells recover for 30 min at 37°, and plate 200 and 20 μL (see Note 30).

4. Conclusions

Traditional plasmid vector construction has been commonly used in many fields yet it is still a time-consuming process and is of limited use due to awkward placement of restriction sites and cumbersome mutagenesis methods. Quick-change site-directed mutagenesis is the simplest and most broadly applicable tool to change DNA sequences in plasmid vectors without using restriction enzymes. Modification of the Quick-change kit as described here provides several advantages in plasmid construction by decreasing the cost of materials while increasing the time in plasmid construction.

5. Notes

1. The wild-type *veA* allele results in decreased DNA extraction quality. To get better DNA, culture the fungus in light but harvest prior to sporulation.
2. Pierce the lid of the Eppendorf tubes.
3. After discarding SN, quick spin and take out residual SN with pipette before drying on bench.
4. Tris–HCl (pH 8) makes DNA more stable than sterile water.
5. You can use leftover buffers in commercial plasmid prep kits such as Qiagen mini prep kit or Biorad min prep kit or you can make your own buffers. All of the steps can be done at room temperature using a microcentrifuge.
6. Decant culture SN, and quick spin to remove residual SN by pipetting out.
7. Instead of drying pellet in the tube by turning it upside down, quick spin to remove residual SN by pipetting after decantation of culture SN.
8. To inactivate possible DNase contamination, incubate your sample at 65°C for 10 min and preserve your sample at −20°C until you need it.
9. Make 0.8% agarose gel for separation. This provides you better results because lower % agarose is easy to melt completely.
10. If gel weight is under 150 mg, just add 3 volumes of QG. Even if gel weight is over 150 mg, do not exceed a total volume more than 600 μL. For example, if gel weight is 200 mg, add 400 μL QG.
11. You can incubate at 65°C for a short time (2–3 min) to dissolve the gel completely.
12. Mix well immediately to prevent resolidification of agarose, which can cause clogging of column. Do not centrifuge because centrifugation can result in precipitation of your DNA.
13. Reapply the flow-through to the column and centrifuge at 3,380 × *g* for 15 s. Repeat this step one more time to maximize recovery of DNA (total three times for flow-through).

14. To remove residual ethanol, spin for 1 min with open lid at 3,380 × *g*.
15. To get more recovery of DNA from column, add 65°C elution buffer into the column and immediately open and close the lid several times to make the buffer absorbed well into the column materials.
16. The 3′ end is more significant because that is where DNA synthesis starts; try to end the primer with a couple of G/Cs.
17. *Amount of template.* The stated 50 ng is simply lifted from the standard Quick-change protocol, 100 ng/50 μL, and was not tested experimentally. 0.2–2 μL of miniprep sample was used in the 25 μL reaction. The advantages of higher template concentration are more product with less number of cycles and higher chance of priming when annealing is suboptimal; however, the disadvantages of having too much of the template are introducing more impurities into the reaction (some of which might be inhibitory), increased background if the *Dpn* I treatment is incomplete, and increase in the misprimed side reactions. There is an optimal balance between all of these factors which each practitioner must establish himself or herself.
18. *Quality of template.* Most of the time current modified plasmid preparation method worked well. However, steps to increase quality using the Qiagen mini-prep kit include (1) using 1 mL culture instead of 3 mL if your plasmid is high copy number, (2) using a PB solution volume that covers all areas of the column that are in contact with cell lysate, and (3) using 750 μL PE for washing twice.
19. Good quality of PCR product is very important with high concentration. To achieve this goal, we suggest to use 30 μL elution buffer of 65°C and heat inactivation at 65°C for 10 min. Usually the volume of 250 ng will be contained in 4–6 μL.
20. The Pfu Fusion polymerase works faster and makes severalfold less errors than Pfu Ultra.
21. DMSO is optional but it is helpful for decreasing melting temperature of DNA duplexes with high GC content and it reduces primers' secondary structure. You can use up to 4% DMSO and make sure to mix DMSO well with other components before adding polymerase.
22. One drop of mineral oil is very important to have reproducible result. The mineral oil keeps 25 μL PCR reaction from evaporating.
23. The larger number of cycles gives better results but we have found that 20–36 cycles can work.
24. *Annealing temperature.* Pfu Fusion works in a broad range of annealing conditions including 5–10° below the calculated

primer TM based on GC content (number of GC × 4 + number of AT × 2). However, 56–58°C should work in the vast majority of cases.

25. *Extension time.* The recommended 1.5 min/kbp at 68° definitely works for Pfu Fusion. Round the number of minutes upwards and stay with this value. We have had success with 1 min/kbp in some cases. Pfu Fusion has some strand displacement activity that may require 2 min/kbp at 65°C for some reactions.
26. *Miscellaneous.* You may increase the primer concentration (approximately twofold), increase concentration of dNTPs, and add extra $MgCl_2$ in the reaction all when using high-fidelity Pfu Fusion.
27. While plasmid DNA isolated from almost all of the commonly used *Escherichia coli* strains (*dam+*) is methylated and is a suitable template for mutagenesis, plasmid DNA isolated from the exceptional *dam− E. coli* strains, including JM110 and SCS110, is not suitable.
28. This is a very critical step to digest parental plasmid DNA. Some protocols say that the digestion can be done overnight at 37°C but there is a greater chance for DNA to be degraded and the reaction volume to be reduced by evaporation. This is especially true of small volumes where 37°C incubations in the water bath or the incubator are hard to control and may not be reproducible because of evaporation/condensation, etc. Therefore we suggest a shorter incubation time than overnight. Or alternatively, you can leave your samples in a PCR machine if you want to incubate your samples overnight.
29. Gel filtration with G50 is very important to remove salts for electroporation. Too much salts or too much cells cause an electric arc and kill cells.
30. Sometimes cell recovery after chemical transformation or electroporation is not consistent. We recommend to try the chemical method first as it is cheaper. If this method does not work, you can try electroporation or you may try using LB instead of SOC, or LB with extra $MgSO_4$ and glucose. A plasmid vector with kanamycin resistance marker needs to be incubated longer (1 h) than ampicillin marker (30 min) to help competent cell recovered.

Acknowledgment

The project was supported by PO1GM084077 from the National Institute of General Medical Sciences to N.P.K.

References

1. Sambrook J, Russell DW (2001) Molecular cloning: a laboratory manual. Cold Spring Harbor, New York
2. Williams SG, Cranenburgh RM, Weiss AME, Wrighton CJ, Sherratt DJ, Hanak JAJ (1998) Repressor titration: a novel system for selection and stable maintenance of recombinant plasmids. Nucleic Acids Res 26:2120–2124
3. Cohen SN (1993) Bacterial plasmids: their extraordinary contribution to molecular genetics. Gene 135:67–76
4. Bahl MI, Hansen LH, Sørensen SJ (2009) Persistence mechanisms of conjugative plasmids. Methods Mol Biol 532:73–102
5. Innis MA, Gelfand DH, Sninsky JJ, White TJ (1990) PCR protocol: a guide to methods and applications. Academic, San Diego
6. Arredondo-Peter R, Bonic A, Sarath G, Klucas RV (1995) Commentary: rapid PCR-based detection of inserts from cDNA libraries using phage pools or direct phage plaques and lambda primers. Plant Mol Biol Reptr 13:138–146
7. Erlich HA, Gelfand D, Sninsky JJ (1991) Recent advances in the polymerase chain reaction. Science 252:1643–1651
8. Rosilawati ML, Yasmon A (2011) Rapid detection of ethambutol-resistant *Mycobacterium tuberculosis* directly from sputum samples by radioisotope (32P)-based PCR dot blot hybridization and sequencing methods. Acta Med Indones 43:34–38
9. Palatinszky M, Nikolausz M, Sváb D, Márialigeti K (2011) Preferential ligation during TA-cloning of multitemplate PCR products–a factor causing bias in microbial community structure analysis. J Microbiol Methods 85:131–136
10. Walker A, Taylor J, Rowe D, Summers D (2008) A method for generating sticky-end PCR products which facilitates unidirectional cloning and the one-step assembly of complex DNA constructs. Plasmid 59:155–162
11. Shaya D, Zhao WJ, Garron ML, Xiao Z, Cui Q, Zhang Z, Sulea T, Linhardt RJ, Cygler M (2010) Mechanism of heparinase II investigated by site-directed mutagenesis and the crystal structure with its substrate. J Biol Chem 285:20051–20061
12. Mullaney EJ, Locovare H, Sethumadhavan K, Boone S, Lei XG, Ullah AH (2010) Site-directed mutagenesis of disulfide bridges in *Aspergillus niger* NRRL 3135 phytase (PhyA), their expression in *Pichia pastoris* and catalytic characterization. Appl Microbiol Biotechnol 87:1367–1372
13. Simionatto S, Marchioro SB, Galli V, Luerce TD, Hartwig DD, Moreira ÂN, Dellagostin OA (2009) Efficient site-directed mutagenesis using an overlap extension-PCR method for expressing *Mycoplasma hyopneumoniae* genes in *Escherichia coli*. J Microbiol Methods 79:101–105
14. Tian J, Liu Q, Dong S, Qiao X, Ni J (2010) A new method for multi-site-directed mutagenesis. Anal Biochem 406:83–85
15. Seyfang A, Jin JH (2004) Multiple site-directed mutagenesis of more than 10 sites simultaneously and in a single round. Anal Biochem 324:285–291
16. Kunkel TA (1985) Rapid and efficient site-specific mutagenesis without phenotypic selection. Proc Natl Acad Sci USA 82:488–492
17. Vandeyar MA, Weiner MP, Hutton CJ, Batt CA (1988) A simple and rapid method for the selection of oligodeoxynucleotide-directed mutants. Gene 65:129–133
18. Anderson EM, Halsey WA, Wuttke DS (2003) Site-directed mutagenesis reveals the thermodynamic requirements for single-stranded DNA recognition by the telomere-binding protein Cdc13. Biochemistry 42:3751–3758
19. US Patent Nos. 6,713,285, 6,391,548, 5,948,663, 5,932,419, 5,789,166, 7,132,265, 7,176,004, 5,286,632 and patents pending
20. Geiser M, Cèbe R, Drewello D, Schmitz R (2001) Integration of PCR fragments at any specific site within cloning vectors without the use of restriction enzymes and DNA ligase. Biotechniques 31:88–90
21. van den Ent F, Löwe J (2006) RF cloning: a restriction-free method for inserting target genes into plasmids. J Biochem Biophys Methods 67:67–74
22. Bayram O, Krappmann S, Ni M, Bok JW, Helmstaedt K, Valerius O, Braus-Stromeyer S, Kwon NJ, Keller NP, Yu JH, Braus GH (2008) VelB/VeA/LaeA complex coordinates light signal with fungal development and secondary metabolism. Science 320:1504–1506
23. US Patent Nos. 7,045,328, 6,734,293, 6,489,150, 6,444,428, 6,183,997, 5,948,663, 5,866,395, 5,545,552 and patents pending
24. Nelson M, McClelland M (1992) Use of DNA methyltransferase/endonuclease enzyme combinations for megabase mapping of chromosomes. Methods Enzymol 216:279–303
25. Geier GE, Modrich P (1979) Recognition sequence of the dam methylase of *Escherichia coli* K12 and mode of cleavage of *Dpn*I endonuclease. J Biol Chem 254:1408–1413

Chapter 12

Reconstitution of a Secondary Metabolite Biosynthetic Pathway in a Heterologous Fungal Host

Takayuki Itoh, Tetsuo Kushiro, and Isao Fujii

Abstract

Expression of multiple genes involved in a particular metabolic pathway in a heterologous host facilitates the study of fungal secondary metabolite biosynthesis and production of useful compounds. Two plasmids with different selection markers, *argB* and the pyrithiamine resistance marker, are used to transform *Aspergillus oryzae* allowing for expression of up to three genes simultaneously.

Key words: Secondary metabolite biosynthesis, Protoplast-PEG transformation, *argB* marker, Pyrithiamine resistance marker

1. Introduction

Fungal secondary metabolites include diverse structural classes of compounds, many of which have been extensively utilized for maintaining human health such as antibacterial, antifungal, and anti-hypercholesterolemia agents (1). These molecules are sequentially assembled from a simple precursor through complex biosynthetic pathways that involve numerous enzymes (2). Polyketides and terpenoids are representative classes of secondary metabolites that exhibit a large variety of structures. The enzymes, polyketide synthases and terpenoid synthases, are mostly responsible for constructing the basic carbon skeletons of these classes, which are then followed by numerous tailoring enzymes such as oxygenases, acyltransferases, and glycosyltransferases to complete modifications to afford the final compounds. Biosynthetic studies on fungal secondary metabolites have recently become of a great interest largely due to the completion and/or ongoing genome sequencing projects

Nancy P. Keller and Geoffrey Turner (eds.), *Fungal Secondary Metabolism: Methods and Protocols*,
Methods in Molecular Biology, vol. 944, DOI 10.1007/978-1-62703-122-6_12,

that greatly facilitate the discovery of novel biosynthetic genes and gene clusters (3). In order to analyze the enzymatic function of each gene and to engineer them to produce useful compounds, a reconstitution of a whole biosynthetic pathway in heterologous expression host is an alternative process to express multiple genes in a single host. Although pathway expression is quite common in bacterial systems such as in actinomycetes, few examples have been published for fungal systems. We have succeeded in expressing multiple pathway genes involved in fungal meroterpenoid biosynthesis in the fungal host *Aspergillus oryzae* (4). *A. oryzae* is a GRAS organism harboring all of the advantages of routine transformation and expression techniques coupled with little endogenous secondary metabolite production to complicate expression of heterologous compounds. This expression system is based on the α-amylase promoter and terminator which have been used successfully for fungal polyketide synthase expression (2), and utilization of two plasmids with different selection markers, one employing *argB* marker (5, 6) while other using pyrithiamine resistance marker (7). Up to three genes were simultaneously expressed in a single host using these techniques.

2. Materials

2.1. Host Strain and Culture Medium

1. The host strain *A. oryzae* M-2-3 *argB*⁻ (5) is maintained on a standard potato dextrose agar (PDA) plate at 30°C.
2. CD-MM agar plate: Weigh 0.3 g $NaNO_3$, 0.2 g KCl, 0.1 g KH_2PO_4, 0.05 g $MgSO_4·7H_2O$, 0.002 g $FeSO_4·7H_2O$, 2 g glucose, 21.86 g sorbitol, and 1.5 g Bacto-agar and make up to 100 mL with water. Mix and adjust pH to 5.5 with HCl.
3. CD-MM top agar: The same composition as in CD-MM agar plate except that 0.8 g Bacto-agar is added instead of 1.5 g.
4. DPY medium: Weigh 2 g Dextrin, 1 g Polypeptone, 2.5 g Bacto-Yeast extract, 0.5 g KH_2PO_4, and 0.05 g $MgSO_4·7H_2O$ and make up to 100 mL with water.
5. CD-starch medium: Weigh 0.3 g $NaNO_3$, 0.2 g KCl, 0.1 g KH_2PO_4, 0.05 g $MgSO_4·7H_2O$, 0.002 g $FeSO_4·7H_2O$, 1 g Polypeptone, and 2 g Starch Soluble, and make up to 100 mL with water. Mix and adjust pH to 5.5 with HCl.

2.2. Expression Plasmids

1. Fungal expression plasmid pTAex3 utilizing α-amylase promoter and terminator (5) is converted to Gateway destination vector pTAex3R by inserting Reading Frame Cassette A fragment (Invitrogen) in the *Sma*I site.

2. pPTR I (Takara), harboring a pyrithiamine resistance marker, is modified by inserting *Srf*I recognition sequence (GCCC GGGC) into the *Hin*dIII recognition site and the resultant pPTR-srfI is used for expression of the second gene.

2.3. Transformation Solutions (8)

1. TF Solution 0: 50 mM Maleic acid, pH 7.5. Weigh 1.16 g maleic acid and make up to 200 mL with water. Mix and adjust pH to 7.5 with NaOH. Sterilize by autoclaving and store at room temperature.
2. TF Solution 1: 1% (w/v) Yatalase, 0.6 M $(NH_4)_2SO_4$ in TF Solution 0. Weigh 0.2 g Yatalase (Takara) and 1.58 g $(NH_4)_2SO_4$ and make up to 20 mL with TF Solution 0. Prepare and sterilize by filtration (pore size 0.45 μm) just before use.
3. TF Solution 2: 1.2 M Sorbitol (Glucitol), 50 mM $CaCl_2{\cdot}2H_2O$, 35 mM NaCl, 10 mM Tris–HCl (pH 7.5). Weigh 87.4 g sorbitol, 2.94 g $CaCl_2{\cdot}2H_2O$, and 0.82 g NaCl and add 4 mL of 1 M Tris–HCl (pH 7.5). Make up to 400 mL with water. Sterilize by autoclaving and store at room temperature.
4. TF Solution 3: 60% (w/v) PEG4000, 50 mM $CaCl_2{\cdot}2H_2O$, 10 mM Tris–HCl (pH 7.5). Weigh 120 g PEG4000 and 1.47 g $CaCl_2{\cdot}2H_2O$ and add 2 mL of 1 M Tris–HCl (pH 7.5). Make up to 200 mL with water. Sterilize by autoclaving and store at room temperature.

3. Methods

3.1. Construction of Plasmids

1. PCR amplify the gene to be expressed with N- and C-terminal primers containing nucleotide sequences for Gateway cloning system (Invitrogen). These are attB1 site (5′-GGGG*ACAAG TTTGTACAAAAAAGCAGGCT*-3′) directly upstream of the start codon at the 5′-terminus and attB2 site (5′-GGGG*ACC ACTTTGTACAAGAAAGCTGGGT*-3′) directly following the stop codon at the 3′-terminus.
2. The amplified full-length fragment is purified by agarose gel electrophoresis and extracted using gel extraction kit.
3. The fragment is introduced into pDONR221 (Invitrogen) donor vector using BP recombinase reaction according to the manufacturer's protocol.
4. The resulting entry clone is then used to introduce the gene fragment into pTAex3R destination vector containing *argB* marker using LR recombinase reaction according to the manufacturer's protocol.

5. This will place the full-length gene between the α-amylase promoter and the terminator in pTAex3R.
6. Because pPTR-srfI possesses neither promoter nor terminator, the second gene of interest is first introduced into pTAex3R between the α-amylase promoter and the terminator, and then a fragment containing promoter-gene-terminator as a whole is amplified by PCR, inserted into the pPTR-srfI vector digested with *Srf*I, and dephosphorylated.
7. A GGGC sequence is introduced at the 5′ end of a forward primer which can reconstitute *Srf*I recognition site with GCCC sequence at the 3′ end of the vector digested with *Srf*I. Therefore, multiple biosynthetic genes can be theoretically introduced at the reconstituted *Srf*I site using this method. To facilitate this process, a GGG sequence is introduced at the 5′ end of a reverse primer which will reconstitute *Sma*I recognition site upon ligation with pPTR-srfI. This will prevent loss of the first gene fragment upon *Srf*I digestion when the second gene is introduced. Both forward and reverse primers are phosphorylated at the 5′-ends to ensure ligation.
8. The third gene of interest can be introduced into regenerated *Srf*I recognition site directly upstream of the α-amylase promoter by the same procedure (step 7). The entire process of steps 6–8 is illustrated in Fig. 1.

3.2. Preparation of Protoplasts

1. *A. oryzae* M-2-3 strain cultured in PDA plate at 30°C is inoculated into 20 mL of liquid DPY medium and cultured for 3–4 days at 30°C with 220 rpm shaking.
2. The culture is transferred into 200 mL of liquid DPY medium and continue culturing for 3–5 days at 30°C with 160 rpm shaking until the culture becomes viscous (see Note 1).
3. The cells are filter collected by passing 20–30 mL of the culture through a sterilized plastic syringe filled with cotton (see Note 2).
4. The collected cells are transferred into a polypropylene tube filled with 20 mL of TF Solution 1 and incubated for 3 h at 30°C with 60 rpm shaking in a water bath (see Note 3).
5. The resulting solution is briefly filtered through a syringe filled with cotton and collected in a centrifugation tube (see Note 4).
6. Centrifuge at 200 × *g* for 10 min and discard the supernatant.
7. Add 20 mL of TF Solution 2 and mix by gently inverting the tube.
8. Centrifuge at 200 × *g* for 10 min and discard the supernatant.

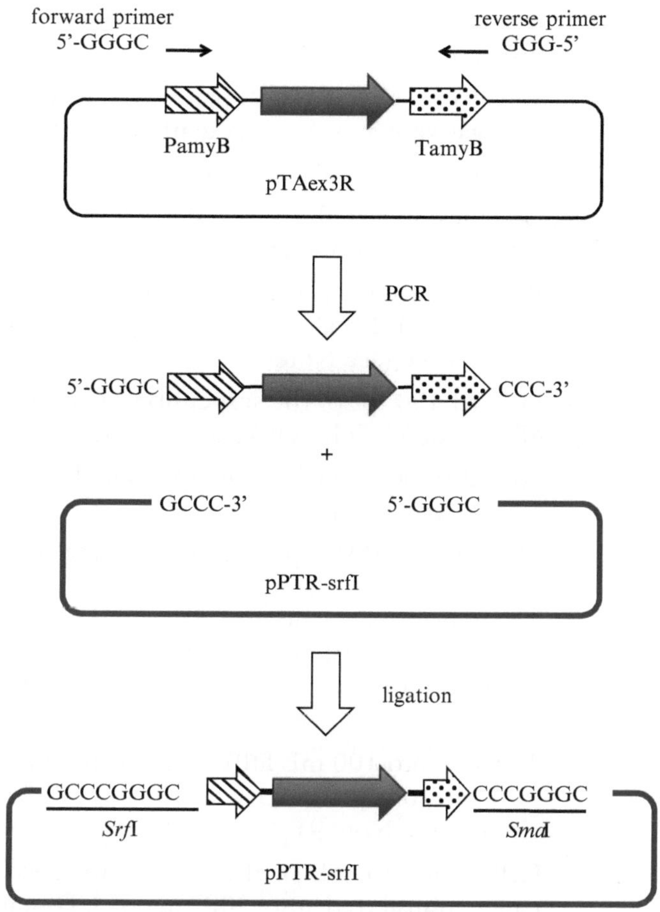

Fig. 1. Construction of pPTR-srfI expression plasmid harboring the gene of interest flanked by the α-amylase promoter and the terminator. A GGGC sequence at the 5′ end of a forward primer can reconstitute *Srf*I recognition site upon ligation into *Srf*I cut pPTR-srfI, while GGG sequence at the 5′ end of a reverse primer will reconstitute *Sma*I recognition site upon ligation.

9. Resuspend the pelletted protoplasts in TF Solution 2 to a final concentration of $1.0–5.0 \times 10^7$/mL.

3.3. Transformation (8)

1. Place 200 μL of protoplast solution in a 15 mL centrifugation tube and add less than 20 μL plasmid solution of concentration more than 0.2 μg/μL (see Note 5).
2. Mix gently and keep on ice for 30 min.
3. Add twice 250 μL and once 850 μL of TF Solution 3 and each time mix gently.
4. Store at room temperature for 20 min.
5. Add 5 mL of TF Solution 2, mix gently, and centrifuge at $200 \times g$ for 10 min.

6. Discard the supernatant, add 500 μL of TF Solution 2, and suspend.
7. Spread on a CD-MM agar plate, overlay with 5 mL of CD-MM top agar kept at 45°C, mix gently, and dry for 30 min.
8. Incubate at 30°C for 4–8 days.
9. Colonies appeared are transferred to a CD-MM agar plate without sorbitol and checked for stability of growth.
10. When the second plasmid pPTR I is used for transformation of *A. oryzae* M-2-3 already harboring pTAex3R, repeat the procedures 1–6 (see Note 6).
11. Discard 450 μL of the suspension, dilute ten times by adding 450 μL of TF Solution 2, and mix (see Note 7).
12. Spread on a CD-MM agar plate supplemented with 0.1 μg/mL of pyrithiamine.
13. Overlay CD-MM top agar and incubate at 30°C for 4–8 days.

3.4. Expression Culture of A. oryzae Transformants

1. The transformant colony of *A. oryzae* M-2-3 grown in CD-MM agar plate is transferred to 20 mL of DPY liquid medium and incubated for 3 days at 30°C at 220 rpm (rotary shaking) (see Note 8).
2. Transfer into 100 mL DPY medium and incubate for 3 days at 30°C at 160 rpm (rotary shaking) until the culture becomes viscous (see Note 9).
3. Cells from 10 mL of the culture are collected onto a filter paper, transferred into 200 mL of CD–starch medium, and incubated for 3 days at 30°C at 160 rpm (rotary shaking).

3.5. Extraction of Products

1. Collect the mycelia through filtration of the culture in vacuo using a filter paper. The filtrate is also used for extraction.
2. Dry the mycelia by pressing it against filter paper and dip into acetone overnight at room temperature for extraction.
3. The extract is concentrated in vacuo, re-extracted with ethylacetate under acidic conditions (pH 3), and again concentrated in vacuo.
4. The extract is further purified using an octadecylsilyl (ODS) column and reverse phase HPLC depending on the nature of the compounds.
5. The filtrate from step 1 is extracted with ethylacetate under acidic conditions (pH 4) and washed with brine.
6. The extract is concentrated in vacuo and further purified using an ODS column and reverse phase HPLC depending on the nature of the compounds.

4. Notes

1. Culture the mycelia until it becomes viscous and not aggregated. If the aggregation is small (less than 1 cm in diameter), mycelia may become viscous if culturing is continued for a longer period; however, if the aggregation is large (more than 2 ~ 3 cm in diameter), start from step 1 and inoculate mycelia from a PDA plate again.
2. Make sure that the moisture is not remaining in the filtered mycelia, as it will affect the efficiency of protoplast formation due to the osmotic pressure.
3. Incubate until there is no longer visible mycelium. If the mycelia does digest, the efficiency of the protoplast formation may be very low.
4. In order not to damage the protoplast cells, do not press the piston of the syringe until the very last. Some moisture may remain on a cotton.
5. The concentration of a plasmid solution should be more than 0.2 μg/μL. Higher transformation efficiency can be achieved when higher concentration is used.
6. It may be possible to perform transformation with both pTAex3R and pPTR I simultaneously; however, the transformation efficiency may become very low. Therefore, a two-step transformation as described is recommended.
7. This ten times dilution is necessary; otherwise the original suspension is too concentrated for the colonies to be separated on the plate.
8. The copy number of the plasmid integrated into the genome varies considerably among colonies. Therefore, it is important to examine several colonies and choose the one that gives the largest production of the compound of interest.
9. Culture the mycelia until it becomes viscous. Otherwise, optimal production of the compound may not be achieved.

Acknowledgments

This work was supported by a Grant-in-Aid for Japan Society for the Promotion of Science (JSPS) Fellows to T.I. from JSPS, a Grant-in-Aid for Young Scientists (B) (No. 21710222) from JSPS and The Mochida Memorial Foundation for Medical and Pharmaceutical Research to T.K., and a Grant-in-Aid for Scientific Research on Priority Areas "Applied Genomics" to I.F. from the Ministry of Education, Culture, Sports, Science and Technology of Japan.

References

1. Hoffmeister D, Keller NP (2007) Natural products of filamentous fungi: enzymes, genes, and their regulation. Nat Prod Rep 24:393–416
2. Fujii I (2010) Functional analysis of fungal polyketide biosynthesis genes. J Antibiot 63:207–218
3. Bergmann S, Schümann J, Scherlach K, Lange C, Brakhage AA, Hertweck C (2007) Genomics-driven discovery of PKS-NRPS hybrid metabolites from *Aspergillus nidulans*. Nat Chem Biol 3:213–217
4. Itoh T, Tokunaga K, Matsuda Y, Fujii I, Abe I, Ebizuka Y, Kushiro T (2010) Reconstitution of a fungal meroterpenoid biosynthesis reveals the involvement of a novel family of terpene cyclases. Nat Chem 2:858–864
5. Gomi K, Iimura Y, Hara S (1987) Integrative transformation of *Aspergillus oryzae* with a plasmid containing the *Aspergillus nidulans argB* gene. Agric Biol Chem 51:2549–2555
6. Fujii T, Yamaoka H, Gomi K, Kitamoto K, Kumagi C (1995) Cloning and nucleotide sequence of the ribonuclease T_1 gene (*rntA*) from *Aspergillus oryzae* and its expression in *Saccharomyces cerevisiae* and *Aspergillus oryzae*. Biosci Biotechnol Biochem 59: 1869–1874
7. Kubodera T, Yamashita N, Nishimura A (2000) Pyrithiamine resistance gene (*ptrA*) of *Aspergillus oryzae*: cloning, characterization and application as a dominant selectable marker for transformation. Biosci Biotechnol Biochem 64:1416–1421
8. Kitamoto K (2002) Molecular biology of the koji molds. Adv Appl Microbiol 40: 634–635

Chapter 13

Multi-Copy Genetic Screen in *Aspergillus nidulans*

Hee Soo Park and Jae-Hyuk Yu

Abstract

With the completion of genomes of various *Aspergillus* species, large-scale genome-wide expression studies can be carried out. Genomics, however, is more powerful and efficient when combined with genetics. A multi-copy-based gain-of-function screen is a complementary method to loss-of-function genetic screen and can identify novel genes that may not be easily identifiable through loss-of-function-type screens. Particularly, gain-of-function genetic screens would identify novel activators or repressors of fungal development and secondary metabolism. Here, we describe a working protocol for the identification of novel regulators in *Aspergillus nidulans*.

Key words: Genetics, *Aspergillus*, Multi-copy screening, Sporulation, Secondary metabolism

1. Introduction

Loss-of-function genetic screen is the traditional approach to identify genes of interest based on inactivation of gene function (1). Chemical mutagenesis can be used to introduce random mutation in certain gene. 4-nitroquinoline 1-oxide (NQO) is an aromatic hydrocarbon that can intercalate between DNA base pair causing various loss-of-function mutations (2). However, loss-of-function approaches often miss genes if the inactivation does not result in clear and detectable phenotypes. Gain-of-function mutagenesis is a complementary method to loss-of-function approach and can identify novel genes. Multi-copy genetic screen is one of high-throughput screening method (3). Recently, several studies found novel regulators of fungal development and/or secondary metabolism employing multi-copy genetic screen (4–6). Generally in *Aspergillus nidulans*, *pyrG*-deficient strains and a wild-type genomic library constructed in the vector pRG3-AMAI, which

Nancy P. Keller and Geoffrey Turner (eds.), *Fungal Secondary Metabolism: Methods and Protocols*,
Methods in Molecular Biology, vol. 944, DOI 10.1007/978-1-62703-122-6_13,

contains the *pyr4*[+] allele of *Neurospora crassa* and an autonomous replicating sequence called AMA1, are being used (7, 8). Here, we describe a detailed protocol for multi-copy genetic screen in *A. nidulans* using pRG3-AMA1 wild-type library.

2. Materials

2.1. *Aspergillus* Transformation

1. KOH solution: 1.1 M KOH solution. Dissolve 6.17 g KOH in 70 mL. Adjust volume to 100 mL with additional dH_2O (see Note 1).
2. KCl_Citric acid solution: 1.1 M KCl and 0.1 M citric acid monohydrate. Dissolve 8.2 g KCl and 2.1 g citric acid monohydrate in 40 mL dH_2O. Adjust pH to 5.8 with 1.1 M KOH solution. After adjusting the pH, transfer the solution into a 100 mL graduated cylinder and make the volume up to 100 mL with dH_2O. Sterilize by autoclaving and store at room temperature (see Note 2).
3. 2× Protoplasting solution: KCl_Citric acid solution with 12.5% (w/v) VinoTeste Pro. Dissolve the 1.25 g of VinoTaste Pro (Novozyme) in 10 mL of KCl_Citric acid solution. Sterilize the solution by filtering this solution through a sterile syringe-driven filter (0.22 μm pore size) (see Note 3).
4. Sucrose cushion solution: 1.2 M sucrose. Dissolve 41.08 g sucrose in 80 mL of dH_2O. Adjust volume to 100 mL with additional dH_2O. Sterilize by autoclaving and store at 4°C.
5. KCl solution: 0.6 M KCl. Dissolve 4.47 g KCl in 80 mL of dH_2O. Adjust volume to 100 mL with additional dH_2O. Sterilize by autoclaving and store at 4°C.
6. KCl_$CaCl_2$ solution: 0.6 M KCl and 50 mM $CaCl_2$. Dissolve 4.47 g KCl and 0.74 g $CaCl_2 \cdot 2H_2O$ in 80 mL of dH_2O. Adjust volume to 100 mL with additional dH_2O. Sterilize by autoclaving and store at 4°C.
7. STC buffer solution: 10 mM Tris–HCl (pH 7.5), 1.2 M sorbitol, 10 mM $CaCl_2$. Dissolve 10 mL 1 M Tris–HCl, 218.6 g sorbitol, and 0.47 g $CaCl_2 \cdot 2H_2O$ in 800 mL of dH_2O. Adjust volume to 1 L with additional dH_2O. Sterilize by autoclaving and store at 4°C.
8. 1 M Tris–HCl solution: Dissolve 15.76 g Tris–HCl per 100 mL dH_2O. Store at room temperature.
9. 1 M Tris base solution: Dissolve 12.11 g Tris base per 100 mL dH_2O. Store at room temperature.
10. PEG solution: 20 mM Tris–HCl (pH 7.5), 25% (w/v) PEG, 0.6 M KCl, and 50 mM $CaCl_2$. Dissolve the following in

20 mL dH_2O: 4.47 g KCl, 0.74 $CaCl_2 \cdot 2H_2O$, 0.802 mL 1 M Tris–HCl, 0.196 mL 1 M Tris base, and 25 g PEG (average molecular weight 3,350). Adjust volume to 100 mL with additional dH_2O. Sterilize by autoclaving. Store at room temperature (see Note 4).

11. Minimal medium (MM) with 0.5% yeast extract: 5 g yeast extract, 10 g D-glucose, 50 mL 20× nitrate salt solution, and 1.0 mL trace element solution per 1 L of dH_2O. Adjust pH to 6.5 with NaOH. For a *pyrG*⁻ strain, add 1 g uridine (0.1%w/v, Acros Organic) and 1 g uracil (0.1% w/v, USB Products) prior to autoclaving. Sterilize by autoclaving. Store at room temperature.
12. 20× Nitrate salt solution: Dissolve the following in 800 mL dH_2O, 120 g $NaNO_3$, 10.4 KCl, 10.4 g $MgSO_4 \cdot 7H_2O$, and 30.4 g KH_2PO_4. Adjust volume to 1 L with additional dH_2O.
13. Trace element solution: Dissolve the following in 800 mL dH_2O, 22 g $ZnSO_4 \cdot 7H_2O$, 11 g H_3BO_3, 5 g $MnCl_2 \cdot 4H_2O$, 5 g $FeSO_4 \cdot 7H_2O$, 1.6 g $CoCl_2 \cdot 6H_2O$, 1.6 g $CuSO_4 \cdot 5H_2O$, 1.1 g $(NH_4)_6Mo_7O_{24} \cdot 4H_2O$, and 50 g $Na_2EDTA \cdot 2H_2O$. Adjust pH to 6.0 with NaOH after each component dissolves and add dH_2O to a final volume of 1 L. Store at room temperature (see Note 5).
14. The selection medium: 10 g D-glucose, 50 mL 20× nitrate salt solution, 1.0 mL trace element solution, 218.6 g sorbitol (1.2 M) per 1 L of dH_2O. Add 15 g agar (1.5%, v/w) prior to autoclaving. Adjust pH to 6.5 with NaOH.
15. Top agar: 10 g D-glucose, 50 mL 20× nitrate salt solution, 1.0 mL trace element solution, 218.6 g sorbitol (1.2 M) per 1 L of dH_2O. Add 7 g agar (0.7%) prior to autoclaving. Adjust pH to 6.5 with NaOH.
16. Sterile Miracloth (EMD).
17. Transfer pipettes.
18. 0.22 μm Syringe filter.
19. The recipient strain should be a *pyrG*-deficient strain (*pyrG*⁻ mutant) that is auxotrophic for uridine and uracil.
20. The pRG3-AMA1-based wild-type genomic library is available from the Fungal Genetics Stock Center, Kansas City, MO.

2.2. Genomic DNA Isolation

1. Minimal medium with 0.5% yeast extract: Add and dissolve 5 g yeast extract, 10 g D-glucose, 50 mL 20× nitrate salt solution, and 1.0 mL trace element solution per 1 L of dH_2O. Adjust pH to 6.5 with NaOH. Sterilize by autoclaving. Store at room temperature.
2. Sterile test tubes (8–10 mL).

3. Breaking buffer: 50 mM Tris–HCl, pH 8.0, 50 mM EDTA (pH 8.0), 5% Triton X-100 (v/v), 8% sucrose (w/v). Dissolve the following in 50 mL dH_2O, 5 mL 1 M Tris–HCl (pH 8.0), 10 mL 0.5 M EDTA, 5 mL Triton X-100, and 8 g sucrose and add dH_2O to a final volume of 100 mL. Sterilize by autoclaving. Store at 4°C.
4. Phenol:Chloroform:Isoamyl alcohol (25:24:1).
5. Zirconia/silica bead (0.5 mm, Research Products International Corp).
6. 2 mL Screw cap microcentrifuge tubes with O-ring cap.
7. TE buffer with RNase A: 10 mM Tris–HCl (pH 7.6) and 1 mM EDTA sterilized by autoclaving. Add 2 mg Ribonuclease A per 100 mL TE buffer. Store at 4°C.
8. Mini-Beadbeater (Bio Spec Products Inc.).

2.3. Plasmid Rescue and Insert Sequencing

1. *Escherichia coli* MAX Efficiency® DH10B™ Competent Cells (Invitrogen).
2. Gene Pulser/Micropulser cuvettes (0.1 cm gap, Bio-Rad).
3. Gene Pulser X cell Microbial system (Bio-Rad).
4. LB medium: Dissolve 25 g LB broth into 1 L dH_2O. To make LB plate add 15 g agar. Sterilize by autoclaving. Allow the medium to cool to 65°C before adding 2 mL of 50 mg/mL ampicillin to a final concentration of 100 μg/mL.
5. S.O.C Medium: Dissolve the following in 70 mL dH_2O, 2 g Bacto® Tryptone, 0.5 g Bacto® Yeast extract, 1 mL 1 M NaCl, 0.25 mL 1 M KCl, 1 mL $MgCl_2$, 1 mL 1 M $MgSO_4$, and 2 mL 1 M Glucose, and add dH_2O to a final volume of 100 mL. The pH should be 7.0. Sterilize by autoclaving.
6. Ampicillin: 50 mg/mL ampicillin. Dissolve 500 mg ampicillin into 10 mL of dH_2O.
7. QIAquick PCR Purification Kit.

3. Methods

3.1. Transformation with Multi-Copy Genomic Libraries

1. Prepare fresh spore suspension of a recipient strain carrying a *pyrG*⁻ allele.
2. Inoculate 1×10^8 spores into 20 mL MM with 0.5% yeast extract and incubate with shaking at 150 rpm on a shaker for 12–14 h at 30°C.
3. Harvest the hyphae by filtering through the sterile Miracloth and wash the hyphae with about 20 mL MM with 0.5% yeast extract.

4. Scrape hyphae from the Miracloth with a sterile spatula, transfer the hyphae into a sterile 50 mL centrifuge tube, and resuspend the hyphae in 10 mL MM with 0.5% yeast extract.
5. Add 10 mL 2× protoplasting solution.
6. Incubate with 100 rpm in a shaker at 30°C for 2 h.
7. Remove any undigested hyphal material by filtration through sterilized Miracloth.
8. Overlay the protoplasting mixture very gently with 10 mL of sucrose cushion solution.
9. Centrifuge for 10 min at 1,800 × *g* at 4°C using HS-4 rotor (Soval).
10. Using a sterile transfer pipette take the protoplasts at the top of the sucrose cushion.
11. Place the collected protoplasts in a sterile 50 mL centrifuge tube, gently mix with an equal volume of KCl solution, and centrifuge for 10 min at 1,800 × *g* at 4°C using HS-4 rotor.
12. Carefully decant the supernatant and resuspend the pellet in 1 mL of KCl solution and transfer the suspension to 1.5 mL microcentrifuge tubes.
13. Centrifuge at 2,400 × *g* for 3 min at room temperature.
14. Decant the supernatant carefully and resuspend the pellet in 1.0 mL KCl_$CaCl_2$ solution.
15. Centrifuge at 2,400 × *g* for 3 min at room temperature.
16. Decant the supernatant carefully and resuspend the pellet in 0.5 mL KCl_$CaCl_2$ solution.
17. Centrifuge at 2,400 × *g* for 3 min at room temperature.
18. Decant the supernatant. Resuspend the protoplast pellet in a volume of 100 μL of KCl_$CaCl_2$ solution at a concentration of 10^8–10^9 protoplasts per 1 mL. Store on ice.
19. To 100 μL protoplast suspension add ~5 μg of the pRG3-AMA1 wild-type library (the plasmids in library contain *pyr4*$^+$) in less than 15 μL TE buffer.
20. Add 50 μL of PEG solution, and mix gently by turning the tube on its side and rotating it. Incubate for 25 min on ice.
21. Add 1 mL of PEG solution and incubate at room temperature for 25 min.
22. Add 10 mL of STC buffer and mix.
23. Spread about 0.5 mL mixture and plate on the selection medium.
24. Add 5 mL top agar (0.7% agar) and mix (see Note 6).
25. Incubate the plates overnight at 30°C.

26. Flip the plates upside down and incubate at 37°C for an additional 2–3 days.
27. Screen for the transformants of interest, e.g., reduced sporulation, accumulation of unusual pigments, restricted growth, etc. Transfer the transformants of interest into regular solid MM and check phenotypes, including secondary metabolism, further.

3.2. Genomic DNA Isolation

1. Inoculate a loopful of fungal spores (or hyphae) in 2 mL liquid MM with 0.5% yeast extract and supplements in an 8 mL test tube and incubate at 37°C slanted for 14–16 h (stationary culture).
2. Collect the mycelial mat, squeeze it with paper towel to remove the medium as much as possible, and transfer the squeezed sample into a 2-mL O-ring screw cap microcentrifuge tube.
3. Add 500 μL of breaking buffer, 500 μL of phenol: chloroform:isoamyl alcohol (25:24:1), and 300 μL of zirconia/silica beads into tubes.
4. Tightly close the cap and insert tubes in a Mini-Beadbeater in a cold room and homogenize at the highest speed for 2 min.
5. Take out the tubes and centrifuge for 10 min at 16,000 × *g* at 4°C or room temperature.
6. Transfer 300 μL of the aqueous (upper) phase into a new 1.5 mL microcentrifuge tube.
7. Add two volumes of ice-cold absolute ethanol (kept at −20°C) and gently mix.
8. Centrifuge the precipitated DNA for 10 min at maximum velocity and decant the supernatant.
9. Add 1 mL of 70% ethanol to the tubes with the DNA pellets.
10. Centrifuge at 16,000 × *g* for 5 min. Decant the supernatant.
11. Centrifuge at 16,000 × *g* for 1 min and carefully remove the supernatant as much as possible.
12. Let the pellet air-dry for 5–10 min.
13. Resuspend the pellet in 50 μL of TE buffer with RNaseA.
14. Incubate for 1 h in the 37°C incubator.
15. Measure DNA concentration spectroscopically and store DNA samples at −20°C until use.

3.3. Plasmid Rescue and Insert Sequencing

1. Thaw 20 μL of *E. coli* MAX Efficiency® DH10B™ Competent Cells on ice.
2. Add 1 μL of isolated *A. nidulans* genomic DNA to one tube containing 20 μL competent cells (see Note 7).
3. Label 0.1 mm electroporation cuvette for each sample and place on ice.

4. Carefully pipette and transfer the mixture of cells and DNA to a chilled cuvette. Tap the suspension to the bottom (see Note 8).
5. Place the cuvette in the electroporator shock-pod and close the shock-pod.
6. Press the "pulse" button (see Note 9).
7. Add 1 mL of S.O.C. medium and transfer the solution to a 15 mL snap-cap tube (e.g., Falcon tube).
8. Shake at 225 rpm at 37°C for 1 h.
9. Spread 100 μL of the dilution on prewarmed LB plates containing 100 μg/mL ampicillin.
10. Incubate plates overnight at 37°C.
11. Pick 2–5 colonies and inoculate 5 mL LB broth containing 100 μg/mL ampicillin overnight at 37°C.
12. Purify the plasmid DNA using QIAquick PCR Purification Kit.
13. Sequence the plasmid DNA with the primer set OMN33 and OMN35 (see Note 10).
14. Search the full genome sequence (the Broad Institute, http://www.genome.wi.mit.edu/annotation/fungi/aspergillus/index.html) with the insert sequences.
15. Reintroduce the plasmids into the recipient strain for confirmation of the resulting phenotypes.
16. If more than one gene resides within the insert, functional validation of each gene is necessary. This can be done by introducing multiple copies of each gene to the recipient strain.

4. Notes

1. Eye protection and care should be used to prevent contact of the solution with eye and skin. The KOH solution should be made fresh before to prevent reaction with CO_2.
2. KCl–Citric acid solution can be stored for short periods or sealed carefully (e.g., with Parafilm) to prevent evaporation.
3. 2× Protoplasting solution should be made less than 1 h before use.
4. If PEG is precipitated then it must be warmed and mixed thoroughly before use to allow the PEG to dissolve. PEG precipitate will destroy protoplasts and reduce transformation frequencies.
5. After autoclaving, the solution will be green but the color of the solution will change to purple with time.

6. After autoclaving, allow top agar to cool to ~50°C.
7. Solutions containing the DNA to be transformed must contain a minimum of salts.
8. Avoid formation of bubbles.
9. Electroporate using the following electroporator conditions: 1.8 kV, 200 Ω, and 25 μF.
10. Primers to sequence the insert in pRG3-AMA1:
    (a) OMN33 5′-AAATAAGCTTGCATGCGC-3′.
    (b) OMN35 5′-GCCAGTGAATTCGAGCTC-3′.

Acknowledgments

The authors express special thanks to Drs. Nak-Jung Kwon and Ni Min for their technical assistance. This work was supported by the National Science Foundation (IOS-0950850), USDA CSREES Hatch (WIS04667), and UW Food Research Institute grants to JHY. HSP was in part supported by the UW FRI Foster Award.

References

1. Käfer E, Mayor O (1986) Genetic analysis of DNA repair in *Aspergillus nidulans*: evidence for different types of MMS-sensitive hyperrec mutants. Mutat Res 161:119–134
2. Osman F, Tomsett B, Strike P (1993) The isolation of mutagen-sensitive nuv mutants of *Aspergillus nidulans* and their effects on mitotic recombination. Genetics 134:445–454
3. Kondou Y, Higuchi M, Matsui M (2010) High-throughput characterization of plant gene functions by using gain-of-function technology. Annu Rev Plant Biol 61: 373–393
4. Ni M, Yu JH (2007) A novel regulator couples sporogenesis and trehalose biogenesis in *Aspergillus nidulans*. PLoS One 2:e970
5. Yu J-H (2010) Regulation of development in *Aspergillus nidulans* and *Aspergillus fumigatus*. Mycobiology 38:229–237
6. Shaaban MI, Bok JW, Lauer C, Keller NP (2010) Suppressor mutagenesis identifies a velvet complex remediator of *Aspergillus nidulans* secondary metabolism. Eukaryot Cell 9:1816–1824
7. Efimov VP, Morris NR (1998) A screen for dynein synthetic lethal in *Aspergillus nidulans* identifies spindle assembly checkpoint genes and other genes involved in mitosis. Genetics 149:101–116
8. Osherov N, Mathew J, May GS (2000) Polarity-defective mutants of *Aspergillus nidulans*. Fungal Genet Biol 31:181–188

Chapter 14

Identification of Protein Complexes from Filamentous Fungi with Tandem Affinity Purification

Özgür Bayram, Özlem Sarikaya Bayram, Oliver Valerius, Bastian Jöhnk, and Gerhard H. Braus

Abstract

Fungal molecular biology has benefited from the enormous advances in understanding protein–protein interactions in prokaryotic or eukaryotic organisms of the past decade. Tandem affinity purification (TAP) allows the enrichment of native protein complexes from cell extracts under mild conditions. We codon-optimized tags and established TAP, previously not applicable to filamentous fungi, for the model organism *Aspergillus nidulans*. We could identify by this method the trimeric Velvet complex VelB/VeA/LaeA or the eight subunit COP9 signalosome. Here, we describe an optimized protocol for *A. nidulans* which can also be adapted to other filamentous fungi.

Key words: *Aspergillus nidulans*, Tandem affinity purification, Filamentous fungi, Protein complexes, Interaction partners, Fungal biochemistry

1. Introduction

Proteins are essential for cellular enzyme activities, signal transduction pathways, membrane transport functions or as structural elements. Advanced biochemical methods combined with molecular biology and mass spectrometry are required to identify protein–protein interaction in vivo to understand the function of a given protein in its protein context. The *t*andem *a*ffinity *p*urification (TAP) method represents an advanced biochemical application that was first described for *Saccharomyces cerevisiae* (1, 2) and was further applied to prokaryotes, plants, and mammalian cells including human for in vivo studies (3–7). It is based on a recombinant fusion tag containing two tags; two tandemly repeated *Staphylococcus aureus* protein A epitopes that bind to IgG sepharose during the

Nancy P. Keller and Geoffrey Turner (eds.), *Fungal Secondary Metabolism: Methods and Protocols*,
Methods in Molecular Biology, vol. 944, DOI 10.1007/978-1-62703-122-6_14, © Springer Science+Business Media, LLC 2012

first enrichment step, and a *c*almodulin *b*inding *p*eptide (CBP) that binds to calmodulin affinity resin in the second purification step. Protein A domains are connected with CBP by a *t*obacco *e*tch *v*irus (TEV) protease cleavage site. The final eluate after TAP tag purification is precipitated by using *t*richloro*a*cetic *a*cid (TCA). This precipitated protein eluate can be directly submitted to mass spectrometry (MS) identification. Further enrichment can be achieved prior to MS, when the eluate is first separated by SDS-PAGE. Protein bands are excised after staining and digested with trypsin and analysed in MS. The identification of putative interacting partners of the protein of interest should be confirmed by other methods, including the *y*east *t*wo-*h*ybrid (YTH) (8, 9) or *b*imolecular *f*luorescence *c*omplementation (BIFC) (9, 10) methods.

Our initial attempts to construct TAP fusion proteins for the filamentous fungus *Aspergillus nidulans* based on the original TAP tag resulted in negligible expression thus hampering co-purification of interacting protein candidates. This problem was circumvented by using site-directed mutagenesis to change rarely used codons of the TAP tag according to the *A. nidulans* codon usage (11). We established a modified TAP method which was successfully applied to identify several protein complexes from the filamentous fungus *A. nidulans* including the COP9 signalosome, the NUDF/NUDC/BNFA complex at spindle pole bodies, and the velvet complexes VelB-VeA-LaeA and VosA-VelB (11–15).

In this study, we describe the adaptation of the TAP purification protocol in detail including all required steps, such as protein extract preparation, purification procedure, visualization of proteins in the SDS gels by two different methods, trypsin digest, and final mass spectrometry identifications. We also provide some solutions to the problems encountered during the application of the method (see Notes). This protocol provides the fungal researchers with new possibilities for the identification of protein complexes in *A. nidulans* and other filamentous fungi.

2. Materials

2.1. Cultivation Media of A. nidulans and Harvesting Mycelia

1. Air pump (2.4 bar max) (KNF).
2. Aspergillus liquid minimal medium with supplements.
3. Büchner funnel.
4. DMSO (dimethyl sulfoxide).
5. Falcon tubes (15 ml).
6. Miracloth (Calbiochem).
7. Mixer mill MM400 (Retsch).

8. 0.96% NaCl containing 1% DMSO (v/v) and 1 mM PMSF (v/v).
9. Phenylmethanesulphonylfluoride (PMSF).
10. Two liter flasks with indents (5× for each strain).
11. Vacuum flask.

2.2. TAP Protocol

1. Acetone.
2. AcTEV™ protease (Invitrogen).
3. BSA (Bovine Serum Albumin).
4. β-Mercaptoethanol.
5. Biofuge Fresco (Heraeus).
6. Calcium chloride.
7. Calmodulin Affinity Resin (Stratagene).
8. Complete, EDTA-free protease inhibitor set (Roche).
9. Dithiothreitol DTT.
10. E-64 (irreversible cysteine protease inhibitor) (Calbiochem).
11. Ethanol.
12. Ethylenediaminetetraacetic acid (EDTA).
13. Ethyleneglycoltetraacetic Acid (EGTA).
14. Falcon tubes (50 and 15 ml).
15. Glycerol.
16. IgG Sepharose™ 6 Fast Flow (GE Healthcare).
17. Imidazol.
18. Magnesium acetate.
19. NaCl.
20. Nonidet P40.
21. Phenylmethanesulphonylfluoride (PMSF).
22. Phosphatase Inhibitor Cocktail Set II (Calbiochem).
23. Poly-Prep Chromatography Columns (Bio-Rad).
24. Rotating platform.
25. Sorvall SS34 centrifuge tubes (50 ml) (Nalgene).
26. Sorvall RC 5B plus centrifuge (Thermo Scientific).
27. Sigma 4K15C centrifuge for falcon tubes (Sigma).
28. Sterilized Erlenmeyer flasks (Schott).
29. Trichloroacetic acid (TCA).
30. Tris base.
31. Tris–Hydrochloride.

2.2.1. Buffers and Stock Solutions for TAP Procedure

Prepare and keep all the reagents at room temperature, unless any specific temperature is indicated. Protease inhibitors, phosphatase inhibitors, and reducing agents must be added to the corresponding buffers 20–30 min prior to application. Some of the components loose their effectiveness within a short time period (see PMSF).

1. *1 M Tris–HCl pH 7.5 (500 ml)*: fill a bottle with 400 ml ddH_2O and add 63.5 g Tris–Hydrochloride as well as 11.8 g Tris base and dissolve it. If the pH shifts up or down, adjust the pH to 7.5 by adding 5N or 10N HCl or NaOH. Make up to 500 ml and autoclave.
2. *1 M Tris–HCl pH 8.0 (500 ml)*: add 44.4 g Tris–HCl and 26.5 g Tris base and prepare as in previous step.
3. *1 M DTT (25 ml):* dissolve 3.85 g DTT in 25 ml ddH_2O and aliquot in 500 μl volumes into 1.5 ml vials and store at -20°C.
4. *1 M imidazole (10 ml)*: dissolve 0.6808 g imidazole in 10 ml ddH_2O and keep at 4°C.
5. *1 M CaCl2 (250 ml)* dissolve 27.74 g $CaCl_2$ in 100 ml ddH_2O, complete to 250 ml, and autoclave.
6. *1 M Magnesium acetate (250 ml)*: dissolve 53.6 g MgOAc in 200 ml ddH_2O. Make up to 250 ml, and autoclave (see Note 1).
7. *500 mM EDTA* (500 ml): 93.6 g EDTA to 300 ml of ddH_2O, mix and adjust the pH to 8.0 (with 5 M NaOH) and dissolve. Complete to 500 ml and autoclave.
8. *100 mM EGTA (100 ml)*: add 38.3 g EGTA in 50 ml ddH_2O, adjust the pH as described for EDTA and complete the volume to 100 ml, and autoclave.
9. *10% BSA*: dissolve 5 g BSA in 50 ml ddH_2O by stirring (50 ml falcon tube), filter sterilize and freeze as 10 ml stocks at -20°C.
10. *100 mM PMSF (100 ml)*: dissolve 1.75 g PMSF in 100 ml EtOH. Make aliquots in 2 ml vials and store at –20°C.
11. *1 mM E-64*: 1 mg E-64 in 2.8 ml DMSO, aliquot each 100 μl into 1.5 ml vials and keep at –20°C.
12. *1 l B250 buffer: 250 mM NaCl (w/v), 100 mM Tris–HCl pH 7.5 (v/v), 0.1% NP-40 (v/v)*: add 14.6 g NaCl in 100 ml of Tris–HCl (1 M pH 7.5 stock), 100 ml glycerol, 1 ml NP-40 (10% stock), 2 ml EDTA (500 mM stock), then complete the volume to 1 l and autoclave. Store at 4°C. Prior to preparation of the crude extracts (10–20 min), add the following elements to 100 ml B250 on the ice: 1 ml PMSF (100 mM stock), two complete EDTA-free protease inhibitor tablets (prior to adding, two pills should be dissolved in 1 ml ddH_2O), 100 μl

DTT (1 M stock) as well as 1 ml Phosphatase Inhibitor Cocktail Set II.

13. *1 l Wash Buffer 250 (WB250): 250 mM NaCl (w/v), 40 mM Tris–HCl pH 8.0 (v/v), 0.1% NP-40 (v/v)*: add 14.60 g NaCl, 40 ml of Tris–HCl (1 M pH 8.0), 1 ml NP-40, complete to 1 l and autoclave and store at 4°C. Before the washing step (20–30 min), add the same components to the 100 ml WB250, which were added to B250.
14. *1 l Wash Buffer 150 (WB150): 150 mM NaCl (w/v), 40 mM Tris–HCl pH 8.0 (v/v), 0.1% NP-40 (v/v)*: similar to WB250 preparation, this time 8.76 g NaCl is added. Autoclave and store at 4°C.
15. *TEV cleavage buffer: TCB (is the same as WB150 and contains 0.5 mM EDTA additionally)*: add 100 μl DTT (1 M stock), 50 μl EDTA (500 mM EDTA stock), 100 μl (1 mM E-64 stock), to 100 ml WB150 on the ice in a cool room and mix by inverting.
16. *Calmodulin Binding Buffer (CBB): 150 mM NaCl (w/v), 40 mM Tris–HCl pH 8.0 (v/v), 1 mM MgOAc, 2 mM CaCl2*: CBB composition is the same as WB150 (without NP-40). Add 100 μl MgOAc (1 M stock), 200 μl $CaCl_2$ (1 M stock), 100 μl imidazole (1 M stock), 70 μl β-mercaptoethanol in 100 ml WB150.
17. *Elution Buffer (EB)*: add following constituents in 8 ml WB150, 2 ml (100 mM EGTA), 10 μl (1 M MgOAc), 10 μl (1 M Imidazol), 7 μl (β-Mercaptoethanol), and 20 μl (10% NP-40 stock).
18. *100% TCA*: dissolve 100 g TCA in 30 ml ddH_2O and complete the volume to 100 ml.
19. *Harvest solution*: 0.96% NaCl, 1% DMSO, 1 mM PMSF

2.3. SDS Page and Staining Protocols

Any SDS polyacrylamide recipe around 12% can be used for running TAP purified samples. However, if the protein of interest is a low molecular weight protein, then it is recommended to use a gradient gel (4–15%) to get a better view of the purification process. Because the gradient gel does not allow the small molecular weight proteins to leave off the SDS page while allowing the bigger size of proteins to run homogenously.

1. Acetic Acid.
2. Ammoniumpersulfate.
3. Brilliant Blue G-Colloidal Concentrate (Sigma).
4. Glycine.
5. Methanol.
6. Protein molecular weight standard (MBI Fermentas).

7. Rotiphorese (30% Acrylamide).
8. Small size lab shaker.
9. Silver nitrate.
10. Sodium thiosulfate.
11. Sodium carbonate.
12. Sodium dodecyl sulfate (SDS).
13. Tetramethylethylendiamine (TEMED).
14. Tetra cell system for SDS page (Bio-rad).
15. *Fixing solution*: 40% EtOH, 10% Acetic acid.
16. *Wash solution*: 30% EtOH.
17. *Developing solution:* 3% Sodium carbonate, 0.05% formaldehyde, 0.0004% sodium thiosulfate.

2.4. Trypsin Digest

1. Acetonitrile, Rotisolv HPLC (Carl Roth GmbH).
2. Ammoniumhydrogencarbonate ≥99% p.a. (Carl Roth GmbH).
3. Formaldehyde 37% (Carl Roth).
4. Formic acid 98–100% (Merck).
5. Iodoacetamide, ultra pure (Sigma).
6. Trypsin, sequencing grade modified (Promega).
7. Water, Lichrosolv (Merck).
8. Digestion solution: 25 mg Trypsin resolved in 80 μl Promega Resuspension Buffer, diluted 1:20 with 50 mM AHC solution.

3. Methods

3.1. Culturing of the Fungal Strains and Crude Extract Preparation

Culturing conditions are important for the conclusions of the TAP purification. Therefore, before starting to TAP purification, it is recommended to perform a time course expression study (immunoblotting) to see at which developmental step the protein of interest is expressed. Growing temperature of the cultures is very important. Because some protein fusions might not be functional at 37°C, thus, culturing the organism at lower temperatures such as 25 and 30°C might help to get better results. While preparing buffer for crude extract preparation, properties of the protein of interest have to be seriously considered. Metalloproteins and membrane proteins might require extra care regarding the buffer components. For interaction studies of metalloproteins, usage of EDTA must be avoided in the buffer systems because it binds the metal ions of the metalloproteins, which can cause the disruption of the natural conformation of the metalloprotein. Therefore, this might lead to loss of interacting proteins. Sometimes it may be difficult to solubilize the membrane bound proteins into the crude extract, hence, increased concentration of nonionic detergents such as NP-40 and Triton X can be applied during the protein extraction procedure.

1. Inoculate 5× 800 ml liquid media with *A. nidulans* strain containing tap tagged construct as well as the wild-type control into *Aspergillus* minimal media (1×10^6 per 1 ml medium) and grow the culture for 20–24 h at 37°C/250 rpm (see Note 2).
2. Place miracloth filter on the Büchner funnel sitting at the top of a vacuum flask connected to the air pump. Filter the mycelia through the miracloth and wash with the harvest solution (see Note 3). Mycelia should appear to be completely dry.
3. Grind the mycelia in liquid nitrogen with either mortar/pestle or a pulverizing machine and transfer the ground mycelia into 15 ml falcon tubes pre-kept in liquid nitrogen and store at −80°C. For each TAP tag experiments, a minimum 6× 15 ml ground mycelia is required. Here ground mycelia can be stored at −80°C up to 1 year.
4. Take three SS34 tubes (50 ml volume) for each strain and put 2× 15 ml ground mycelia into each SS34 tube (6× 15 ml ground mycelia in total, 15 ml ground mycelia correspond to 6–7 g), which ends up 36–42 g per strain. Here prepare your buffer components (B250) (see Note 4).
5. Add protease and phosphatase inhibitors as well as reducing agent DTT to the buffer B250. Use 5 ml pipet to reach the bottom of tubes containing the mycelia and give 5 ml B250 for each SS34. Give another 5 ml B250 to the top but this time mix them again with a plastic or metal stick. Add additional 2–3 ml buffer if required (see Note 5). Close the caps of the tubes and vortex them until buffer absorbed by the mycelia. Mixture will move slowly during vortexing, which is quite normal. Then, keep tubes on ice for 5–10 min. 100 ml buffer is needed for 6× SS34 tubes preparation of the extracts (6× 10–12 ml) and washing sepharose (2× 5 ml).
6. Centrifuge the ice-kept sorval tubes ≥20,000 × g for 30 min at 4°C. Meanwhile take the BSA-treated 50 ml falcon tubes from rotator and decant the BSA (see Note 6). Add 5 ml B250 into these falcon tubes and resuspend 300–350 μl of IgG sepharose (cut the pipette tip in order to pipet the sepharose easily) and centrifuge them at 8,00 × g for 1 min at 4°C. Remove the supernatant. After centrifugation of the sorval SS34 tubes, transfer the brownish supernatant (crude extract) into the falcon tubes that contain washed IgG sepharose. For each strain three SS34 tubes will yield around 45–50 ml crude extract. Here, the concentration of the protein extract can be determined by using various methods.

3.2. TAP Purification and SDS Page Running

1. Incubate crude extract/IgG sepharose mixture on a rotating platform at 4°C for 2–3 h (see Note 7). A summary of the TAP workflow is given in Fig. 1.
2. Add necessary components (protease/phosphatase inhibitors, DTT) to the washing buffers W250 and W150 towards the end of incubation time. Only add PMSF just before using the

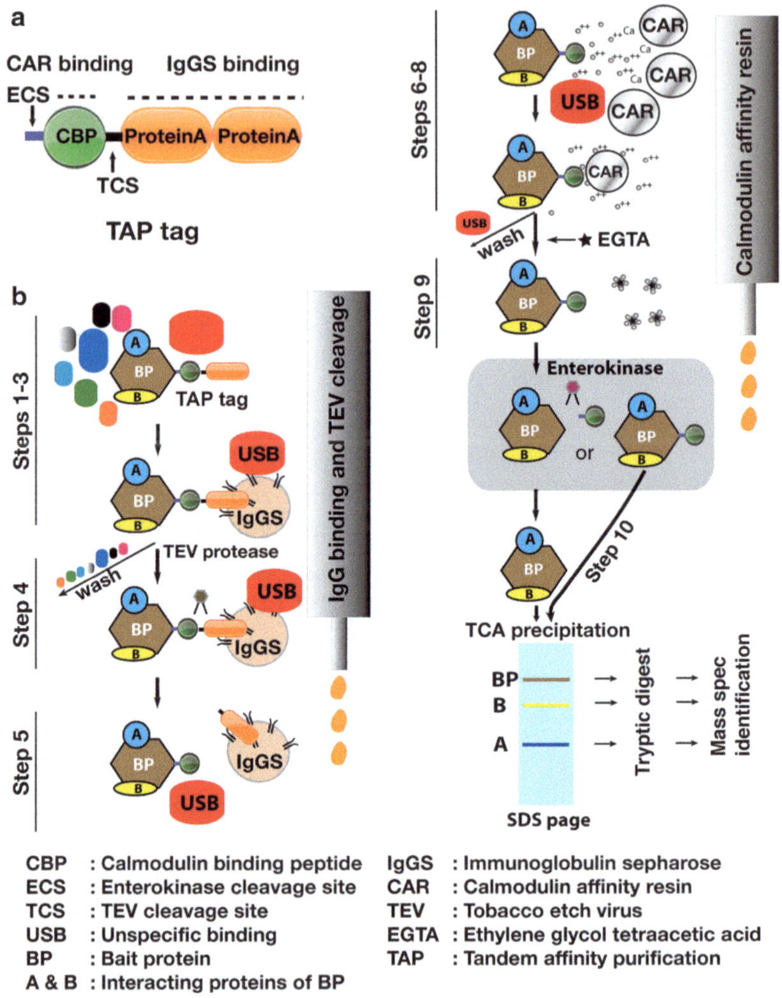

Fig. 1. Workflow of the tandem affinity protocol. (**a**) Domains found in the C-terminal TAP tag. Protein of interest is fused to C-TAP in the following order: (ECS)-CBP-(TCS)-Protein A domain. (**b**) Application steps of the protocol. Steps 1–3: Binding of the protein of interest TAP fusion to the IgG sepharose. Step 4: washing the proteins bound to the IgG unspecifically. Step 5: TEV cleavage of the CBP domain and fusion protein from the IgG bound protein A domains. Step 6–7: Binding of CBP to the calmodulin affinity resin. Step 9: EGTA elution of the CAR bound fusion protein and interaction partners. Step 10: TCA precipitation of the eluate and running of the proteins on the SDS gel. A further purification step with enterokinase treatment can be performed to cleave CBP peptide off the target protein. This step is not essential to get interaction partners, but may be required for functional studies such as enzyme assay with the purified complex.

buffer (its half life in aqueous solutions is very short, 110 min at pH 7.0 and 35 min at pH 8.0 (16)).

3. Pour the crude extract/IgG sepharose suspension into chromatography columns and let the extract flow through with gravity. Accumulation of the sepharose at the bottom of the columns might slow down the flowing speed of the liquid phase. If it is available, a 20 ml injector might help to apply some pressure on the top of the columns to facilitate the flowing of unbound proteins.
4. Wash the IgG sepharose with 2× 10 ml W250 buffer, once with 10 ml W150 and TEV cleavage buffer (see Note 8).
5. After TCB wash, close the bottom of chromatography column with the yellow plug and add 1 ml TCB as well as 20 μl (200 U) of AcTEV™ enzyme and close the upper cap and tightly secure the both caps by sealing with parafilm. Put the chromatography columns on a rotator and rotate the columns at 4°C for 3–4 h or alternatively O/N (see Note 9).
6. Prior to the calmodulin binding step, prepare CBB and EB. Take new columns and fill them with 3 ml CBB and plug the bottom of the column. Afterwards, pipet 300 μl of calmodulin affinity resin (CAR) in CBB and open the plug to let CBB flow through. Close the bottom of columns with yellow cap and add 6 ml of CBB buffer as well as 7 μl 1 M $CaCl_2$ in order to neutralize the effect of EDTA present in TCB.
7. After TEV protease treatment of the old column, remove the upper cap of old columns and place them on the top of the new columns. Remove the lower cap of old column and transfer the TEV protease treated suspensions into the new columns containing the CAR and CBB buffer. Wash the walls of the old columns with an additional 1 ml TEV buffer in order to wash down the proteins that stick to the sepharose and wall of the columns. Close the upper cap of the CBB columns and rotate them at 4°C for 1 h.
8. At the end of incubation, first open the upper cap and then lower cap thus allowing the CBB to flow through. Wash the columns with 3× 1 ml CBB (containing 0.02% NP-40) and close the lower cap.
9. Prepare 1.5 ml vials for each column, first add 500 μl EB buffer into the columns and wait 2–3 min, then open the caps and eluate the proteins into the vials. Repeat the elution step again by adding 500 μl EB and keep the vials in the ice.
10. Start the TCA precipitation step by adding 333 μl of 100% TCA to the 1 ml eluate (makes 25% TCA in total), vortex them periodically every 8–10 min by keeping the vials in the ice for 30 min. Centrifuge the vials at 10,000 ×*g* at 4°C for 25 min and carefully decant the supernatant. Wash the unseen pellet

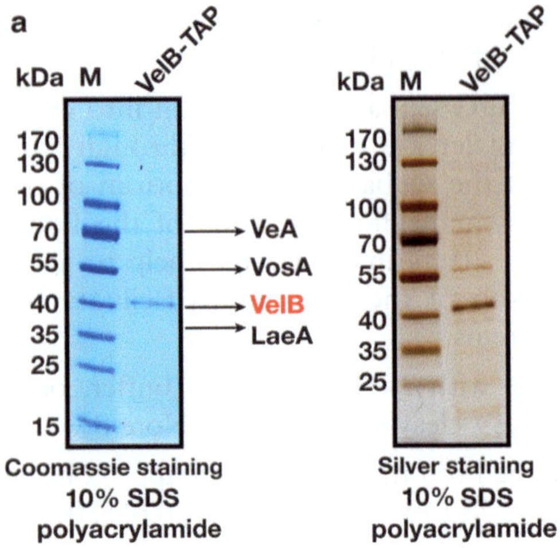

b

Gene Identity	Description	P Score*1	SPPV*2
AN0363.3	VelB	170.26	4.45E-12
AN1959.3	VosA	168.24	9.61E-11
AN1052.3	VeA	260.27	1.57E-12
AN0807.3	LaeA	30.20	1.40E-5

*1 Protein score *2 Sequest protein probablity value

Fig. 2. Identification of the interaction partners of the VelB-TAP protein from sexually induced cultures of *A. nidulans*. (**a**) Brilliant blue G-stained and silver stained 10% SDS-polyacrylamide gel electrophoresis of TAP for VelB protein. Strains were grown 20 h vegetatively and shifted to the plates and kept in the darkness for 12 h to induce sexual development. (**b**) Peptides identified from the bands belong to AN0363.3, AN1959.3, AN1052.3, AN0807.3 locus products. Protein score represents the total scores of XC (cross-correlation values).

with 1 ml cold acetone and centrifuge again at 10,000×g at 4°C for 5 min. Decant supernatant and carefully remove the rest by using a pipet without disturbing the pellet. Acetone must be stored at –20°C and prior to use keep on ice.

11. Put the vials in a speed vac and dry them at 35–37°C for 5 min. If a speed vac is not available, keep the cap of vials open at 37°C for 10–15 min. Add 30 µl of laemli buffer (3×) to each vial and boil at 95°C for 10 min. Here the samples can be stored at –20°C up to 2–3 months and the next step can be performed on the following day or in the following weeks. Before running them in SDS page, shortly heat up at 95°C (1 min) (see Note 10).
12. Prepare the SDS polyacrylamide gel (10%), load and run the samples in the SDS gel (Fig. 2).
13. Stain with the Coomassie brilliant blue according to the protocol provided by the provider (below) (see Note 11).

3.3. Brilliant Blue Staining Protocol After Modifications (17)

1. Fix the SDS polyacrylamide gel in fixer solution (7% glacial acetic acid and 40% methanol (v/v)) with shaking at RT for 1 h. Alternatively, here fixation can be carried out O/N.
2. Take a 50 ml falcon tube and fill with 40 ml Brilliant blue G-colloidal concentrate and add 10 ml methanol. Mix well by vortexing and shaking, remove the fixer solution and add staining solution and incubate on a shaker at RT for 1–2 h.

3. Remove the dye solution and incubate the gel in destaining solution (10% acetic acid and 25% methanol (v/v)) for 30–40 s.
4. Rinse the gel with washing solution (25% methanol (v/v)) for 1 min with shaking and add new washing solution. Scan the gel in a scanner (Fig. 2a). Here the stained SDS gel can be kept in washing solution up to 24 h if it is needed (see Note 12).

3.4. Silver Staining Protocol After Modifications (18)

1. Fix the gel in fixing solution for 2 h or O/N and wash the gel in wash solution 3× 15 min.
2. Incubate the gel in thiosulfate solution (0.02% sodium thiosulfate) for 1 min with shaking and wash the gel in dH_2O (3× 20 s).
3. Incubate the gel in silver nitrate solution (0.2% Silver nitrate) on a shaker for 20 min and wash in dH_2O (2× 20 s).
4. Place the gel in developing solution until the bands become slightly visible and wash in dH_2O (2× 1 min). Further incubate the gel in stop solution (0.5% Glycine) for 5 min followed by washing in dH_2O for 30 min (Fig. 2a).

3.5. Tryptic In-Gel Digest According to Ref. (19), HPLC and Database Search

1. Analyse the bands under laminar flow hood. This must be a human keratin-free environment, therefore, wear vinyl gloves and cut out the bands with a new scalpel and put the bands into new vials in order to digest with trypsin. Alternatively, to get a better evaluation, the entire lane can be cut into six to seven pieces and each piece has to be cut into smaller pieces (see Note 12). For storage of gel pieces cover them individually with deionized water in vials.
2. Remove the water and add 30 μl acetonitrile (or more, until the gel pieces are covered), leave at room temperature for 10 min.
3. Remove acetonitrile (acetonitrile dehydrates the gel pieces, therefore volume will be bigger than 30 μl). Dry the gel pieces in a speed vac at 60°C for 10 min.
4. Reduce proteins by adding 150 μl 10 mM DTT in 100 mM ammonium hydrogencarbonate (AHC) to gel pieces and incubate for 1 h at 56°C.
5. Remove DTT solution and alkylate the proteins at cysteine residues by adding 150 μl 55 mM iodoacetamide (in 100 mM AHC) to gel pieces and incubate for 45 min at RT in the dark.
6. Remove iodoacetamide solution and wash gel pieces with 150 μl 100 mM AHC for 10 min at room temperature.
7. Remove AHC and add 150 μl acetonitrile to gel pieces, leave at room temperature for 10 min.

8. Remove acetonitrile and repeat steps 6 and 7 (without iodoacetamide) once more.
9. Remove acetonitrile and dry gel pieces in the speed vac at medium temperature level for 10 min.
10. Add 30 μl trypsin digestion solution to gel pieces and leave the vials on ice for 45 min. Calculate the amount of trypsin digestion solution (TDS) for the samples. For example, digestion of 12 samples requires at least 400 μl TDS. Some samples with many pieces of polyacrylamide gel might require more than 30 μl TDS. Therefore, 20 μl trypsin is added into 380 μl 50 mM AHC solution to make 400 μl TDS. The rest of the trypsin in resuspension buffer can be stored at −20°C for a couple of months.
11. Remove remainders of unsoaked trypsin digestion solution and add 30 μl 50 mM AHC for O/N incubation at 37°C.
12. Centrifuge vials at 10,000 × *g* at RT for 1 min and collect supernatants in fresh vials (from here on supernatants containing tryptic peptides are collected and not discarded any more).
13. Add 30 μl 20 mM AHC to gel pieces and leave samples for 10 min at RT.
14. Combine supernatants with previous eluate, add further 30 μl 50% acetonitrile/5% formic acid to gel pieces, leave them at RT for further 20 min.
15. Centrifuge the vials at 10,000 × *g* at RT and combine supernatants in collection tubes.
16. Repeat the steps 14 and 15 twice.
17. Completely dry collected supernatants in the speed vac at 60°C (dried peptide pellets can be stored at 4°C until LCMS analysis).
18. Resolve peptide pellet in 5–10 μl (or in an appropriate volume) sample buffer (95% LC-water/5% LC-acetonitrile/0.1% formic acid) for LCMS analysis.
19. Run LCMS method as described (20) and collect MS and MS/MS data during the chromatographic separation.
20. Extract the peak list from the raw data files and perform the database search for protein identification with either Mascot or Sequest analysis programs (Fig. 2b). This step can be varied depending on the model and type of the MS device.

3.6. Data Analysis

It is likely that there are common impurities, which are copurified with the TAP fusion protein. These are most often translation elongation factors, various ribosomal proteins, heat shock proteins (chaperons), and various metabolic enzymes of primary metabolism. One of the drawbacks of TAP system is that it is not suitable for protein complexes interacting in a transient manner. In those cases, some cross-linkers, e.g., formaldehyde, can be applied to increase the potential of the method to copurify transitory interaction partners. For starters, it is always suggested to run a wild-type control to evaluate the specificity of the interactions in the TAP purified lane in comparison to wild-type lane, which always helps to distinguish between the interaction partners and false positives.

3.7. Conclusions

Understanding the molecular mechanisms governing the various developmental processes of the filamentous fungi and *Aspergillus* species undoubtedly requires advanced biochemical methods. In this context, TAP method is a very useful system to discover the novel connections of protein of interest. Here, we describe an optimized version of TAP protocol for the *Aspergillus* species, which can also be applied to other filamentous fungi outside of *Aspergillus* genus with slight modifications.

4. Notes

1. Long time storage of MgOAc on the bench can lead to some precipitations, shake this solution before using.
2. Here rich/full media can also be used especially for slowly growing strains. Using rich media will increase the yield of mycelia and will contribute to the final concentration of protein extract. If TAP purification results in too many heat shock proteins, growing the strains at 30°C might help to reduce heat shock protein copurifications.
3. This step is very crucial for regulatory proteins that are degraded more rapidly, here DMSO penetrates the cells together with PMSF instantly and inhibits the proteases while the fungal cells are alive. Adding first PMSF in 0.96% NaCl can cause some precipitations, therefore first add DMSO, and then PMSF.
4. Mycelia transfer into sorval SS34 and steps until the TCA precipitation (step 9) is advised to be performed in a cold room facility if available (at 4°C). Do not let the frozen mycelia melt down before adding the protein buffer (B250) and all tubes used during TAP purification must be pre-cooled by keeping them in ice in advance (it is even better to leave all tubes needed for purification at 4°C the night before).
5. This bottom and top buffering is necessary to mix up the low volume of buffer with high amount of mycelia in order to

obtain highly concentrated protein extract. Adding excessive amount of buffer (15–20 ml for each SS34 tubes) will make vortexing much easier, however, protein content of the crude extract will be more diluted, which is not recommended for the first IgG binding step.

6. 50 ml Falcon tubes should be treated with 10% BSA O/N before starting with the TAP purification protocol. This step helps to protect the poorly expressed proteins by saturating the walls of the falcon tube.
7. Longer than 4 h incubation with IgG sepharose should be avoided for the sake of proteins. Otherwise, proteins might undergo degradation process.
8. More washing might be required if many proteins have been identified in mass spectrometry. In such cases, three to four times washing can be applied after IgG binding step.
9. TEV enzyme cleavage can take longer than normal if accessibility of the TEV cleavage site of the fusion protein is partly limited due to the folding state. Then incubation time can be prolonged to O/N or cleavage reaction can be performed at 16°C for 2 h, which might also increase the degradation risk of the fusion protein.
10. Because of the remnants of the TCA, color of the loading dye (laemli) might turn from blue into orange. This does not affect the fidelity of the experiment.
11. Here as an alternative silver staining can also be performed. It is more sensitive than the coomassie to visualize the weak bands (Fig. 2). However, silver staining might interfere with mass spectroscopy identifications (19).
12. Even if there are no bands visible on the coomassie stained polyacrylamide gel, cut whole lane into 6–7 pieces and perform tryptic digest. In some cases with poorly expressed proteins, this approach gave substantial identifications of the interacting proteins.

Acknowledgments

This work has been funded by grants from the Deutsche Forschungsgemeinschaft (DFG), the Volkswagen-Stiftung, and the Fonds der Chemischen Industrie. Özlem Sarikaya Bayram is supported by the excellence stipend of Göttingen Graduate School for Neurosciences and Molecular Biosciences (GGNB) and Bastian Jöhnk by the German-Mexican DFG Research Unit 1334.

References

1. Puig O, Caspary F, Rigaut G, Rutz B, Bouveret E, Bragado-Nilsson E, Wilm M, Seraphin B (2001) The tandem affinity purification (TAP) method: a general procedure of protein complex purification. Methods 24:218–229
2. Rigaut G, Shevchenko A, Rutz B, Wilm M, Mann M, Seraphin B (1999) A generic protein purification method for protein complex characterization and proteome exploration. Nat Biotechnol 17:1030–1032
3. Butland G, Zhang JW, Yang W, Sheung A, Wong P, Greenblatt JF, Emili A, Zamble DB (2006) Interactions of the Escherichia coli hydrogenase biosynthetic proteins: HybG complex formation. FEBS Lett 580:677–681
4. Rohila JS, Chen M, Cerny R, Fromm ME (2004) Improved tandem affinity purification tag and methods for isolation of protein heterocomplexes from plants. Plant J 38: 172–181
5. Rohila JS, Chen M, Chen S, Chen J, Cerny R, Dardick C, Canlas P, Xu X, Gribskov M, Kanrar S, Zhu JK, Ronald P, Fromm ME (2006) Protein-protein interactions of tandem affinity purification-tagged protein kinases in rice. Plant J 46:1–13
6. Burckstummer T, Bennett KL, Preradovic A, Schutze G, Hantschel O, Superti-Furga G, Bauch A (2006) An efficient tandem affinity purification procedure for interaction proteomics in mammalian cells. Nat Methods 3: 1013–1019
7. Gregan J, Riedel CG, Petronczki M, Cipak L, Rumpf C, Poser I, Buchholz F, Mechtler K, Nasmyth K (2007) Tandem affinity purification of functional TAP-tagged proteins from human cells. Nat Protoc 2:1145–1151
8. Gyuris J, Golemis E, Chertkov H, Brent R (1993) Cdi1, a human G1 and S phase protein phosphatase that associates with Cdk2. Cell 75:791–803
9. Hoff B, Kuck U (2005) Use of bimolecular fluorescence complementation to demonstrate transcription factor interaction in nuclei of living cells from the filamentous fungus Acremonium chrysogenum. Curr Genet 47: 132–138
10. Hu CD, Chinenov Y, Kerppola TK (2002) Visualization of interactions among bZIP and Rel family proteins in living cells using bimolecular fluorescence complementation. Mol Cell 9:789–798
11. Busch S, Schwier EU, Nahlik K, Bayram O, Helmstaedt K, Draht OW, Krappmann S, Valerius O, Lipscomb WN, Braus GH (2007) An eight-subunit COP9 signalosome with an intact JAMM motif is required for fungal fruit body formation. Proc Natl Acad Sci USA 104:8089–8094
12. Helmstaedt K, Laubinger K, Vosskuhl K, Bayram O, Busch S, Hoppert M, Valerius O, Seiler S, Braus GH (2008) The nuclear migration protein NUDF/LIS1 forms a complex with NUDC and BNFA at spindle pole bodies. Eukaryot Cell 7:1041–1052
13. Bayram O, Krappmann S, Ni M, Bok JW, Helmstaedt K, Valerius O, Braus-Stromeyer S, Kwon NJ, Keller NP, Yu JH, Braus GH (2008) VelB/VeA/LaeA complex coordinates light signal with fungal development and secondary metabolism. Science 320:1504–1506
14. Braus GH, Irniger S, Bayram O (2010) Fungal development and the COP9 signalosome. Curr Opin Microbiol 13:672–676
15. Sarikaya Bayram O, Bayram O, Valerius O, Park H, Irniger S, Gerke J, Ni M, Han K, Yu JH, Braus GH (2010) LaeA control of velvet family regulatory proteins for light-dependent development and fungal cell-type specificity. PLoS Genet 6:e1001226
16. James GT (1978) Inactivation of the protease inhibitor phenylmethylsulfonyl fluoride in buffers. Anal Biochem 86:574–579
17. Neuhoff V, Arold N, Taube D, Ehrhardt W (1988) Improved staining of proteins in polyacrylamide gels including isoelectric focusing gels with clear background at nanogram sensitivity using Coomassie Brilliant Blue G-250 and R-250. Electrophoresis 25:1327–1333
18. Blum H, Hildburg B, Hans JG (1987) Improved silver staining of plant proteins. RNA andDNAinpolyacrylamidegels.Electrophoresis 8:93–99
19. Shevchenko A, Wilm M, Vorm O, Mann M (1996) Mass spectrometric sequencing of proteins silver-stained polyacrylamide gels. Anal Chem 68:850–858
20. Valerius O, Kleinschmidt M, Rachfall N, Schulze F, Lopez Marin S, Hoppert M, Streckfuss-Bomeke K, Fischer C, Braus GH (2007) The Saccharomyces homolog of mammalian RACK1, Cpc2/Asc1p, is required for FLO11-dependent adhesive growth and dimorphism. Mol Cell Proteomics 6:1968–1979

Chapter 15

Correlating Secondary Metabolite Production with Genetic Changes Using Differential Analysis of 2D NMR Spectra

Ry R. Forseth and Frank C. Schroeder

Abstract

For many fungi the number of known secondary metabolites is surprisingly small compared to the astonishingly large number of terpene cyclase, polyketide synthase (PKS), and non-ribosomal peptide synthetase (NRPS) secondary metabolite gene clusters found in their genomes. Correspondingly, the majority of fungal secondary metabolite genes have not yet been associated with the biosynthesis of any known small molecules, and it seems likely that for many more PKS and NRPS known small molecule products represent but a fraction of the entire spectrum of metabolites produced by the associated pathways. Comparative metabolomics based on differential analysis by 2D NMR spectroscopy (DANS) in conjunction with LC–MS analyses is emerging as a highly effective tool for pursuing small molecule structures and biosynthetic pathways associated with orphan PKS and NRPS gene clusters. Here we describe the use of DANS paired with LC–MS analyses for the comparison of the metabolomes of various fungal strains including wild-type (WT), PKS/NRPS overexpressing, and/or corresponding PKS/NRPS knock-out (KO) strains.

Key words: DANS, NMR spectroscopy, Nonribosomal peptide synthetase, Polyketide synthase, Metabolomics, Mass spectrometry, Biosynthesis

1. Introduction

A number of recent examples have demonstrated the utility of comparative metabolomics using 2D NMR spectroscopy and mass spectrometry for identifying the biogenic small molecules (1) whose biosynthesis is dependent on a specific genetic background (2, 3) or environmental conditions (4, 5). In fungi, biosynthesis of many if not most secondary metabolites appears to depend on an integrated network of regulatory processes that responds to specific environmental stimuli (6, 7). As a result, most fungal PKS and NRPS genes are not active under standard laboratory conditions and correspondingly their small molecule products have not been

Nancy P. Keller and Geoffrey Turner (eds.), *Fungal Secondary Metabolism: Methods and Protocols*, Methods in Molecular Biology, vol. 944, DOI 10.1007/978-1-62703-122-6_15, © Springer Science+Business Media, LLC 2012

identified. Variation of fungal growth conditions can induce activation of additional biosynthetic pathways, and it was shown that 2D NMR-based comparative metabolomics can be used to detect and identify associated changes in secondary metabolite production (4). This first example for DANS was based on comparison of a specific type of high-resolution 2D NMR spectra (a well-known version of $^{1}H,^{1}H$ correlation referred to as dqfCOSY or 2qfCOSY) of entire fungal metabolite extracts. Whereas previously 2D NMR spectra had been used almost exclusively for the characterization of isolated, pure natural products, this example demonstrated that differential comparison of such spectra provides a highly effective means to map and characterize changes in fungal metabolism. dqf-COSY spectra are particularly well-suited for this type of analysis because of their good dynamic range and information-rich peak fine structure (8) which enables interpretation even in cases of significant peak overlap and includes detailed proton–proton coupling information (see Note 1) (9).

Subsequent studies established DANS as a general method for correlating secondary metabolite production with genetic changes. The first example where DANS was utilized to link secondary metabolite gene cluster expression and specific metabolic output came from the study of the *Bacillus subtilis pksX* gene cluster, a very large hybrid polyketide/nonribosomal peptide synthase. The antibiotic metabolites produced by the *pksX*-encoded pathway had eluded structure elucidation due to their rapid degradation under ambient laboratory conditions and upon chromatography (10). Therefore a strategy was developed in which 2D NMR spectra of metabolite extracts derived from a *pksX*-expressing strain (*pksX*$^{+}$) and a *pksX* knock-out strain (*pksX*$^{-}$) were compared. Initial comparison of dqf-COSY spectra corresponding to extracts derived from *pksX*$^{+}$ and pksX^{-} led to the identification of *partial structures* (i.e., fragments of a larger chemical structure) belonging to the sought-after *pksX*-dependent metabolites. Additional 2D NMR spectroscopic analysis including: ($^{1}H,^{13}C$)-HMQC, ($^{1}H,^{13}C$)-HMBC, ($^{1}H,^{15}N$)-HMBC, and ROESY spectra, in conjunction with high-resolution positive-ion electrospray MS of the crude extracts, enabled the identification of the highly labile hexaene bacillaene and several-related metabolites as the products of the PksX megacomplex.

DANS methodology has also been applied to higher organisms. In pursuit of endogenous small molecule signals used by the model organism *Caenorhabditis elegans*, Pungaliya et al.'s DANS comparison of WT worms and *daf-22* mutant worms, a strain known to be defective in the production of the signaling molecules of interest, led to the characterization of a family of glycosides of the dideoxy sugar ascarylose (ascarosides) with potent biological activity (11). In this study the use of chemically synthesized ascarosides served as standards for parallel LC–MS analysis of crude worm extracts, which confirmed the identity of the *daf-22*-dependent *C. elegans* metabolites identified via DANS.

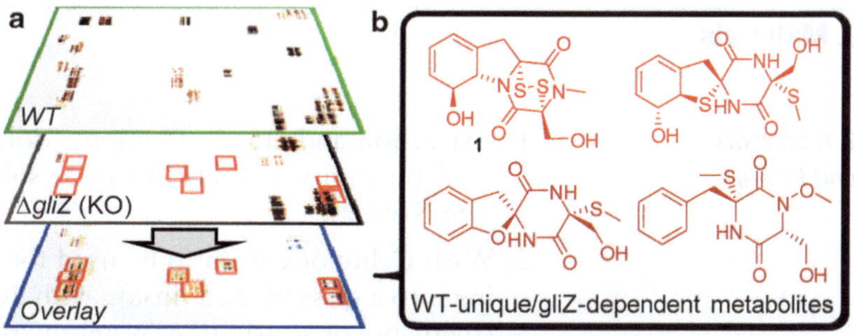

Fig. 1. Schematic overview for the use of DANS to identify *gliZ*-dependent metabolites in *A. fumigatus* (3). (**a**) Spectroscopic signals present in both WT and *ΔgliZ* (KO) spectra, representing metabolites not strongly affected by the *ΔgliZ* mutation, are eliminated in the overlay, whereas signals only present in WT (i.e. they are *gliZ*-dependent) remain (*red boxes*). (**b**) Further analysis of the COSY signals highlighted by DANS, with additional 2D NMR and HPLC–MS analysis, established full chemical structures of *gliZ*-dependent metabolites, including both known and novel compounds.

A recent study of the *Aspergillus fumigatus* metabolome demonstrated that DANS can reveal large numbers of new metabolites even for previously well-studied pathways (3). Gliotoxin (1, Fig. 1b), a major product of the *gli* non-ribosomal peptide synthetase (NRPS) gene cluster, is strongly associated with virulence of *A. fumigatus.* Despite identification of all genes of the *gli* cluster, the pathway of gliotoxin biosynthesis had remained elusive, in part because few potential intermediates have been characterized. DANS-based comparison of the WT metabolome with that of *ΔgliZ* (Fig. 1), a knock-out strain devoid of the gene encoding the transcriptional regulator of the *gli* cluster, revealed nine novel *gliZ*-dependent metabolites including unexpected structural motifs. Their identification provided detailed insight into the biosynthesis of gliotoxin and related compounds and may benefit studies of the role of the *gli* cluster in *A. fumigatus* virulence.

DANS-based comparative metabolomics paired with HPLC–MS analyses promises to greatly improve detection of changes in fungal metabolite production and to simplify identification of the structures of differentially produced metabolites. Potential applications include comparison of fungi grown under different culture conditions as well as comparison of different genotypes or ecotypes. However, great care must be taken when designing DANS studies to ensure that metabolite extracts are obtained using strictly reproducible conditions and remain uncontaminated. Below we outline a detailed protocol and note some pitfalls for DANS analysis of fungal samples, discussing specific aspects of extract preparation, sample enrichment, data acquisition, processing and interpretation, and replication. As a specific example, we present a general outline for DANS/LC–MS comparison of a wild-type strain with an NRPS or PKS knock-out strain.

2. Materials

2.1. Solvents and Filters

1. Extraction and chromatography: Solvents must be of highest available purity. Use HPLC-grade solvents with purity ≥99.9% (see Note 2).
2. Wash Celite or cotton to be used for extract filtration by placing into a glass frit and rinsing with 20 ml of extraction solvent (methanol or acetonitrile/water mixtures) per 1 ml of filtration material.

2.2. NMR Spectroscopy

1. NMR solvents: Use methanol-d_4 (see Note 3), acetonitrile-d_3, acetone-d_6, or DMSO-d_6 of 99.8% or greater purity.
2. NMR tubes: Use high quality 5 mm tubes (see Note 4).
3. NMR spectrometer and probes: Use of NMR spectrometers with a ^1H resonance frequency of 600 MHz or higher is recommended; however, high-resolution dqfCOSY spectra acquired at 900 MHz can provide more than twice the resolution of spectra acquired at 600 MHz. Use inverse-detection or HCN triple resonance probes (see Note 5).

2.3. Fractionation and HPLC–MS

1. Fractionation of extracts for DANS: Use a programmable, fully automated flash chromatography system (e.g., Teledyne ISCO CombiFlash), equipped with either normal-phase silica gel or reverse-phase C18 (octadecyl-functionalized) silica columns.
2. LC–MS: Use a standard reversed phase column (e.g., 4.6 × 150 mm, 5 μm particle diameter) and acetonitrile–water (alternatively methanol–water) solvent gradients (see Note 3). Addition of 0.1% acetic acid can improve chromatographic separation for some compound classes. Analyze all samples under carefully standardized conditions, using both positive- and negative-ion electrospray ionization.
3. Use a rotary evaporator with vacuum control and temperature-controlled water bath for evaporation of extracts and fractions.

3. Methods

Fungal secondary metabolism can be affected by minor, often unnoted changes in culture conditions. Therefore, we recommend that all analyses are carried out using samples from at least two independent biological replicates.

3.1. Preparation of Fungal Cultures

1. Wild-type and knock-out strains (as well as any additional strains to be compared, for example a PKS/NRPS overexpressor) must be cultured at the same time under identical conditions. Culture parameters must be carefully controlled in order to ensure comparability with later replicates. These include temperature, humidity, light levels, media (make one batch and partition out for individual strains), incubation time, and inoculum concentration, in addition to other parameters specific to the type of culture (e.g., liquid/solid).
2. The number of individual culture replicates for each strain required for DANS varies depending on media type (rich/poor) and fungal species. If possible, enough cultures should be set up, pooled, and processed to produce at least 50–100 mg of metabolite extract for each strain.

3.2. Culture Extraction

1. Lyophilize fungal cultures (liquid culture or homogenized solid/agar based), break up the residue into a fine powder, and extract with organic solvent. Suitable solvent systems for extraction of compounds of a wide polarity range include (from least to most polar): a mixture of ethyl acetate and methanol (9:1, v:v), acetonitrile (3), methanol, or a mixture of acetonitrile and water (9:1, v:v) (see Note 3). Use 100 ml solvent per 2 g of lyophilized culture. Extract for at least 2 h with effective stirring. Process all cultures on the same day, using equivalent volumes of the same batch of extraction solvents.
2. Filter the extract over pre-washed Celite (5 g per 100 ml extract) or cotton (see Note 6). Rinse filter with 100 ml of extraction solvent.
3. Evaporate filtered extracts in vacuo using a rotary evaporator equipped with a room temperature water bath. Do not heat the extracts above room temperature. Use the same size of evaporation flask and the same vacuum settings for all extracts.

3.3. Optional Pre-fractionation/ Polarity Segregation

Crude fungal extracts usually represent highly complex mixtures of small molecules (complexity can be assessed by acquiring a dqf-COSY spectrum for a crude fungal extract, see below) and contain compounds of a wide polarity range. Some compounds in such crude extracts may not be well soluble in the NMR solvent, thus preventing their detection. Furthermore, the presence of poorly soluble components in metabolite samples often degrades the quality of the NMR spectra. The procedures described below employ reproducible chromatographic enrichment to yield samples of reduced chemical complexity and increased polarity uniformity, helping to prevent precipitate formation, and improving NMR spectral line shapes as well as dynamic range. Two alternative procedures are described, for samples derived from extraction procedures that use either relatively polar or nonpolar solvents. These protocols separate the wild-type and knock-out mutant extracts (as well as any additional samples) into sets of three "pools" each.

3.3.1. Samples from Polar Extraction

1. Dissolve wild-type fungal sample completely in suitable organic solvent (e.g., the solvent used for extraction) in a round-bottomed flask. Use at least 5 ml of solvent per 100 mg of extract or more if needed for complete dissolution.
2. Add 1 g (per 100 mg of extract dissolved) of Celite or octadecyl-functionalized silica gel to the solution.
3. To adsorb the solute extract onto the solid phase, remove solvent in vacuo using rotary evaporation. To ensure reproducible chromatography amongst sets of fungal extracts, remove residual solvent after rotary evaporation by subjecting the sample to reduced pressure (0.1 mmHg, ~1 h).
4. Remove the extract-imbued solid phase from flask and set aside. Add 10 ml of additional solvent and 1 g of Celite or octadecyl-functionalized silica gel to flask and repeat step 3 to collect residual extract stuck to the sides of the flask.
5. Condition a 100 g reverse-phase C-18 flash chromatography column (for 80–500 mg of extracts) by changing solvent composition from 100% organic solvent to 1–5% organic solvent/99–95% water (or water with 0.1% acetic acid) over 10 column volumes, holding at 99–95% water for five additional column volumes.
6. Load the solid phase adsorbed sample onto the chromatography system as described by the manufacturer.
7. Fractionate the sample using a solvent-gradient based chromatography method. A reverse-phase gradient (acetonitrile–water or methanol–water) that works well for fungal extracts was recently described (3).
8. Combine individual fractions to give three pools corresponding to high (e.g., 1–5% organic solvent), medium (e.g., 6–54% organic solvent), and low (e.g., 55–100% organic solvent) polarities. UV chromatograms acquired by the chromatography system during a test separation run can assist with determining which fractions to combine.
9. Repeat steps 1–8 for knock-out extract (and any additional samples to be compared via DANS) using identical solvents, media, and fractionation parameters. Comparison of UV chromatograms acquired for different extracts can be utilized to assess reproducibility of elution times, helping to determine which fractions should be included in a given pool.

3.3.2. Samples from Nonpolar Extraction

1. Dissolve wild-type fungal sample completely in suitable organic solvent (e.g., the solvent used for extraction) in a round-bottomed flask. Use at least 5 ml of solvent per 100 mg of extract or more if needed for complete dissolution.
2. Add 2 g (per 100 mg of extract dissolved) of normal-phase silica gel to solution.

3–4. Same as polar extract enrichment procedures above.

5. Condition a 40 g Normal-Phase Silica Gel flash chromatography column (for extracts 100–1,000 mg) by changing solvent gradient composition from 50% more polar solvent (e.g., methanol, acetonitrile, ethyl acetate, with optional 0.1% acetic acid) to 100% less polar organic solvent (e.g., dichloromethane, hexanes) over 8 column volumes, followed by conditioning with the same solvent for 4 additional column volumes.

6. Same as polar extract enrichment procedures above.

7. Fractionate the sample using a solvent-gradient based chromatography method. Example, chromatographic method for fractionation of *Aspergillus flavus* ethyl acetate:methanol (9:1) extracts: start with 100% dichloromethane for 2.5 column volumes, followed by a linear increase of methanol over 20 column volumes up to 30% methanol, which is held for an additional 10 column volumes.

8. Combine individual fractions to give three pools. For example when using the chromatographic method in step 7, combine fractions with 0–8% methanol as pool#1 (least polar), fractions with 8–16% methanol as pool#2 (medium polar), and 17–30% methanol as pool#3 (most polar). UV chromatograms acquired by the chromatography system during a test separation run can help guide pooling.

9. Same as polar extract enrichment procedures above.

3.4. NMR Sample Preparation

The following steps assume use of a 5 mm sample tube and 0.6 ml of NMR solvent. If using smaller samples volumes, sample and solvent amounts must be adjusted proportionally.

1. Suspend 50 mg of extract in 0.4 ml of deuterated NMR solvent in a 1 ml vial and place on a shaker for 30 min. Centrifuge to remove insoluble material and transfer supernatant into a fresh 1 ml vial (see Note 7).
2. Evaporate all solvent in vacuo using a rotary evaporator (using a proper adaptor for the 1 ml vial). Redissolve the residue in 0.6 ml of NMR solvent (see Note 8). If any turbidity remains at this stage, reevaporate all solvent and repeat step 1. Treat all samples to be compared in exactly the same way.
3. Place NMR samples into NMR tubes, taking care that solvent amount for all samples is identical. Store samples in the dark at room temperature for at least 12 h prior to NMR spectroscopy.
4. Acquire ^{1}H spectra for all samples. Inspect lines shape and peak position for known polar components (e.g., media components such as nicotinamide). If the spectra show considerable variation in proton chemical shift values, especially for signals representing chemical species with acidic or basic functional groups, the pH of the sample should be adjusted such that

peaks corresponding to the same chemical species in the various samples have the same chemical shift values (see Note 9).

3.5. Acquisition and Processing of 2D NMR Spectra

1. Acquisition parameters: For acquisition of dqfCOSY spectra at 600 MHz, use 0.8 s acquisition time, 80 complex increments per 1 ppm of sweep width, a relaxation delay of 1.6 s, and 8–32 scans per increment (see Note 10). Phase cycled dqfCOSY sequences must be used, not gradient-selected versions (see Note 11). For field strengths greater than 600 MHz, proportionally increase the number of complex increments per 1 ppm of sweep width.
2. Processing of NMR spectra: All spectra are processed in two different ways: (1) for comparative analysis (spectral overlay) and (2) for detailed follow-up analysis. Method 1 (for spectral overlay): shorten time domain to 0.25 s, zero-fill to 4096 data points in F2, apply a squared sine bell window function, and perform first Fourier transform. Next, zero-fill to 4096 data points in F1, apply a squared sine bell window function, and apply second Fourier transform. Switch display to absolute value-mode and use real part of the spectrum for overlay (see below). Method 2 (for detailed follow-up analysis): use full-time domain, zero-fill to 8192 data points in F2 (see Note 12), apply cosine window function, and perform first Fourier transform. Next, zero-fill to 4096 data points in F1, apply a cosine window function, and apply second Fourier transform. Phase correct spectra as appropriate.
3. DANS overlay of dqfCOSY spectra using graphics software: Overlays of large dqfCOSY spectra are possible with several NMR processing software packages, e.g., Bruker's Topspin. However, differences between spectra can often be visualized better using overlays of spectral images using graphics software such as Adobe Photoshop. For the purpose of detecting differences between a wild-type and a knock-out strain, the dqfCOSY spectra are arranged as follows. Display spectra as color maps in the NMR processing software (e.g., MNOVA, Topspin) using a color scheme that correlates color saturation and/or brightness with signal intensity. Place bitmaps from the spectra to be compared (e.g., spectra of wild-type and knock-out strain) into two sets of layers in a Photoshop document as follows. On top of a white background layer, two layers are placed containing absolute value-mode dqfCOSY spectra for wild-type and knock-out strain. For these two layers, the blending mode is set to "difference." On top of this stack of layers, another two layers are placed, again containing the wild-type and knock-out spectra. For these three layers, "multiply" is used as the blending mode. Finally, a "curves" adjustment layer is placed as the topmost layer. This curves layer is used to invert colors and to adjust gamma to increase or decrease contrast.

4. Comparative analysis: In order to compare spectra, visibility of layers is toggled off and on as needed. For example, to visualize all signals from compounds present in wild-type, the layer containing the wild-type spectrum in "difference" mode is turned on and visibility of all other layers containing spectra is turned off. An example for resulting spectral image is in Fig. 2a.

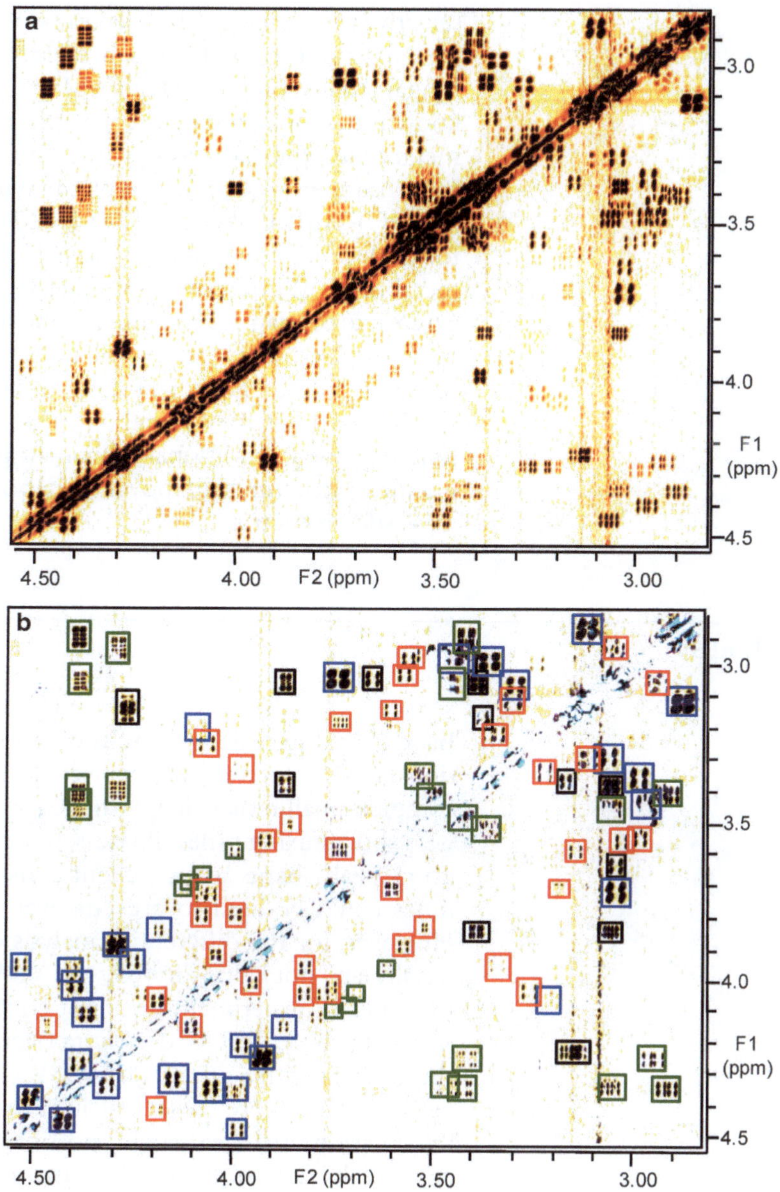

Fig. 2. Actual DANS spectra from analysis of *A. fumigatus gliZ*-dependent metabolites (3). (**a**) Section of the WT dqfCOSY spectrum in absolute-value mode. (**b**) DANS overlay of WT and Δ*gliZ* (KO) spectra, showing signals representing known gliotoxin derivatives (*blue*), novel *gliZ*-dependent compounds (*red*), and previously described fungal metabolites (*black*). Several crosspeaks representing metabolites present in both WT and Δ*gliZ* were not fully suppressed in the overlay due to chemical shift variation (*green*).

Additionally turning on visibility of the layer containing the knock-out spectrum in "multiply" mode results in suppression or color changes for all signals, except those only present in wild-type and absent in the knock-out spectrum (Fig. 2b) (see Note 13).

5. Construct partial structures based on interpretation of the identified differential dqfCOSY cross peaks, using the phase-sensitive processing of the spectra (see step 2) (8, 12). Compare partial structures with known metabolites of the investigated fungal strains, and use HPLC–MS to confirm presence of any known metabolites. For the identification of potentially novel metabolites that are present in high concentrations (>1% of total sample), acquire supplemental (^{1}H,^{13}C)-HSQC and (^{1}H,^{13}C)-HMBC spectra (12) of wild-type and mutant sample for additional comparative analysis. If the abundance of the metabolites of interest is lower (<1% of the total) or in cases of extensive overlap in the dqfCOSY spectra, additional chromatographic fractionation will be required to obtain satisfactory dqfCOSY, (^{1}H,^{13}C)-HSQC, and (^{1}H,^{13}C)-HMBC data. If the presence of new metabolites is confirmed, isolate compounds using additional chromatography (see Note 14). Confirm structures of novel metabolites by high-resolution mass spectrometry.

4. Notes

1. Instead of dqfCOSY spectra, several other types of two-dimensional NMR spectra could be used for the differential analysis of complex small molecule mixtures, for example TOCSY spectra with short mixing times. Phase-sensitive TOCSY spectra exhibit good peak shape and excellent sensitivity, and short mixing times could be used to prevent over-crowding of the spectra. However, for the differential analysis of fungal extracts, which usually are not mass-limited, the slightly higher sensitivity of the TOCSY experiment does not confer any significant advantage. We chose dqfCOSY spectra over TOCSY spectra because the fine structure of the dqfCOSY cross peaks allows for the extraction of more structural information (coupling constant information) than would be possible to derive from TOCSY cross peaks. This difference becomes relevant in the second step of the differential analysis, in which signals, identified as representing compounds of interest, are analyzed in greater detail. For this second step, the dqfCOSY method thus provides a clear advantage. Nonetheless, for mass-limited samples, differential analysis based on TOCSY spectra could be useful.

2. Care should be taken to avoid or reduce contact of the sample and solvents to plastic materials/surfaces, as many plastics contain soluble components that can result in significant contamination of samples and degradation of NMR spectral quality.
3. Methanol (when used as extraction, chromatography, or NMR solvent) can react with labile metabolites (e.g., methylsulfanyl derivatives, carboxylic acids, unsaturated aldehydes, and ketones), resulting in artifact formation. Acetonitrile/water mixtures should be used if the presence of methanol-sensitive compounds is suspected. Additionally, methanol and acetonitrile/water mixtures will extract media carbon sources such as glycerol, glucose, or other mono- and disaccharides. Signals corresponding to these components will dominate NMR spectra acquired for crude methanol extracts, requiring their eventual removal from the sample via chromatography (see Subheading 3.3).
4. The use of 5 mm NMR tubes and corresponding NMR probe heads offers excellent dynamic range at good sensitivity, facilitating detection of even very minor components of a metabolome. In cases where extract quantity is limited, smaller sample tubes/probe heads should be used.
5. The use of cryogenic (cold) probes can be advantageous for mass-limited samples; however, in cases where sample size is not limiting, room temperature probes may provide better results with fewer artifacts when acquiring dqfCOSY spectra. For (^{1}H,^{13}C)-HSQC and (^{1}H,^{13}C)-HMBC, cryogenic probes are generally preferable (8).
6. Of the two filtration materials, Celite is more effective at removing proteins and other macromolecular components from the sample, which often cause line shape problems and/or precipitate formation during NMR acquisition. However, cotton tends to be less chemically abrasive than Celite and should be used when labile chemical species are suspected.
7. Appropriate choice of NMR solvent is critical for spectral quality and dynamic range. Methanol-d_4 dissolves most organic compounds, but leads to deuterium exchange of many types of protons (hydroxyl, amine, amide, and also β-keto carbonyl) and may cause problems for compounds susceptible to nucleophilic addition or substitution (see Note 3). DMSO-d_6 or acetonitrile-d_3 with 5–10% D_2O can be used alternatively, but may not dissolve very lipophilic compounds. Any insoluble small molecules will be retained in the centrifugation pellet.
8. The use of 0.4 ml of solvent in the centrifugation step ensures that the actual NMR sample, prepared using a slightly larger solvent volume of 0.6 ml, will not be completely saturated. This is necessary to avoid formation of precipitates during NMR spectroscopic analysis.

9. The proton chemical shifts of small molecules containing acidic or basic functional groups can vary greatly (>1 ppm) depending on sample pH; this variation can complicate the interpretation of DANS spectra. To adjust the pH of an NMR sample, make a solution of 5% acetic acid-d_4 in an NMR solvent of choice (acidified solution). Choose an internal standard ^{1}H signal to monitor during the process of adjusting the sample pH; for example, the well resolved singlet corresponding to the pH-sensitive proton at position 2 of nicotinic acid or nicotinamide (8.6–8.7 ppm). Titrate the sample with the acid solution (~10 μl/aliquot), monitoring how sample chemical shifts respond to the addition of the acidified solution. Continue the process iteratively until internal standard ^{1}H signal and overall peak shift landscape is consistent for all samples that will be compared in the DANS analysis (11).
10. Use of 8 scans per increment may be sufficient for concentrated samples; however, on most spectrometers artifact suppression is better at 16 scans per increment. For high dynamic range requirements or dilute samples 32 (or more) scans per increment may be required.
11. Use of gradient-selected dqfCOSY sequences (i.e. sequences that use gradients for coherence selection) results in spectra with significantly worse line shape characteristics and lower signal-to-noise and dynamic range. The use of homospoil pulses at the beginning of the relaxation delay d1 is recommended.
12. Digital resolution in F2 in the Fourier-transformed spectrum should be no lower than 1 Hz/data point, and preferably closer to 0.5 Hz/data point. Spectra with large sweep widths, or spectra acquired at high field strength, may thus require zero-filling to 16384 data points in F2 prior to Fourier transformation.
13. The graphical comparison of 2D-NMR spectra as described here can be regarded as a "weighing" of one spectrum against one or more related spectra. Essentially, the "multiply" function weighs the RGB values of a given data point against the RGB values of corresponding data points in the overlaid spectra. The use of the "multiply" blending mode as a simple out-of-the-box weighing mode is convenient; however, other blending modes may in some situations provide better results. For example, simply subtracting one spectrum from another would create a mask that could be used to completely block out any signals occurring in both spectra. However, in our experience color shifts of signals as they result from the "multiply" blending mode are generally better to work with than complete signal extinction, because the preserved multiplicity structures of partially suppressed peaks allows one to confirm that they indeed represent superposition(s) of equivalent cross

peaks in the compared spectra, as opposed to mutual extinction of unrelated cross peaks that accidentally occur at similar chemical shift values in the compared spectra.

14. Isolation is not a requirement for structure determination of novel compounds; however, full structural characterization of pure compounds (or significantly enriched samples) is often easier than their identification from spectra of complex mixtures. Nonetheless, there is small but increasing number of metabolites that have never been isolated in pure form and, therefore, were identified based on spectra of complex metabolite mixtures (1).

Acknowledgments

This work was supported by the National Institutes of Health (Grant No. GM088290 and GM008500) and DuPont Crop Protection.

References

1. Forseth RR, Schroeder FC (2011) NMR-spectroscopic analysis of mixtures: from structure to function. Curr Opin Chem Biol 15: 38–47
2. Challis GL (2008) Genome mining for novel natural product discovery. J Med Chem 51:2618–2628
3. Forseth RR, Fox EM, Chung D, Howlett BJ, Keller NP, Schroeder FC (2011) Identification of cryptic products of the gliotoxin gene cluster using NMR-based comparative metabolomics and a model for gliotoxin biosynthesis. J Am Chem Soc 133:9678–9681
4. Schroeder FC, Gibson DM, Churchill AC, Sojikul P, Wursthorn EJ, Krasnoff SB, Clardy J (2007) Differential analysis of 2D NMR spectra: new natural products from a pilot-scale fungal extract library. Angew Chem Int Ed Engl 46:901–904
5. Shiloach J, Reshamwala S, Noronha SB, Negrete A (2010) Analyzing metabolic variations in different bacterial strains, historical perspectives and current trends-example *E. coli*. Curr Opin Biotechnol 21:21–26
6. Georgianna DR, Fedorova ND, Burroughs JL, Dolezal AL, Bok JW, Horowitz-Brown S et al (2010) Beyond aflatoxin: four distinct expression patterns and functional roles associated with *Aspergillus flavus* secondary metabolism gene clusters. Mol Plant Pathol 11:213–226
7. Yin W, Keller NP (2011) Transcriptional regulatory elements in fungal secondary metabolism. J Microbiol 49:329–339
8. Claridge TDW (2009) High-resolution NMR techniques in organic chemistry. Elsevier, Amsterdam, Boston, p xiv, 383
9. Taggi AE, Meinwald J, Schroeder FC (2004) A new approach to natural products discovery exemplified by the identification of sulfated nucleosides in spider venom. J Am Chem Soc 126:10364–10369
10. Butcher RA, Schroeder FC, Fischbach MA, Straight PD, Kolter R, Walsh CT et al (2007) The identification of bacillaene, the product of the PksX megacomplex in *Bacillus subtilis*. Proc Natl Acad Sci USA 104:1506–1509
11. Pungaliya C, Srinivasan J, Fox BW, Malik RU, Ludewig AH, Sternberg PW et al (2009) A shortcut to identifying small molecule signals that regulate behavior and development in *Caenorhabditis elegans*. Proc Natl Acad Sci USA 106:7708–7713
12. Bross-Walch N, Kuhn T, Moskau D, Zerbe O (2005) Strategies and tools for structure determination of natural products using modern methods of NMR spectroscopy. Chem Biodivers 2:147–177

Chapter 16

Chromatin Immunoprecipitation Analysis in Filamentous Fungi

Stefan Boedi, Yazmid Reyes-Dominguez, and Joseph Strauss

Abstract

Chromatin immunoprecipitation (ChIP) is used to map the interaction between proteins and DNA at a specific genomic locus in the living cell. The protein–DNA complexes are stabilized already in vivo by reversible crosslinking and the DNA is sheared by sonication or enzymatic digestion into fragments suitable for the subsequent immunoprecipitation step. Antibodies recognizing chromatin-linked proteins, transcription factors, artificial tags, or specific protein modifications are then used to pull down DNA–protein complexes containing the target. After reversal of crosslinks and DNA purification locus-specific quantitative PCR is used to determine the amount of DNA that was associated with the target at a given time point and experimental condition. DNA quantification can be carried out for several genomic regions by multiple qPCRs or at a genome-wide scale by massive parallel sequencing (ChIP-Seq).

Key words: Chromatin, Protein–DNA interactions, Antibody, Quantitative PCR, Filamentous fungi

1. Introduction

Chromatin, the complex of histone and non-histone proteins associated with nuclear DNA, is the natural substrate of all DNA-dependent processes such as replication, recombination, repair or transcription. Experimental approaches to study chromatin dynamics and modifications such as chromatin accessibility or "Chromatin immunoprecipitation" (ChIP) assays have been developed and applied also in filamentous fungi (1–4). The ChIP technique was pioneered by Varshavsky and colleagues (5, 6) and is used to detect and to quantify interactions between DNA and associated proteins in vivo. It aims to determine whether specific proteins such as histones or transcription factors are located at a given genomic region under a given experimental condition. The original method

Nancy P. Keller and Geoffrey Turner (eds.), *Fungal Secondary Metabolism: Methods and Protocols*, Methods in Molecular Biology, vol. 944, DOI 10.1007/978-1-62703-122-6_16,

is based on analysis of DNA that is reversibly crosslinked to, and is co-precipitating with, protein complexes. In order to stabilize these interactions during the IP step, intact cells are usually treated with formaldehyde in their growth medium. Native ChIP procedures have been developed which omit the crosslinking step, but have so far not been applied in filamentous fungi. Time and concentration of formaldehyde crosslinking depends on the organism and needs to be optimized empirically for each medium and species.

Briefly, chromatin and other DNA-associated proteins (in cell lysates or intact cells) are temporarily crosslinked, then genomic DNA is sheared by sonification or enzymatic restriction to obtain smaller DNA fragments of similar size (usually between 300 and 500 bp). Antibodies recognizing epitopes in the protein of choice (target protein) are then added to the mix to immunoprecipitate DNA–protein complexes. Subsequently, the crosslinks are reversed, protein components are digested and DNA is isolated. Finally, the amount of precipitated DNA is determined by quantitative PCR in comparison to input DNA. The values indicate how much of the antibody target associates with the analyzed DNA locus at a certain time point under a given experimental condition. If many loci should be analyzed in parallel, a library of co-precipitated fragments can be generated and analyzed by sequencer-based fragment analysis (7) or multiple qPCRs. A genome-wide map of binding events can be generated by hybridizing the co-precipitated fragments to microarray probes (ChIP-Chip) (8) or by massive-parallel sequencing of libraries generated from precipitated fragments (ChIP-Seq) (9, 10). Systematic identification of DNA targets on a genome-wide level with extremely high resolution by ChIP-Seq has revolutionized mapping of DNA–protein interactions or the characterization of specific histone modifications associated with metabolic or developmental processes (11, 12). In filamentous fungi, these genome-wide techniques have been applied so far only for *Neurospora crassa* (13, 14).

In addition to determining genomic locations of target protein binding, ChIP analysis also provides the possibility to test for condition-specific protein modifications. Antibodies directed against epitopes such as protein phosporylation, acetylation, methylation, or ubiquitination are commercially available and have been used partially also in filamentous fungi (Table 1). Crucial to all ChiP experiments are controls that reveal the background of DNA co-precipitated by unspecific antibody binding (see below). When specific modifications of a protein are analyzed additional input controls are necessary. These controls include a separate ChIP with an antibody that recognizes a different epitope in the analyzed protein, optimally an epitope that is not subject to posttranslational modifications (PTMs). For example, to test for histone H3 N-terminal modifications such as methylation or phosphorylation, an antibody recognizing the conserved C-terminus of histone H3 is often used (see Table 1).

Table 1
List of antibodies used for ChIP in filamentous fungi

Antibodies against protein tags	Company	Product number	References	Organism	Comment
S-tag antibody	Abcam	ab19321	(16)	*A. nidulans*	ChIP antibody directed against S-tagged KapK. Subsequent amplification of the precipitated DNA at the NirA binding site to examine in vivo interaction of these two proteins involved in nitrate assimilation
HA tag antibody	Abcam	ab9110	(17)	*A. nidulans*	ChIP antibody directed against HA-tagged GATA factor AreA to determine protein binding to promoter sites of genes involved in nitrate assimilation
anti-GFP	Abcam	ab290	(14)	*N. crassa*	Identification of centromeric DNA by ChIP-seq of DNA associated with tagged versions of centromere foundation proteins and one kinetochore protein

Antibodies against histones, histone marks and non-histone proteins	Company	Product number	References	Organism	Comment
anti-H3K4me2	Abcam	ab32356	(18)	*A. nidulans*	Determination of histone H3K4 dimethylation associated with monodictyphenone and emodin production in background of a *cclA* gene deletion
	Upstate	07-030	(14)	*N. crassa*	Determination of histone H3K4 dimethylation at centromeric loci identified by ChIP-seq
anti-H3K4me3	Abcam	ab1012	(18)	*A. nidulans*	Examination of histone H3K4 trimethylation associated with monodictyphenone and emodin production in background of a *cclA* gene deletion
	Active Motif	AR-0169	(14)	*N. crassa*	Determination of histone H3K4 trimethylation at centromeric loci identified by ChIP-seq

(continued)

Table 1 (continued)

Antibodies against histones, histone marks and non-histone proteins	Company	Product number	References	Organism	Comment
anti-H3K9me3	Abcam	ab8898	(18)	*A. nidulans*	Examination of histone H3K9 trimethylation associated with monodictyphenone and emodin production in background of a *cclA* gene deletion
			(19)	*A. nidulans*	Investigation of the role of H3K9 histone methyltransferase ClrD in the regulation of sterigmatocystin cluster genes
			(20)	*F. graminearum*	Determination of histone H3K9 trimethylation within the aurofusarin and the *Tri5* gene clusters upon deletion of the heterochromatin protein 1 homolog
	Active Motif	39161	(14)	*N. crassa*	Determination of histone H3K9 trimethylation at centromeric loci identified by ChIP-seq
anti-5me-C	Diagenode	MAb-5MECYT-100	(14)	*N. crassa*	Correlation between DNA methylation and heterochromatic marks of centromeric DNA
Anti-histone H3 (acetyl K9)	Abcam	ab4441	(21)	*A. nidulans*	Examination of histone H3 acetylation associated with SAGA-dependent activation of secondary metabolism gene clusters
Anti-acetyl-histone H3 (Lys14)	Upstate	07-353	(4)	*N. crassa*	Quantification of specific histone H3K14 acetylation underlying blue light response in *N. crassa*
Anti-histone H3 (acetyl K14)	Abcam	ab46984	(21)	*A. nidulans*	Examination of histone H3 acetylation associated with SAGA-dependent activation of secondary metabolism gene clusters

Anti-acetyl-histone H3 (binding to acetylated K9 and K14)	Upstate	06-599	(4)	*N. crassa*	Quantification of histone H3 acetylation underlying blue light response in *N. crassa*
			(22)	*A. nidulans*	Investigation of the effect of histone acetylation in the proline assimilation cluster
			(17)	*A. nidulans*	Determination of acetylation activities mediated by the GATA factor AreA involved in nitrate assimilation
			(18)	*A. nidulans*	Examination of histone H3 acetylation associated with monodictyphenone and emodin production in background of a *cclA* gene deletion
			(19)	*A. nidulans*	Determination of acetylation levels of H3 at the promoter site of AflR, the specific activator of secondary metabolite cluster genes, in background of different regulator deletions
			(21)	*A. nidulans*	Examination of histone H3 acetylation associated with SAGA-dependent activation of secondary metabolism gene clusters
			(1)	*N. crassa*	Analysis of histone H3 acetylation state at *frq* locus encoding a factor involved in regulation of clock and light-controlled genes
Anti-hyperacetylated histone H4 (Penta)	Upstate	06-946	(23)	*A. parasiticus*	Correlation between AflR binding to promoters of aflatoxin biosynthesis genes and levels of acetylated histone H4
Anti-acetyl-histone H4	Upstate	06-866			
Anti-tetra-acetylated histone H4 (K5, K8, K12, K16)	Upstate	Not specified in publication	(1)	*N. crassa*	Examination of histone H4 acetylation state at *frq* locus encoding a factor involved in regulation of clock and light-controlled genes
Anti-dimethyl R26 histone H3	Abcam	ab8047	(23)	*A. parasiticus*	Did not detect the predicted size protein in *A. parasiticus* whole-cell protein extracts in Western blot analysis. No longer included in Abcam product range
Histone H4 (symmetric di methyl R3)	Abcam	ab5823	(23)	*A. parasiticus*	Did not detect the predicted size protein in *A. parasiticus* whole-cell protein extracts in Western blot analysis. Correlation between AflR binding to promoters of aflatoxin biosynthesis genes and levels of methylated histone H4
Antibody specific to AflR	Non-commercial		(24, 23)	*A. parasiticus*	Correlation between AflR binding to promoters of aflatoxin biosynthesis genes and levels of acetylated histone H4

(continued)

Table 1 (continued)

Antibodies against histones, histone marks and non-histone proteins	Company	Product number	References	Organism	Comment
HP1 antibody	Abcam	ab24726	(19)	*A. nidulans*	Examination of heterochromatin protein 1 homolog occupancy at promoters of different genes involved in sterigmatocystin production
Anti-histone H3	Upstate	06-755	(4)	*N. crassa*	ChIP antibody directed against N terminus of histone H3 to quantify modifications of the histones detected by an additional ChIP antibody relative to histone presence.
anti-Histone H3—C term	Abcam	ab1791	(22) (17) (18) (19) (21) (20)	*A. nidulans* *A. nidulans* *A. nidulans* *A. nidulans* *A. nidulans* *F. graminearum*	ChIP antibody directed against C-terminus of histone H3 to quantify modifications of the histone detected by an additional ChIP antibody relative to histone presence
anti-WC-1 anti-WC-2	Non-commercial		(1)	*N. crassa*	Antibody directed against transcription factors White Collar-1 (WC-1) and White Collar-2 (WC-2), respectively, to determine binding of the WC proteins to elements within the promoter of *frq*, an additional factor involved in regulation of clock and light-controlled genes
			(13)	*N. crassa*	WC-1 and WC-2 dimerize to form the White Collar Complex (WCC), a regulator of gene expression in response to light and the circadian clock (25, 26). A genome-wide ChiP-Seq approach was performed to identify WCC targets in downstream regulatory pathways
anti-CSW-1	Non-commercial		(1)	*N. crassa*	Localization of CSW-1, which is required for chromatin remodeling, at promoter sites of the *frq* gene, that encodes a factor involved in regulation of clock and light-controlled genes

In contrast to in vivo footprinting which exclusively reveals direct DNA–protein interactions (15) ChIP detects all proteins interacting with DNA, either through their own DNA-binding domain or in a complex with DNA-binding factors. This indirect interactions need to be taken into account for interpretation of the ChIP results. On the other hand, these indirect interactions also provide the opportunity to test by ChIP protein–protein interactions between a target protein and a DNA-binding factor, when the DNA recognition sequence of the DNA-binding protein is known. This method has, for example, been used to map the interactions between the nuclear export in KapK with its cargo protein NirA, a DNA-binding transcription factor mediating nitrate induction (2).

Crucial for the success of ChIP experiments are the available antibodies. The specificity of these antibodies should be determined prior to ChIP experiments. This step is usually carried out in Western analysis of crude protein extracts of an isogenic mutant strain not expressing the target protein, if such a mutant is viable. When a tag should be used to precipitate the target protein the control strain should be isogenic without expressing the tagged version of the target protein. Control strains for ChIP against PTMs of the target usually carry replacements of amino acids that receive the tested modification, although for many proteins such replacements lead to loss of cell viability. In all these cases, where such mutations are lethal the use of a second-site epitope in the target protein is advisable for a control (e.g., histone H3 C-terminus). After Western analysis of crude protein extracts a preliminary ChIP analysis is carried out to test the antibody for its specificity under ChIP conditions. The experimentators should note that ChIP conditions represent native interactions between the epitope of a protein and the antibody used for IP, whereas Western analysis is usually carried out with cell extracts containing denatured proteins. To clarify if the antibody also specifically interacts with the protein of choice under non-denaturating conditions, a Western analysis of immunoprecipitated protein samples can be performed. If a signal is detected only in IP samples with the specific antibody and not in the control samples (only resin without antibody, mutants strain, strain harboring untagged version of target protein) the specificity of the antibody is satisfying for ChIP.

The presented step-by-step protocol is an optimized procedure for ChIP in filamentous fungi. It can be directly applied to *Aspergillus nidulans* and *Fusarium graminearum*. Moreover, in Table 1 we have summarized all antibodies that have been successfully used so far for ChIP in filamentous fungi. It should be noted that these organisms grow apically in submerged cultures and on solid media and thus their cells have different age and may be in different metabolic states at the time of harvesting. Aerial cultures on solid media also represent cells of different age and after longer incubation differentiation into conidia or sexual structures occurs. The precise time required for entry into differentiation stages depend on the individual species. For interpretation of ChIP results, these crucial facts need to be considered.

2. Materials

2.1. Reagents Stocks

1. 1 M HEPES–KOH pH 7.5.
2. 5 M NaCl.
3. 1 M Tris-HCl pH 8.
4. 0.5 M EDTA pH 8.
5. 20% SDS.
6. 20% Triton X-100.
7. 4 M LiCl.
8. Ig epal CA630.
9. 1% Na-deoxycholate.
10. 1 M $NaHCO_3$.
11. 100× Fungal protease inhibitor mix (Promega).
12. Salmon sperm DNA/protein A agarose slurry (Upstate).
13. Promega BSA assay.
14. 1 M Tris–HCl pH 6.5.
15. Proteinase K (1 mg/mL, MBI).

2.2. Reagents Setup

1. *Sonication buffer* 50 mM HEPES–KOH pH 7.5, 140 mM NaCl, 1 mM EDTA, 1% Triton X-100, 0.1% Na-deoxycholate, 1× Fungal protease inhibitor.
2. *ChIP dilution buffer* 1.1% Triton X-100, 1.2 mM EDTA, 16.7 mM Tris-HCl pH 8, 167 mM NaCl, 1× fungal protease inhibitor.
3. *Low salt wash buffer* 150 mM NaCl, 0.2% SDS, 0.5% Triton X-100, 2 mM EDTA, 20 mM Tris-HCl pH 8.
4. *High salt wash buffer* 500 mM NaCl, 0.2% SDS, 0.5% Triton X-100, 2 mM EDTA, 20 mM Tris-HCl pH 8.
5. *LiCl salt wash buffer* 0.25 mM LiCl, 0.5% Na-deoxycholate, 1 mM EDTA, 10 mM Tris-HCl pH 8, 0.5% Ig epal CA 630.
6. *TE buffer* 1 mM EDTA, 10 mM Tris-HCl pH 8.
7. *Resuspension buffer* SDS 1%, $NaHCO_3$ 0.1 M.

3. Methods

This protocol has been optimized for *A. nidulans* cells from liquid cultures, additional specifications for solid cultures are provided. Changes in the protocol that specifically concern *F. graminearum* and *Fusarium fujikoroi* liquid cultures are indicated by (a). Bold numbering from 1 to 27 indicates the sequential steps of the procedure.

3.1. Chemical Crosslinking of Mycelia

1. Add formaldehyde to a final concentration of 1% to the submerged culture and shake the flaks for 15 min at the same temperature at which growth was performed (the "desired temperature"). For aerial cultures, grow mycelia on cellophane and peel off the cellophane with the mycelia from the solid medium plate and transfer the mycelium together with the cellophane to a Petri-dish containing 20 mL liquid minimal media and 1% of formaldehyde. Slightly shake the mixture for 15 min at the desired temperature. Important: If you need to couple ChIP data with transcriptional data you may reserve part of the mycelia for RNA extraction before formaldehyde treatment (see also Notes 1–3 for general considerations in experimental setup).
2. Add glycin to a final concentration of 125 mM and incubate for 5 min at the desired temperature. Filtrate mycelia from submerged cultures and freeze them immediately in liquid nitrogen. The mycelia from the aerial cultures should have detached by now from the cellophane. Filtrate the mycelia and freeze it immediately in liquid nitrogen.

3.2. Sonication

3. Grind 100 mg of mycelia with mortar and pestle in liquid nitrogen, transfer it to a 2 mL Eppendorf tube and add 1 mL of sonication buffer.
4. As stated above, sonication conditions need to be optimized empirically for each type of equipment and organism. Any sonication equipment may be used, such as sonication probes or water baths with sonication function, just make sure the temperature within the sonicated sample stays low by constantly keeping the samples on ice. The company DIAGENODE designed a cooled sonicator for ChIP and the conditions for sonication of *A. nidulans* and *F. graminearum* cells are provided here as an example. Using the Bioruptor™200 from DIAGENODE the samples are sonicated in cycles of 2 min "on" and 1 min "off" during a total treatment time of 15 min at the maximal power. The samples should be always on ice. The sonication time and power should be determined empirically

when using different sonicators. Keep the samples on ice after the sonication procedure (see also Note 4).

5. Centrifuge for 1 min at 15871 ×*g* at 4°C. Recover the supernatants in new Eppendorf tubes and keep them on ice.
6. To pre-clear the samples, add 40 µL Salmon sperm DNA/protein A agarose slurry (Upstate) and turn the tubes on a rotary shaker for 1 h at 4°C. Spin for 30 s at 15871 ×*g*. Transfer the supernatant to a new tube.
7. Measure the protein content (e.g., by Bradford assay).
8. Dilute the sample with ChIP dilution buffer to a concentration of 200 µg protein/mL in a final volume of 1.2 mL.

3.3. Antibody Incubation

9. Take 1 mL of the quantified chromatin sample and add 2 µg of antibody. A 1:100 dilution in respect to 200 µg of protein is recommended (see also Note 5).
10. Incubate over night at 4°C on a rotary shaker.
11. Reserve 100 µL of sonicated and quantified chromatin for use as non-precipitated input control and freeze it at −80°C.

3.4. Precipitation of the Protein–Antibody Conjugate

12. Any protein-A conjugate carrier may be used at this step (e.g., protein A-sepharose) but in our experience it is more convenient to use paramagnetic particles as protein A carrier as the separation between pellet and supernatant is more easily achieved with these beads. Add 30 µL of Protein A beads for immunoprecipitation (e.g., Dynabeads®, Invitrogen). The chromatin–Dynabeads complexes are incubated for 1 h at 4°C on a rotary shaker.
13. Place the samples on the magnetic rack and carefully remove and discard the liquid phase. It is important not to disturb the chromatin–beads complex and therefore vacuum pumps should be avoided to remove the supernatant.
14. Add 1 mL of low salt washing buffer, invert the tubes gently and place them again on the magnetic rack. Remove and discard the remaining liquid.
15. Add 1 mL of high salt wash buffer, gently invert the tubes twice and place them again on the magnetic rack. Remove and discard the remaining liquid.
16. Add 1 mL of LiCl wash buffer, gently invert the tubes twice and place them again on the magnetic rack. Remove and discard the remaining liquid. When you work with paramagnetic protein A beads *skip* this last step because the *LiCl washing buffer* interferes with the magnetic function of the beads.
17. Wash the chromatin–Dynabeads complex twice in 1 mL of TE buffer; remove and discard the TE buffer.

3.5. Resuspension of the Chromatin

18. Add 50 μL of fresh Resuspension buffer and incubate at 65°C for 15 min.
    (a) *For Fusarium species* use a tenfold diluted Resuspension buffer (resulting in SDS 0.1%, $NaHCO_3$ 0.01 M)
19. Centrifuge at 15871 ×*g* for 2 min. Transfer the *SUPERNATANT* to a new tube. Repeat this step (final volume recovered 100 μL).

3.6. Reverse Crosslinking and Recovery of DNA

20. Add 4 μL NaCl (5 M) and incubate for 4 h at 65°C.
21. Treat 100 μL of the input sample that was reserved at step 11 with 4 μL NaCl (5 M), 4 h at 65°C. These incubation steps (20 and 21) can be extended overnight and afterwards samples can be stored at –20°C.
22. 100 μL samples of the resuspendend chromatin are incubated with 2 μL EDTA (0.5 M), 4 μL 1 M Tris–HCl pH 6.5, and 2 μL Proteinase K (1 mg/mL, MBI) for 1 h at 45°C. This step serves to reverse crosslinking and to digest the associated proteins.
23. DNA is purified from the samples most easily with a PCR purification kit (QIAgen). Other purification procedures may be used after an optimization procedure. Using these kits, the samples are eluted in 100 μL of an 1:10 dilution of EB (QIAgen). The DNA can be stored for up to 2 months at -20°C.
    (a) *For Fusarium species*, the DNA is purified in our hands best by the Cetyltrimethylammonium bromide (CTAB) method instead of PCR purification kit (reason unknown). Start with the addition of 400 μL CTAB buffer (100 mM Tris-HCl pH 8, 20 mM EDTA pH 8, 1.4 M NaCl, 2% (w/v) CTAB, 2% (v/v) β-Mercaptoethanol freshly added). Then the samples are incubated by shaking for 1 h at 65°C. 400 μL CI buffer (Chloroform:Isoamylalcohol 24:1) are added and after mixing of the two phases by vigorously inverting the tubes the samples are centrifugated for 10 min at 15871 ×*g*. 450 μL of the upper phase are transferred to a new tube and incubated with 20 μg RNAse A for 30 min at 37°C. The DNA is precipitated by the addition of 500 μL Isopropanol and centrifugation for 10 min at 15871 ×*g*. The liquid is poured off and the DNA adhering to the tube wall is washed by the addition of 1 mL 75% (v/v) ethanol. After centrifugation for 10 min at 15871 ×*g* the liquid is poured off. Remaining drops are spun down and removed by the use of a pipette. The pellets are dried for 10 min at 65°C (with open lids) and the DNA is eluted by the addition of 50 μL of an 1:10 dilution of EB (QIAgen).

The samples are now ready to be analyzed. Quantitative PCR is the most common method to quantify the relative abundance of a target DNA region. This relative abundance generally directly correlates with the abundance of a given antibody epitope (protein or PTM) at the tested genomic location, but exceptions exist to this general rule (see also Note 6).

3.7. Quantification of the co-precipitated DNA

24. Dilute the input control DNA 100-fold in water.
25. Set up real-time PCRs in duplicate with the diluted input DNA. For primer selection, see Note 7.
26. The concentration of specific DNA fragments can vary considerably depending on which protein was immunoprecipitated. Therefore, empirical testing over several dilutions of the final eluates (step 23) may be necessary to find the optimal template concentration for qPCR. For general transcription factors as such as the nitrogen regulator AreA or histone proteins 5 μL of a non-diluted and 1:10 diluted aliquot of the final eluate has proven as initially suitable concentrations for qPCR (unpublished observations). For pathway-specific regulators such as the nitrate activator NirA, where much less targets are present in the genome, only non-diluted samples of the eluates have resulted in successful amplification (unpublished observation). All quantitative PCR protocols can be used for ChIP qPCR and suppliers may also provide specially adapted programs for read-out of ChIP experiments. For the protocol described here, DNA quantification is carried out on a myIQ cycler (Biorad) using Platinum® SYBR® Green qPCR SuperMix-UDG. Processing of ChIP samples for microarray analysis or massive parallel sequencing is described in the literature (8, 10, 11, 12, 13, 14).

4. Notes

1. *Experimental conditions and repetitions.* Two main forms of fungal culture can be used to study different aspects of metabolism and/or development: submerged cultures or aerial cultures. Since ChIP contains many experimental steps and slight variations in each step may lead to significantly different results in experimental repetitions, it is advisable to work with several technical as well as biological repeats. In order to assess the variability among identical experimental setups at least two biologically independent experiments should be performed and both should be analyzed by three technical repeats. However, not too many samples should be processed at the

same time because incubation times with reagents at the different steps should be as equal as possible.

2. *Optimization of the amount of input material.* In order to assure that crosslinking of the mycelia is homogeneous, it is important that the amount of biomass is in the range given below and as similar as possible in all experiments. For the aerial cultures, spores are inoculated on the top of a cellophane disk that is placed onto the medium of choice (minimal medium, complete medium). For the submerged cultures is important to inoculate the same amount of spores in each flask.
3. *Optimization of chemical crosslinking.* In order to capture the DNA–protein interaction in a cellular context, formaldehyde is used as a crosslinking agent. The time of crosslinking for both submerged and aerial cultures needs to be empirically determined when setting up ChIP for the first time for each organism. This step may be problematic, as the extent of crosslinking may vary from sample to sample. It is crucial that all the samples are in contact with formaldehyde for the same amount of time. Longer or shorter periods of individual samples within an experimental series will result in significant variations of ChIP results.
4. *Optimization of sonication efficiency.* Shearing of the chromatin is a critical step in successful implementation of ChIP. Shearing efficiency can be simply tested by agarose gel electrophoresis. It is recommended that the chromatin is sheared to a size range between 300 and 1,000 bp, with the majority of fragments around 500 bp. If the chromatin is not sheared to this extent, the resolution of the ChIP assay will decrease. For example, if primers are designed for a binding site and control site primers are binding within a region of 1,200 bp, inefficient shearing will result in the detection of binding at both loci. If the sonication produces bigger fragments, try to increase the first power of the sonicator and then to increase the time. When submerged cultures and aerial mycelial samples are compared (e.g., ChIP to elucidate binding of sporulation regulators) it needs to be assured that the average fragment sizes of the shared chromatin are in the same range. We have observed that aerial culture samples require a longer sonication time or increased sonication power.
5. *Optimization of the antibody (Ab) amount used for immunoprecipitation.* The crosslinked DNA–protein complex is specifically immunoprecipitated using an Ab against the protein-of-interest or a tag fused to the protein of choice. The quality and specificity of the Ab contributes immensely to the outcome of the ChIP procedure. Prior to the ChIP experiment it is advisable to test the specificity of the Ab by conducting a Western

analysis of immunoprecipitated samples. This step ensures that the Ab also recognizes the epitope under non-denturating conditions. Optimally, this analysis should include controls such as precipitating with the resin only or precipitating in a genetic background in which the target protein is not present or is not tagged.

6. *Accessibility of the antibody to target epitope.* Consider that the accessibility of the Ab to a protein residing within a large protein complex can vary considerably depending on the composition of a potentially dynamic complex at different conditions, in different strains or at different genomic locations. Thus, measuring low amounts of target protein under a certain condition may not always mean that the protein amount is low but that the accessibility if the Ab to its epitope is low. Also, the presence of one protein modification (e.g., histone H3 phosphorylated at serin 10, H3P10) could have an influence on the interaction of the Ab with the neighboring epitope (e.g., trimethylated H3K9). As Ab preparations differ considerably in specificity and titer, the amount of Ab used to immunoprecipitate the crosslinked protein–DNA complexes needs to be empirically determined every time a new batch of Ab is used. All antibodies which were reported in the published literature to be suitable for ChIP in filamentous fungi are summarized in Table 1.

7. *Optimization of oligonucleotides.* Primers for qPCR should be designed in a way that the amplicon is between 50 and 150 bp long. Double-check whether the targeted regions have extensive secondary structures as this will significantly decrease the efficiency of the PCR. In order to increase the specificity of detection it is advisable to choose primer T_m above 60°C.

Acknowledgments

ChIP development and optimization in *A. nidulans* was supported by grants from the Austrian Science Fund P19731-B11 and Vienna Science and Technology Fund—WWTF—Project LS 09-042. The adaptation to *Fusarium* species was carried out in the frame of the Austrian Science Fund "Special Research AreA" Project SFB–S10003.

References

1. Belden WJ, Loros JJ, Dunlap JC (2007) Execution of the circadian negative feedback loop in Neurospora requires the ATP-dependent chromatin-remodeling enzyme CLOCKSWITCH. Mol Cell 25:587–600
2. Bernreiter A, Ramon A, Fernández-Martínez J, Berger H, Araújo-Bazan L, Espeso EA, Pachlinger R, Gallmetzer A, Anderl I, Scazzocchio C et al (2007) Nuclear export of the transcription factor NirA is a regulatory checkpoint for nitrate induction in *Aspergillus nidulans*. Mol Cell Biol 27:791–802
3. Gonzalez R, Scazzocchio C (1997) A rapid method for chromatin structure analysis in the filamentous fungus *Aspergillus nidulans*. Nucleic Acids Res 25:3955–3956
4. Grimaldi B, Coiro P, Filetici P, Berge E, Dobosy JR, Freitag M, Selker EU, Ballario P (2006) The *Neurospora crassa* white collar-1 dependent blue light response requires acetylation of histone H3 lysine 14 by NGF-1. Mol Biol Cell 17:4576–4583
5. Solomon MJ, Varshavsky A (1985) Formaldehyde-mediated DNA-protein crosslinking: a probe for *in vivo* chromatin structures. Proc Natl Acad Sci USA 82:6470–6474
6. Solomon MJ, Larsen PL, Varshavsky A (1988) Mapping protein-DNA interactions in vivo with formaldehyde: evidence that histone H4 is retained on a highly transcribed gene. Cell 53:937–947
7. Basheer A, Berger H, Reyes-Dominguez Y, Gorfer M, Strauss J (2009) A library-based method to rapidly analyse chromatin accessibility at multiple genomic regions. Nucleic Acids Res 37:e42
8. Pillai S, Chellappan SP (2009) ChIP on chip assays: genome-wide analysis of transcription factor binding and histone modifications. Methods Mol Biol 523:341–366
9. Collas P, Dahl JA (2008) Chop it, ChIP it, check it: the current status of chromatin immunoprecipitation. Front Biosci 13:929–943
10. Ma W, Wong WH (2011) The analysis of ChIP-Seq data. Methods Enzymol 497:51–73
11. Barski A, Zhao K (2009) Genomic location analysis by ChIP-Seq. J Cell Biochem 107: 11–18
12. Jothi R, Cuddapah S, Barski A, Cui K, Zhao K (2008) Genome-wide identification of in vivo protein-DNA binding sites from ChIP-Seq data. Nucleic Acids Res 36:5221–5231
13. Smith KM, Sancar G, Dekhang R, Sullivan CM, Li S, Tag AG, Sancar C, Bredeweg EL, Priest HD, McCormick RF et al (2010) Transcription factors in light and circadian clock signaling networks revealed by genomewide mapping of direct targets for *Neurospora* white collar complex. Eukaryot Cell 9: 1549–1556
14. Smith KM, Phatale PA, Sullivan CM, Pomraning KR, Freitag M (2011) Heterochromatin is required for normal distribution of *Neurospora crassa* CenH3. Mol Cell Biol 31:2528–2542
15. Wolschek MF, Narendja F, Karlseder J, Kubicek CP, Scazzocchio C, Strauss J (1998) In situ detection of protein-DNA interactions in filamentous fungi by *in vivo* footprinting. Nucleic Acids Res 26:3862–3864
16. Bernreiter A, Ramon A, Fernandez-Martinez J, Berger H, Araujo-Bazan L, Espeso EA, Pachlinger R, Gallmetzer A, Anderl I, Scazzocchio C et al (2007) Nuclear export of the transcription factor NirA is a regulatory checkpoint for nitrate induction in *Aspergillus nidulans*. Mol Cell Biol 27:791–802
17. Berger H, Basheer A, Bock S, Reyes-Dominguez Y, Dalik T, Altmann F, Strauss J (2008) Dissecting individual steps of nitrogen transcription factor cooperation in the *Aspergillus nidulans* nitrate cluster. Mol Microbiol 69:1385–1398
18. Bok JW, Chiang YM, Szewczyk E, Reyes-Dominguez Y, Davidson AD, Sanchez JF, Lo HC, Watanabe K, Strauss J, Oakley BR et al (2009) Chromatin-level regulation of biosynthetic gene clusters. Nat Chem Biol 5:462–464
19. Reyes-Dominguez Y, Bok JW, Berger H, Shwab EK, Basheer A, Gallmetzer A, Scazzocchio C, Keller N, Strauss J (2010) Heterochromatic marks are associated with the repression of secondary metabolism clusters in *Aspergillus nidulans*. Mol Microbiol 76:1376–1386
20. Reyes-Dominguez Y, Boedi S, Sulyok M, Wiesenberger G, Stoppacher N, Krska R, Strauss J (2011) Heterochromatin influences the secondary metabolite profile in the plant pathogen *Fusarium graminearum*. Fungal Genet Biol 49(1):39–47
21. Nutzmann HW, Reyes-Dominguez Y, Scherlach K, Schroeckh V, Horn F, Gacek A, Schumann J, Hertweck C, Strauss J, Brakhage AA (2011) Bacteria-induced natural product formation in the fungus *Aspergillus nidulans* requires Saga/Ada-mediated histone acetylation. Proc Natl Acad Sci USA 108: 14282–14287
22. Reyes-Dominguez Y, Narendja F, Berger H, Gallmetzer A, Fernandez-Martin R, Garcia I,

Scazzocchio C, Strauss J (2008) Nucleosome positioning and histone H3 acetylation are independent processes in the *Aspergillus nidulans prnD-prnB* bidirectional promoter. Eukaryot Cell 7:656–663

23. Roze LV, Arthur AE, Hong SY, Chanda A, Linz JE (2007) The initiation and pattern of spread of histone H4 acetylation parallel the order of transcriptional activation of genes in the aflatoxin cluster. Mol Microbiol 66:713–726
24. Roze LV, Miller MJ, Rarick M, Mahanti N, Linz JE (2004) A novel cAMP-response element, CRE1, modulates expression of nor-1 in *Aspergillus parasiticus*. J Biol Chem 279: 27428–27439
25. Ballario P, Vittorioso P, Magrelli A, Talora C, Cabibbo A, Macino G (1996) White collar-1, a central regulator of blue light responses in *Neurospora*, is a zinc finger protein. EMBO J 15:1650–1657
26. Linden H, Macino G (1997) White collar 2, a partner in blue-light signal transduction, controlling expression of light-regulated genes in *Neurospora crassa*. EMBO J 16:98–109

Chapter 17

Targeted Proteomics for Metabolic Pathway Optimization

Tanveer S. Batth, Jay D. Keasling, and Christopher J. Petzold

Abstract

A crucial part of optimization of metabolically engineered organisms is producing balanced levels of pathway proteins. Typically, protein levels are monitored by Western blot analysis; however, application to multiple enzyme pathways can be difficult without unique antibodies for each enzyme in the pathway. Furthermore, it can be time consuming, and cost prohibitive during exploratory stages of pathway design when many different proteins must be monitored simultaneously. We present here a targeted proteomics approach that uses selected-reaction monitoring (SRM) mass spectrometry to quantify multiple proteins in a sample. SRM methods provide high selectivity and high sensitivity to enable rapid quantification of multiple proteins in an engineered pathway regardless of sequence or organism of origin.

Key words: Targeted proteomics, Selected-reaction monitoring mass spectrometry, Protein quantification, Metabolic pathway optimization

1. Introduction

Successful metabolic engineering relies on analytical tools that identify bottlenecks or other problems that limit production. Traditionally, Western blot analysis has been used for selected detection and quantification of proteins. Western blot analysis is useful for many assays of the same protein yet, applying it to quantitative analysis of multi-protein pathways is expensive and challenging. Alternatively, targeted proteomics via selected-reaction monitoring (SRM) mass spectrometry can be used when quantifying multiple proteins in a sample (1). The SRM [also referred to as multiple reaction monitoring (MRM)] technique has been used for small molecule drug metabolism and pharmacokinetic (DMPK) studies since the first generation of triple quadrupole mass spectrometers became commercially available in the early 1990s (2), but has only recently been applied to peptides. Here, we demonstrate

Nancy P. Keller and Geoffrey Turner (eds.), *Fungal Secondary Metabolism: Methods and Protocols*,
Methods in Molecular Biology, vol. 944, DOI 10.1007/978-1-62703-122-6_17, © Springer Science+Business Media, LLC 2012

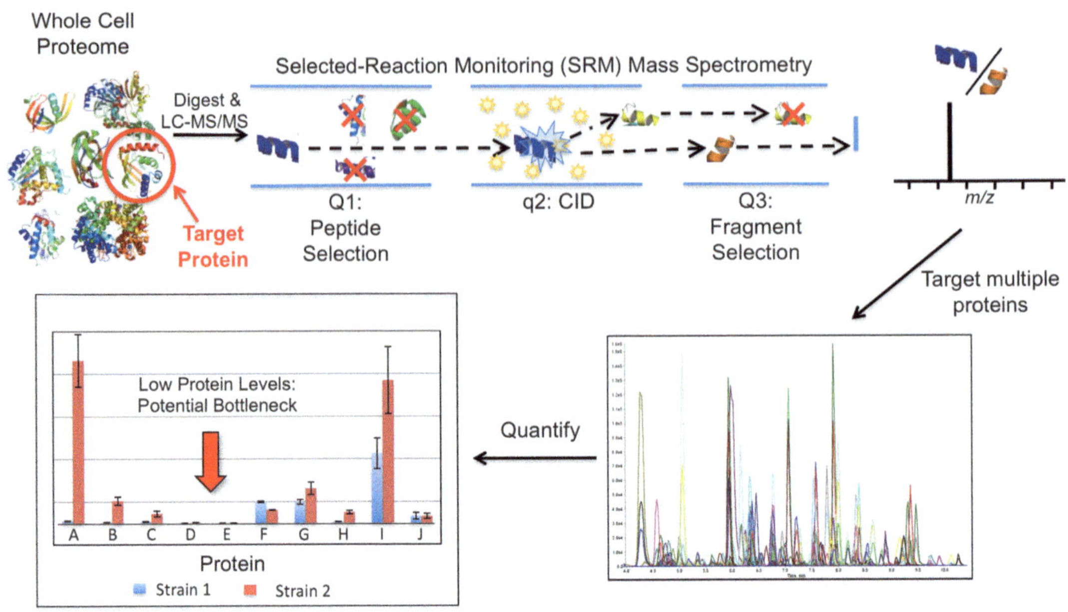

Fig. 1. General schematic of targeted proteomics workflow.

that coupled to liquid chromatographic (LC) separation, SRM-based analysis provides rapid detection and quantification of target proteins (Fig. 1). By using two points of mass selection, the peptide mass and a peptide-specific MS/MS fragment ion, the background signal and noise are greatly reduced. Furthermore, since the entire mass range is not scanned, a significant increase in sensitivity of the target proteins is typically observed for whole cell lysate samples. Over 100 proteins can be detected and quantified in a single 1 h LC-MS/MS run by using scheduled SRM methods (3). This method enables rapid identification of pathway proteins that are poorly produced for subsequent engineering, saving both time and expense. This analysis method is applicable to any system for which a sequenced genome is available and is easily implemented on LC-triple quadrupole mass spectrometers. Efforts are underway to establish a compendium of peptides and transitions for multiple organisms including *Saccharomyces cerevisiae*, human, and mouse (http://www.srmatlas.org) (4). Thus, quantification of proteins native to these organisms, and samples prepared by the methods described below, will be routine.

2. Materials

2.1. Specialized Chemicals

C18 Macro Spin Columns were purchased from Harvard Apparatus (Holliston, MA). For targeted proteomics experiments, mass spectrometric-grade trypsin was obtained from Sigma-Aldrich

(St. Louis, MO). LC-MS Grade water from J.T. Baker (Phillipsburg, NJ) was used to prepare all buffers and solutions. When making buffers and solutions, be sure to follow all appropriate waste disposal regulations guidelines and procedures.

2.2. Buffers and Solutions

1. Resuspension buffer: Dissolve powdered ammonium bicarbonate (AMBIC) in 90% LC-MS grade water/10% LC-MS grade methanol to a final concentration of 100 mM.
2. Trypsin was prepared according to manufacturer's instructions by dissolving lyophilized trypsin in 1 mM hydrochloric acid.
3. Yeast lysis buffer: Dissolve 3.96 g of urea in 5 ml of 1 M triethylammonium bicarbonate (TEAB) and bring the final volume up to 10 ml with LC-MS grade water to give a final concentration of 6 M urea in 500 mM TEAB. Prepare this solution fresh for each use.
4. Reducing agent *tris* (2-carboxyethyl)phosphine (TCEP) should be made at a concentration of 200 mM in LC-MS grade water. Small aliquots (20 μl) should be made to preserve the reducing power of the buffer. While TCEP is preferred, other reducing agents such as dithiothreitol (DTT) or beta-mercaptocthanol (BME) can be also used.
5. Protein alkylation solution: add 36 mg of iodoacetic acid (IAA) to 1 ml of 100 mM sodium hydroxide (NaOH). This will give a 200 mM IAA solution. Other blocking groups such as iodoacetamide or methyl methanethiosulfonate (MMTS) can be used as well.

3. Methods

3.1. *Escherichia coli* Sample Preparation

1. When culturing *E. coli* biomass for proteomics, 15 OD × ml of cells is sufficient for multiple analyses. Fifteen OD × ml of cells is more than enough protein even for high-flow chromatography columns (e.g., 2.1 mm i.d.), but allows for facile lysis and protein extraction (Fig. 2a).
2. Pellet cells by centrifugation at 8,000 × *g* (4°C) for 5 min. After centrifugation, the supernatant should be decanted and cells flash frozen in liquid nitrogen for storage at −80°C until use (Fig. 2b, see Note 1).
3. The cells should be thawed on ice and vortexed to homogenize the cells without adding any buffers or solvents. Take an aliquot (100 μl) of the liquefied cell pellet and pipette into a fresh 1.7 ml tube (see Note 2).
4. Perform protein extraction by using chloroform/methanol precipitation (see Note 3) as described previously (5) with the

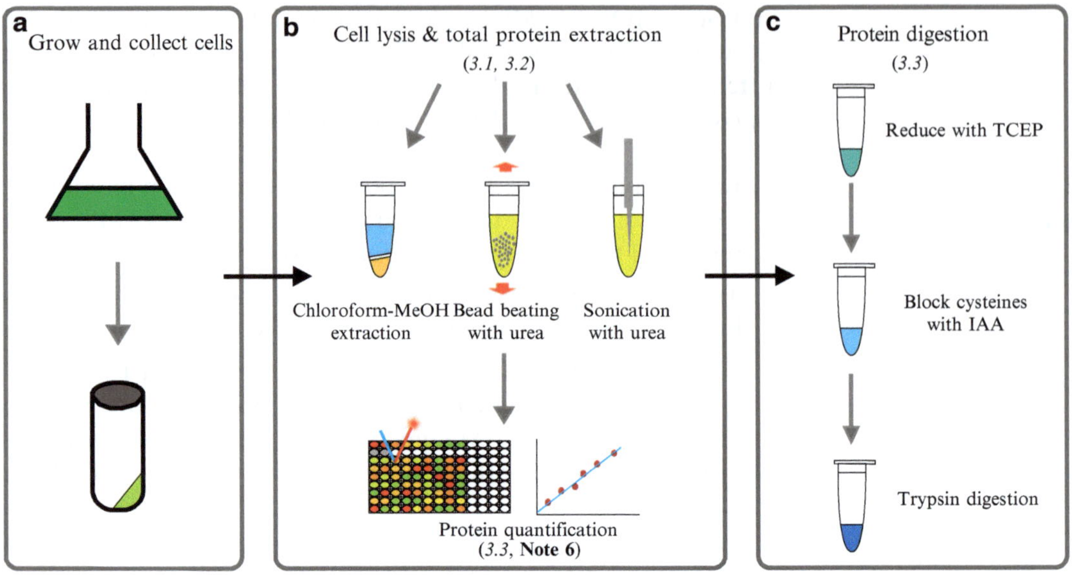

Fig. 2. Workflow schematic for targeted proteomic sample preparation. (**a**) Cell culture and collection for subsequent analysis. (**b**) Cell lysis and total protein extraction from the bacterial and yeast cultures. Several cell lysis methods can be employed to extract protein followed by protein quantification to detect the amount of protein. (**c**) Total protein digestion procedure. The detailed methods can be found in the sections listed in parentheses.

exception that protein pellets be allowed to dry in a vacuum concentrator for 30 min. The procedure is as follows

(a) Pipet 100 μl of cells into a 1.7 ml eppendorf tube.

(b) Add 400 μl of 100% methanol (MeOH); vortex well.

(c) Add 100 μl of chloroform; vortex well.

(d) Add 300 μl of LC-MS grade water; vortex well.

(e) Spin at maximum speed on benchtop centrifuge for 1 min (the phases should separate at this point).

(f) Remove and discard the upper layer (MeOH) carefully without removing protein layer.

(g) Add 300 μl of MeOH to the tube; vortex well.

(h) Spin at maximum speed for 2 min at room temperature (this causes the protein to form a pellet).

(i) Remove the MeOH and chloroform layers.

(j) Dry the protein pellet in a vacuum concentrator (~30 min).

5. Resuspend in AMBIC buffer (see Note 4) for protein quantification.

3.2. Yeast Sample Preparation

1. Pellet ~20 OD × ml yeast samples by centrifugation at 8,000 × *g* (4°C) for 5 min. Decant the supernatant and discard. Flash-freeze the remaining pellet in liquid nitrogen and store at −80°C until further use.

2. For cell lysis, resuspend the cells in 400–700 μl of yeast lysis buffer (6 M urea in 500 mM TEAB solution). Vortex to break pellet and let sit for ~10 min on ice.
3. Transfer 300–500 μl to a fresh 2 ml tube with screw cap; add equal volume of beads to your sample. It is recommended not to completely fill the tube.
4. Bead beat for three cycles of 30–60 s with 1–2 min on ice between each cycle (see Note 5).
5. Spin tubes in a benchtop centrifuge at maximum speed for 30 min at 4°C. Following centrifugation, transfer supernatant to fresh tubes. Samples can be stored at −20°C until protein quantification.

3.3. Protein Quantification and Digestion

1. The protein concentration can be determined by using several routine methods that are available in kits. The accuracy of most protein concentration measurements can be variable, thus it is important to minimize differences in sample handling and to use several dilutions and several replicates when doing the analysis (see Note 6). There are also several different trypsin protocols (6); the most common one for complex mixtures is described here.
2. Once the concentration of the protein mixture is known, take 100 μg of protein from each sample in separate tubes and dilute each to a final concentration of 0.5–1 mg/ml in 100 mM AMBIC buffer (Fig. 2c, see Note 7).
3. To reduce the disulfide bonds in the proteins, add TCEP to a final concentration of 5 mM and incubated at room temperature for 30 min.
4. Once the disulfide linkages are reduced, block them from reforming by adding the 200 mM IAA solution. The amount of IAA used should be 1 μmol/mg protein (i.e., 2 μl of the 200 mM solution per 100 μg protein). Allow the reaction to proceed at room temperature for 30 min in the dark.
5. After alkylation, add trypsin to a final trypsin:total protein ratio of 1:50 and incubate at 37°C for 12–16 h (see Note 8).
6. Following digestion, store samples in −20°C until analysis (see Note 9). Further clean up of the peptide mixtures may be needed if high concentrations of urea are present in the samples. Desalt the samples through a C18 preparative spin column or by other methods (7) if protein extraction was accomplished by using urea or other buffers containing high salts.

3.4. Targeted Proteomics

Typically, targeted proteomics experiments (8) are conducted with a triple quadrupole mass spectrometer operating in SRM mode coupled to a liquid chromatography system. Peptides are normally separated C18 reverse-phase chromatography and for high sensitivity analysis nano-LC systems perform well; however, for most routine analyses, higher flow LC systems are sufficient and more robust. When operating in SRM mode, the triple quadrupole mass spectrometer selects peptide ions in the first quadrupole, fragments the peptide in the second quadrupole, and selects a sequence-specific fragment in the third quadrupole. The dwell time (typically 20–75 ms) for each transition can be optimized to give the greatest sensitivity while still allowing a large number of transitions to be detected in one LC-MS analysis.

3.5. SRM Transition Development

3.5.1. Transition Selection Criteria

When developing SRM assays and picking transitions, certain peptides should be avoided. The following criteria should be applied when selecting peptide transitions.

1. Transitions corresponding to doubly charged peptides are preferred because they are typically the most intense charge state observed.
2. Peptides with cysteine, methionine should not be selected as these residues are susceptible to posttranslational modifications such as disulfide bonds (cysteine) or oxidation (methionine).
3. Doubly charged peptide ions should be greater than 450 *m/z* (peptide MW > 900 Da), smaller peptides risk background ion redundancy, reducing confidence in the transition.
4. Peptides resulting from missed cleavage sites should be avoided because they give poor quantification.
5. y-Series peptide fragment ions are typically intense and good choices.
6. Fragment ions that are greater than the *m/z* of parent peptide are preferred because low mass fragments are less sequence specific.

3.5.2. Empirical Selection

SRM peptide transitions need to be only developed once for a given mass spectrometer (Fig. 3). Empirical selection consists of analyzing a sample to identify a protein of interest via shotgun (nontargeted) proteomics. Shotgun proteomic analysis identifies peptides by sequencing the peptide via collision-induced dissociation (CID) (9, 10). After a peptide is fragmented via CID the mass spectrum consists of peptide fragments of varying intensity (9, 11). In positive ion mode, y-series fragments usually have the highest ion intensity (see Note 10), however, depending on the collision energy that is used, some y-series fragments will be more intense. From the shotgun proteomic experiments good peptide and fragment ion targets for SRM analysis can be optimized as SRM transitions (see Note 11). Only a limited number of transitions should be tested in a single LC-MS run to permit high dwell times (75–100 ms) for added sensitivity. Once the best peptide/fragment ion transitions are identified and optimized the dwell time can be reduced to increase the number of transitions per method.

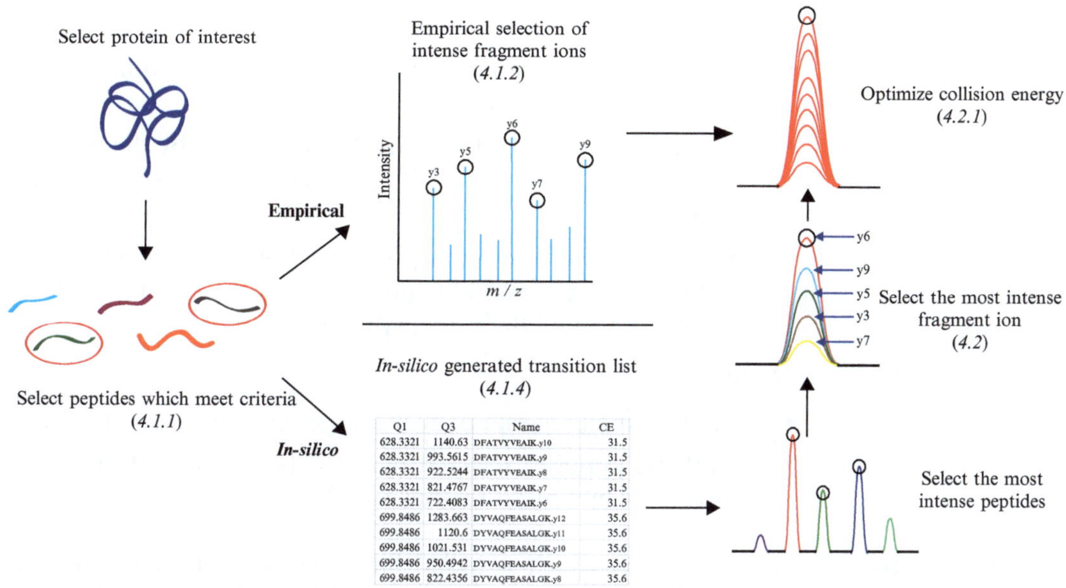

Fig. 3. SRM methods development workflow for empirical and in silico selection of peptide/fragment ion pairs (transitions). Following the selection of the protein of interest the transitions are tested and optimized. The detailed methods can be found in the sections listed in parentheses.

Advantages:

- Able to determine best y-series ions.
- Reduced method development time in determining which peptides ionize preferentially for a given protein.
- Availability of public SRM transition databases such as SRMAtlas ((3, 12), http://www.srmatlas.org).
- Able to determine which peptides are prone to modification that can complicate quantification experiments.

Disadvantages:

- Fails if the peptide is not detected in shotgun proteomic experiments.
- Can be challenging to develop and select transitions for low abundance proteins.
- Increased method development time if performing shotgun proteomic experiment for the first time on a sample containing protein of interest.

3.5.3. Synthetic Peptides

Another faster, yet more expensive approach is the utilization of synthetic peptides (3). Peptides synthesized by using high-throughput peptide synthesis systems or ordered from manufacturers (13, 14) can be infused directly into the mass spectrometer to determine the optimal fragment ions.

Advantages:

- Typically, fast transition optimization without requiring prior observation (see Note 12).
- Able to develop scheduled SRMs (3) quickly by determining LC retention time from the purified peptide standard (see Note 13).
- Absolute quantification is possible by adding set amount of synthetic isotope labeled peptides or by constructing a calibration curve using unlabeled peptides.

Disadvantages:

- Can be expensive especially if a large set of peptides desired.
- Synthetic peptides selected may not be observed from actual sample for biological reasons (see Note 14).

3.5.4. In Silico Selection

Several programs such as MRM Pilot (AB Sciex, Foster City, CA), Skyline (15), and other methods (16–20) can be used to develop a list of transitions. Programs such as Skyline are free to use and manufacturer independent. Similar to empirical selection use of high dwell times is encouraged to increase sensitivity during development.

Advantages:

- User defined in silico protease digest of protein sequence and resulting peptides and fragment masses.
- Peptide and fragment *m/z* generation. Able to select or filer various fragment types (y-series or b-series).
- Able to add and define modifications or heavy-isotope labeling on peptide/residue and determine resulting *m/z* value of peptide and fragment.
- Able to generate transition list and full method export for some manufactures.

Disadvantages:

- Modified peptides are not evident without empirical evidence that can offset quantitative information.
- Unable to determine which transitions yield the best signal and each transition must be tested (see Note 13).
- For larger or multiple set of proteins, multiple different methods across multiple runs will be required to cover each predicted peptide for each protein, increasing method development time.

3.6. SRM Transition Optimization

Once the transitions have been combined into one method the fragment ion choice and collision energy can be optimized. For peptides with several transitions per peptide, there should be overlapping peaks corresponding to multiple y-series ions (see Note 14).

This phenomenon is useful for addressing transitions that produce multiple elution peaks (see Note 15) (21). Selection of 2–3 transitions per peptide is recommended for high quality quantification (see Note 16).

3.6.1. Collision Energy Optimization

After the selecting the most intense transition for a list of peptides, it is recommended to optimize the collision energy (CE) of the transition. Optimizing of peptide collision energies is a key component for optimizing fragment ion signal intensity, this is especially important (and challenging) when dealing with low abundance proteins. Collision energies can vary across instruments, manufacturers, and collision cells, thus a spread of collision energies should be employed to find the CE that produces the most abundant signal. Programs such as Skyline (15) or MRMPilot can generate methods and transition lists that optimize CE as well. The resulting extracted ion chromatograms should produce overlapping peaks corresponding to different collision energies (Fig. 3).

3.6.2. SRM Quantification

As suggested previously, it is advised to pick several different peptides per protein to account for variability that might occur from peptide to peptide. Some peptide transitions will instinctively ionize far greater than others and as a result will have greater peak area. Analysis programs such as *mProphet* (16) are useful for processing large SRM datasets. It is recommended to run samples in replicates, biological and/or technical. The peak areas of all peptides should be summed up to provide a profile for that particular protein of interest. Although relative protein levels are usually good at providing a general translation profile of the proteins in interest, it is recommended not to make any statements on the absolute levels without internal standards or absolute quantification.

4. Notes

1. If the cells are pelleted too tightly thawing them for cell lysis and protein extraction can be inconsistent, this can disrupt homogenization.
2. 1.7 ml Eppendorf tubes are preferred over 2 ml tubes because of the cone shaped bottom, this allows the protein pellet to concentrate in between the chloroform and methanol/water layer.
3. The chloroform/methanol procedure works best on gram-negative and -positive bacteria, the technique does not work well for organisms with more complicated cell walls such as yeast.

4. Adding a small amount of methanol before the AMBIC buffer is recommended as it can further denature and enhance the solubility of hydrophobic proteins. The volume you should add depends on the size of the protein pellet from the extraction but 300 μl of a 100 mM solution with 10% methanol works well for 15 OD × ml of cells. Protein pellets are resuspended by heavy vortexing and by scratching the tubes across a tube rack followed by sonication for 10 min in a sonic bath. A small amount of undissolved solid may not go into solution even after vortexing and tube scratching, be sure to remove that pellet or transfer suspended supernatant to a new tube.
5. Bead beating is highly variable and can be difficult to reproduce, because factors such as bead size, bead beating machine, tube size, sample volume, and timing can cause variability in total amount of protein released. This step may require some optimization depending on the bead beater, bead size, and sample volume. It can be difficult to remove the supernatant and transfer to a fresh tube when using smaller beads, one way to circumvent this is by putting a hole at the bottom of the tube using a hot needle and centrifuging the tube to gather the supernatant.
6. Since the accuracy of the SRM method is dependent on consistent sample loading across samples ensuring that the same amount of protein is present in each load is essential. Addition of an internal standard protein at this point is advised. The DC Protein reagent (BioRad, Hercules, CA) is recommended for protein extracted via the chloroform/methanol protocol. The DC protein assay employs the Lowry method for protein quantification. Lowry-based protein assays employ the reduction of copper ions from amide bonds, thus these methods should not be used with buffers that contain urea because urea will react with copper to give false positives. In this case, a fluorescent assay (22) is recommended. Due to the nature of protein quantification there will be values that might be outliers compared to replications and dilutions. It is also important to verify that absorbance or fluorescence readings for all dilutions are within the standard curve.
7. For yeast sample preparation, be careful to keep the final urea concentration below 1 M for trypsin digestion, as it can inhibit protease activity. Trypsin is active even in large amounts of methanol (up to 50% methanol), thus the addition of up to ~20% MeOH is encouraged to enhance the solubility of hydrophobic peptides. The buffer volume should be adjusted for the addition of reducing buffer, blocking agent, and protease buffer. For nanoLC-MS/MS analysis sample loads of 500 ng–2 μg of total protein are typical, generally, for larger columns and high flow LC systems higher quantities of protein are necessary.

8. There are many different trypsin digestion protocols such as digestion for 4 h in a 37°C water bath or digestion in a 50°C water bath for 2–3 h or even shorter methods which involve the use of microwaves (6, 23). Even though the majority of the peptides (~85%) are formed in less than 4 h the authors recommend overnight digestion at 37°C as a starting point. There are several forms of trypsin available including trypsin that is blocked to reduce self-digestion, thus adjusting the trypsin:total protein ratio may be needed to optimize digestion.
9. Other recommendations include flash freezing tubes with liquid nitrogen or the addition of formic acid, however most trypsin activity is generally stopped at this point. Caution must be paid to final concentration of peptides post elution from spin column and total amount of product and solvent volume.
10. For example, peptide with sequence ARNDCQEG, would result in y-series ion peaks corresponding to mass over charge of ions (+1) such as –RNDCQEG (y7), ––NDCQEG (y6), –––DCQEG (y5) etc.
11. Due to small differences in shotgun and targeted proteomic CID, the most intense y-series fragment ion from shotgun proteomics experiments might be optimal fragment ion for SRM analysis. Thus, testing other y-series ions should be included in the SRM optimization step.
12. Through direct infusion of synthetic peptides, users can quickly determine the most intense fragments and best collision energy without the need to separate on LC or run sample containing peptide.
13. Selecting peptides without prior observations or testing from sample might lead to selecting a peptide which does not ionize well, elute well, or is highly posttranslationally modified (24).
14. Despite the selectivity of SRM transitions, there can be background signal from peptides formed by digestion of native proteins that result in multiple elution peaks for a transition. Synthetic peptides can help identify the chromatography peak corresponding to the peptide of interest (21). Alternatively, the peptide sequence can be validated by using an information-dependent acquisition (IDA) experiment triggered from the SRM signal.
15. Including multiple transitions for a given peptide is useful to identify the retention time of the peptide. This will allow the user to eliminate transitions that generate a double peak in the chromatogram. In a complex mixture there can be a lot of ion redundancy at the peptide *m/z* and the fragment ion as well (21).
16. It is recommended that 2–4 different peptides are selected for each protein for added sensitivity and confidence in the quantification.

Acknowledgments

This work conducted by the Joint BioEnergy Institute was supported by the Office of Science, Office of Biological and Environmental Research, of the U.S. Department of Energy under Contract No. DE-AC02-05CH11231.

References

1. Lange V, Picotti P, Domon B, Aebersold R (2008) Selected reaction monitoring for quantitative proteomics: a tutorial. Mol Syst Biol 4:222
2. Baillie TA (2008) Metabolism and toxicity of drugs. Two decades of progress in industrial drug metabolism. Chem Res Toxicol 21(1): 129–137
3. Picotti P, Rinner O, Stallmach R et al (2010) High-throughput generation of selected reaction-monitoring assays for proteins and proteomes. Nat Methods 7(1):43–46
4. Deutsch EW, Lam H, Aebersold R (2008) PeptideAtlas: a resource for target selection for emerging targeted proteomics workflows. EMBO Rep 9(5):429–434
5. Wessel D, Flügge UI (1984) A method for the quantitative recovery of protein in dilute solution in the presence of detergents and lipids. Anal Biochem 138:141–143
6. Sun W, Gao S, Wang L et al (2006) Microwave-assisted Protein Preparation and Enzymatic Digestion in Proteomics. Mol Cell Proteomics 5(4):769–776
7. Yeung Y-G, Nieves E, Angeletti RH, Stanley ER (2008) Removal of detergents from protein digests for mass spectrometry analysis. Anal Biochem 382(2):135–137
8. Bluemlein K, Ralser M (2011) Monitoring protein expression in whole-cell extracts by targeted label- and standard-free LC-MS/MS. Nat Protoc 6(6):859–869
9. Hunt DF, Yates DR, Shabanowitz J et al (1986) Protein sequencing by tandem mass spectrometry. Proc Natl Acad Sci USA 83:6233–6237
10. Prakash A, Tomazela DM, Frewen B et al (2009) Expediting the development of targeted SRM assays: using data from shotgun proteomics to automate method development. J Proteome Res 8(6):2733–2739
11. Papayannopoulos IA (1995) The interpretation of collision-induced dissociation tandem mass spectra of peptides. Mass Spectrom Rev 14:49–73
12. Picotti P, Lam H, Campbell D et al (2008) A database of mass spectrometric assays for the yeast proteome. Nat Methods 5(11):913–914
13. Wenschuh H, Volkmer-Engert R, Schmidt M et al (2000) Coherent membrane supports for parallel microsynthesis and screening of bioactive peptides. Biopolymers 55(3):188–206
14. Kirkpatrick DS, Gerber SA, Gygi SP (2005) The absolute quantification strategy: a general procedure for the quantification of proteins and post-translational modifications. Methods 35(3):265–273
15. MacLean B, Tomazela DM, Shulman N et al (2010) Skyline: an open source document editor for creating and analyzing targeted proteomics experiments. Bioinformatics 26(7):966–968
16. Reiter L, Rinner O, Picotti P et al (2011) mProphet: automated data processing and statistical validation for large-scale SRM experiments. Nat Methods 8(5):430–435
17. Fusaro VA, Mani DR, Mesirov JP, Carr SA (2009) Prediction of high-responding peptides for targeted protein assays by mass spectrometry. Nat Biotechnol 27(2):190–198
18. Unwin RD, Griffiths JR, Whetton AD (2009) A sensitive mass spectrometric method for hypothesis-driven detection of peptide post-translational modifications: multiple reaction monitoring-initiated detection and sequencing (MIDAS). Nat Protoc 4(6):870–877
19. Sherwood CA, Eastham A, Lee LW et al (2009) MaRiMba: a software application for spectral library-based MRM transition list assembly. J Proteome Res 8(10):4396–4405
20. Mead JA, Bianco L, Ottone V et al (2009) MRMaid, the web-based tool for designing multiple reaction monitoring (MRM) transitions. Mol Cell Proteomics 8(4):696–705
21. Sherman J, McKay MJ, Ashman K, Molloy MP (2009) How specific is my SRM?: the issue of precursor and product ion redundancy. Proteomics 9(5):1120–1123

22. Mackintosh JA, Veal DA, Karuso P (2005) Fluoroprofile, a fluorescence-based assay for rapid and sensitive quantitation of proteins in solution. Proteomics 5(18):4673–4677

23. Wisniewski JR, Zougman A, Nagaraj N, Mann M (2009) Universal sample preparation method for proteome analysis. Nat Methods 6(5):359–362

24. Mallick P, Schirle M, Chen SS et al (2007) Computational prediction of proteotypic peptides for quantitative proteomics. Nat Biotechnol 25(1):125–131

Chapter 18

Investigation of In Vivo Protein Interactions in *Aspergillus* Spores

Kwang Cheol Jeong and Jae-Hyuk Yu

Abstract

Understanding in vivo protein–protein interactions is critical to dissect precise functions of the regulatory proteins of fungal secondary metabolites. As many fungi differentially produce a diverse array of secondary metabolites during their lifecycle, it is important to understand the cell-type specific regulation of secondary metabolism. However, due to the difficulty of sample preparation of biologically active proteins in fungal spores, protein–protein interaction studies have been generally restricted. While some outstanding studies revealed protein–protein interactions of selected regulators, including the *velvet* proteins in vegetative cells, a detailed protocol for investigating the protein–protein interactions in the fungal spores has not yet been reported. Here, we describe a working protocol for the purification and identification of interacting protein partners of the spores of *Aspergillus nidulans* employing the VelB protein as an example.

Key words: *Aspergillus*, Protein interaction, Novel regulator, FLAG-tag, Spores

1. Introduction

One of the most intriguing aspects of the fungal lifecycle is the production of secondary metabolites that are regulated by internal and external inputs (signals). Understanding the gene regulation mechanism of secondary metabolite production has been a long-term goal in fungal science, and a number of gene regulation mechanisms are now known (1–5). Recent studies have revealed key upstream regulators called the *velvet* regulators (VeA, VelB, VosA, and VelC) that not only govern morphological development but also regulate secondary metabolism in fungi (6–8). Moreover, LaeA has been shown to be a global regulator of secondary metabolism as it plays an essential role in the production of various secondary metabolites including sterigmatocystin, lovastatin,

Nancy P. Keller and Geoffrey Turner (eds.), *Fungal Secondary Metabolism: Methods and Protocols*, Methods in Molecular Biology, vol. 944, DOI 10.1007/978-1-62703-122-6_18, © Springer Science+Business Media, LLC 2012

penicillin, gliotoxin, and hyphal pigments in *Aspergillus* (9). In the nucleus the LaeA protein interacts with VeA the founding member of the *velvet* proteins, and VeA interacts with VelB, forming a trimeric complex (6, 7). In addition, VelB can form a homodimer or heterodimer with VosA (8). Importantly, the formation of velvet–LaeA complexes are thought to be highly dynamic and the resulting complexes would control different sets of biological processes, e.g., trehalose biogenesis or secondary metabolism, depending on the developmental stages and cell types.

Understanding protein–protein interactions leads to an advanced understanding of biological functions of a given protein by providing *in vivo* biochemical evidence. Several studies have shown protein–protein interaction data in the vegetative fungal cells using S-tag or tandem affinity purification (TAP) tag (7, 8, 10). Fluorescence protein tagging also has been broadly used for the localization of proteins in cells (6). However, the protein–protein interaction studies in fungal spores have been highly restricted in due to the difficult steps involved in the sample preparation of biologically active proteins. Moreover, the methods mentioned above still have some disadvantages, including high level of nonspecific background affinity, and/or disrupting the activity of other native proteins and protein–protein interactions. Here, we describe a detailed protocol for protein–protein interactions in asexual (conidia) and sexual (ascospores) spores of *Aspergillus nidulans* using FLAG-tag. The VelB protein is used as an example.

2. Materials

2.1. Spore Harvest and Lysis

1. Spore harvest buffer: PBS with 0.02% Triton X-100. Dissolve the following in 800 ml distilled H_2O. Add 8 g of NaCl, 0.2 g of KCl, 1.44 g of Na_2HPO_4, 0.24 g of KH_2PO_4 and adjust pH to 7.4. Add 0.2 ml of Triton X-100. Adjust volume to 1 l with additional distilled H_2O. Sterilize by autoclaving.
2. Spore wash buffer: 1× PBS without Triton X-100.
3. Spore lysis buffer: 50 mM Tris-HCl (pH 7.5), 150 mM NaCl, 5 mM EDTA, 15 mM EGTA, 0.02% Triton X-100, 20% Glycerol, 1 mM PMSF, and 1× Protease inhibitor cocktail. Prepare working solution from stock solution by dilution (see Note 1).
4. PMSF (200 mM) stock solution: dissolve 0.035 g of PMSF in 1 ml of 100% ethanol. Store at −20°C.
5. Protease inhibitor cocktail (25×): Dissolve one tablet of Protease inhibitor cocktail (Roche Cat #116974981001) in 2 ml of distilled water.

6. Zirconia/silica bead (0.5 mm, Research Products International Corp, Cat No. 9834).
7. 2 ml Screw cap microcentrifuge tubes with O-ring cap.
8. Mini-Beadbeater (Biospec, Cat. No. 607).
9. Dounce homogenizer (Norman D Erway Glass).

2.2. Pull-Down Using Anti-FLAG M2 Affinity Agarose

1. Anti-FLAG: M2 affinity agarose (Sigma Cat #A2220) (see Note 2).
2. 4× SDS-PAGE sample buffer: 0.3 M Tris–HCl (pH 6.8), 10% SDS, 25% B-mercaptoethanol, 45% glycerol, 0.1% bromophenol blue.
3. Sarmix® MR1 tube rotator (Sarstedt).

2.3. SDS Polyacrylamide Gel Components

1. *N,N,N',N'*-Tetramethyl-ethylenediamine (TEMED) (Sigma, T9281). Store at 4°C.
2. Resolving gel: Mix 3.33 ml of 30% Acrylamide/Bis-acrylamide (37.5:1), 3.75 ml of 1 M Tris-HCl (pH 8.8), 2.8 ml of distilled water, 100 μl of 10% SDS, 50 μl of 10% Ammonium sulfate (APS), and 5 μl of TEMED.
3. Stacking gel: Mix 1.33 ml of 30% Acrylamide/Bis-acrylamide (37.5:1), 1.25 ml of 1 M Tris-HCl (pH 6.8), 7.3 ml of distilled water, 100 μl of 10% SDS, 50 μl of 10% APS, and 10 μl of TEMED.
4. APS: 10% solution in distilled H_2O.
5. SDS-PAGE gel running buffer (10×, 1 l): dissolve 30 g of Tris base, 144 g of glycine, 10 g of SDS and adjust volume to 1 l with additional distilled H_2O.
6. Coomassie staining buffer: Dissolve 1.0 g of Coomassie powder-Brilliant Blue R-250 (Sigma Chemical Company, St. Louis, MO, USA) in 500 ml of distilled water in 1 l beaker and add 400 ml of methanol and 100 ml of acetic acid.
7. Destaining solution: Mix 200 ml of methanol, 100 ml of acetic acid, and 700 ml of distilled water in 1 l baker.

3. Methods

3.1. Preparation of the Protein Samples from the Asexual Spores Under Native Conditions

1. Grow WT and FLAG-tagged strains in appropriate media. Typically five plates of 2-day cultures are enough for 2–3 binding experiments. Scrape spores using cell scrappers from the plate using 10 ml of spore harvest buffer and pass through miracloth to remove mycelia and to collect spores.

2. Centrifuge 10 min at 1,700 × *g* at 4°C using HS-4 rotor (Sorval).
3. Remove the supernatant and add 10 ml of cold spore harvest buffer.
4. Sonicate on ice (use a sonicator equipped with a microtip). Burst spores for 300 s (50% pulse) at 100–150 W to remove rodlets (see Note 3).
5. Centrifuge 10 min at 1,700 × *g* at 4°C using HS-4 rotor and remove the supernatant.
6. Resuspend the spores in 20 ml of cold spore wash buffer and centrifuge 10 min at 1,700 × *g* at 4°C. Remove the supernatant and add 20 ml of cold spore wash buffer. Centrifuge 10 min at 1,700 × *g* at 4°C and remove the supernatant.
7. Resuspend the spores in 2–3 ml of spore lysis buffer and transfer 1.5 ml sample into a 2 ml O-ring screw cap microcentrifuge tube.
8. Add 200–300 μl of zirconia/silica beads to the tubes and break the cells.
9. Break the cells for 1 min using a bead beater and then keep the tube on ice for 10 min to cool it down. Heat generated during bead beating will denature proteins. Break the cells for an additional minute.
10. Centrifuge the lysate at 10,000 × *g* for 10 min at 4°C to precipitate the cell debris. Transfer the supernatant in a new tube and add 1.0 ml of spore lysis buffer to the cell debris pellet and resuspend.
11. Respin once at 10,000 × *g* for 10 min and save the supernatant. Combine the supernatant in 2 ml of tube and centrifuge at 10,000 × *g* for 20 min to remove cell debris.
12. Pipette the clear supernatant into a new 2 ml tube and store on ice.
13. Take 10 μl aliquots and measure protein concentration.
14. Take 30 μl aliquots from each, add 10 μl of 4× Sample Buffer, and store at −20°C.

3.2. Preparation of the Protein Samples from Ascospores Under Native Conditions

1. Grow the WT and FLAG-tagged strains in appropriate media. Grow 15 plates of 7-day culture (wrapped in aluminum foil) for one binding experiment. Scrape the asexual spores and cleistothecium from the plates using spore harvest buffer and pass through miracloth to remove asexual spores.
2. Scrape cleistothecium on the miracloth using a sterile spatula and resuspend in 10 ml of spore harvest buffer.
3. Centrifuge 5 min at 770 × *g* at 4°C using HS-4 rotor (Sorval).
4. Remove the supernatant and add 10 ml of cold spore harvest buffer.

5. Break the cleistothecium using dounce homogenizer and pass through miracloth to remove mycelia and collect ascospores. You will get red ascospores (see Note 4).
6. Centrifuge 10 min at 1,700 × *g* at 4°C using HS-4 rotor and remove the supernatant.
7. Resuspend the ascospores in 1.5 ml of spore lysis buffer and transfer into a 2-ml O-ring screw cap microcentrifuge tube.
8. Add 200–300 μl of zirconia/silica beads to the tubes and break the cells.
9. Break the cells for 1 min using a bead beater and then keep the tube on ice for 10 min to cool it down. Heat generated during bead beating will denature proteins. Break the cells 1 more min.
10. Centrifuge the lysate at 10,000 × *g* for 10 min at 4°C to pellet cell debris. Save the supernatant in new tube and add 1.0 ml of spore lysis buffer and resuspend. Respin one time at 10,000 × *g* for 10 min and save the supernatant. Combine the supernatant in 2 ml of tube and centrifuge at 10,000 × *g* for 20 min to remove cell debris.
11. Pipette the clear supernatant into a new 2 ml tube and store on ice.
12. Take 30 μl aliquots from each, add 10 μl of 4× Sample Buffer, and store at −20°C.
13. Take 10 μl aliquots and measure protein concentration.

3.3. Pull-Down Using Anti-FLAG Affinity Gel Under Native Conditions

1. Add 40 μl of the 50% anti-FLAG M2 affinity agarose slurry to the cleared lysate (total of 10 mg protein) and incubate for 3 h in the cold room on the tube rotator (see Note 5).
2. After incubation, spin down anti-FLAG agarose 2 min at 1,500 × *g*.
3. Take 30 μl aliquots from the supernatant; add 10 μl of 4× Sample Buffer (Unbound fraction).
4. Add 1 ml of spore lysis buffer and incubate on the Nutator 5 min (Wash 1). Centrifuge 2 min at 1,500 × *g*.
5. Take 30 μl aliquots from the supernatant; add 10 μl of 4× Sample Buffer (Wash 1 fraction).
6. Add 1 ml of lysis buffer and incubate on the Nutator (Wash 2). Centrifuge 2 min at 1,500 × *g*.
7. Take 30 μl aliquots from the supernatant; add 10 μl of 4× Sample Buffer (Wash 2 fraction).
8. Add 1 ml of lysis buffer and incubate on the Nutator (Wash 3). Centrifuge 2 min at 1,500 × *g*.
9. Take 30 μl aliquots from the supernatant; add 10 μl of 4× Sample Buffer (Wash 3 fraction).

10. Add 40 μl of 4× sample buffer (Elution fraction).
11. Analyze by SDS-PAGE. Boil samples for 5 min and load 20 μl of samples (see Note 6).

3.4. 10% SDS-PAGE Gel Electrophoresis

1. To make two resolving gels, mix 3.33 ml of 30% acrylamide mix, 3.75 ml of 1 M Tris-HCl (pH 8.8), 2.8 ml of distilled water, 100 μl of 10% SDS, 50 μl of 10% APS, and 5 μl of TEMED in 15 ml of conical tube. Pour the gel solution in Bio-Rad mini gel with 1 mm spacer before acrylamide is solidified. Leave space for stacking gel and add about 200 μl of water saturated isobutanol to prevent gel contacting atmospheric oxygen and level the resolving gel. Let the resolving gel at least 15 min to be solidified. When the resolving gel is solidified, pour off the isobutanol and wash the gel surface with distilled water.
2. To prepare separating gel, mix 1.33 ml of 30% acrylamide mix, 1.25 ml of 1 M Tris-HCl (pH 6.8), 7.3 ml of distilled water, 100 μl of 10% SDS, 50 μl of 10% APS, and 10 μl of TEMED in 15 ml of conical tube. Insert comb (1 mm thickness) in the separating gel gently. Let the gel sit for at least 15 min to solidify.
3. Prepare protein samples to run by heating for 5 min at 95°C for 5 min and then centrifuge for 5 min at maximum speed.
4. Load the samples in the well of stacking gel (10 μl/well) and electrophorese at 15 mA/gel constantly, so if you run two gels use 30 mA until the blue dye front reaches the bottom of the separating gel.
5. After gel running, separate the gel plates using a spatula and submerge the gels in Coomassie stain for 30 min with gentle shaking to stain the protein bands.
6. Pour off Coomassie stain and rinse the gels with destaining solution.
7. Cut bands, which are specifically detected on FLAG-tagged lane using razor blades. Use razor blade for one band to prevent cross contamination of the protein samples.
8. Transfer the bands in 1.5 ml of tubes and analyze the samples using MALDI-TOF mass spectrometer.

4. Notes

1. Prepare fresh working solution each time as PMSF and protease inhibitor cocktail are unstable.
2. As the anti-FLAG affinity gel is stored in 50% glycerol with buffer, make sure to remove glycerol just prior to use and equilibrate it with the lysis buffer. To equilibrate the resins,

take 40 μl of resin and mix it with 1 ml of the lysis buffer, and then rotate on the shaker for 1 h. Centrifuge the resins at low speed (1,500 × *g*) for 2 min and remove the supernatant. Resuspend the resins in 40 μl of lysis buffer. Do not exceed 1,500 × *g*, it may break the agarose.

3. Because the fungal spore rodlet is very hydrophobic, it interrupts protein–protein interaction during binding experiments. Removing the rodlets in the spore surface is the most critical point of getting successful results.
4. It is not necessary to sonicate ascospores as they lack rodlet on the surface.
5. Do not incubate samples more than 3 h because protease inhibitor is no longer active in 3 h, resulting in protein degradation.
6. To find specific proteins that bind to the FLAG-tagged proteins, make sure to load WT samples to differentiate the bands that exist only on the FLAG-tagged fraction. If it can be detected visually in the Coomassie staining, the band can be analyzed by MALDI-TOF.

Acknowledgments

We express special thanks to Hee Soo Park, Kwang-Soo Shin, Young Hwan Kim for their technical assistance. This work was supported by the National Science Foundation (IOS-0950850) and UW Food Research Institute grants to JHY.

References

1. Hoffmeister D, Keller NP (2007) Natural products of filamentous fungi: enzymes, genes, and their regulation. Nat Prod Rep 24: 393–416
2. Shwab EK, Keller NP (2008) Regulation of secondary metabolite production in filamentous ascomycetes. Mycol Res 112:225–230
3. Yu J-H (2006) Heterotrimeric G protein signaling and RGSs in *Aspergillus nidulans*. J Microbiol 44:145–154
4. Yu J-H, Mah J-H, Seo J-A (2006) Growth and developmental control in the model and pathogenic aspergilli. Eukaryot Cell 5:1577–1584
5. Yu J-H (2010) Regulation of development in *Aspergillus nidulans* and *Aspergillus fumigatus*. Mycobiology 38:229–237
6. Ni M, Yu JH (2007) A novel regulator couples sporogenesis and trehalose biogenesis in *Aspergillus nidulans*. PLoS One 2:e970
7. Bayram O, Krappmann S, Ni M, Bok JW, Helmstaedt K, Valerius O, Braus-Stromeyer S, Kwon NJ, Keller NP, Yu JH, Braus GH (2008) VelB/VeA/LaeA complex coordinates light signal with fungal development and secondary metabolism. Science 320:1504–1506
8. Sarikaya Bayram O, Bayram O, Valerius O, Park HS, Irniger S, Gerke J, Ni M, Han KH, Yu JH, Braus GH (2010) LaeA control of velvet family regulatory proteins for light-dependent development and fungal cell-type specificity. PLoS Genet 6:e1001226
9. Bok JW, Keller NP (2004) LaeA, a regulator of secondary metabolism in *Aspergillus* spp. Eukaryot Cell 3:527–535
10. Liu HL, Osmani AH, Ukil L, Son S, Markossian S, Shen KF, Govindaraghavan M, Varadaraj A, Hashmi SB, De Souza CP, Osmani SA (2010) Single-step affinity purification for fungal proteomics. Eukaryot Cell 9:831–833

Chapter 19

Purification of a Vesicle–Vacuole (V) Fraction from *Aspergillus*

Anindya Chanda, Ludmila V. Roze, and John E. Linz

Abstract

Recent studies conducted in our laboratory demonstrate that *Aspergillus parasiticus* synthesizes and stores aflatoxin in transport vesicles and endosomes. Proteomics data suggest that enzymes involved in the synthesis of other secondary metabolites as well as enzymes involved in response to heat, osmotic, and oxidative stress also localize to these subcellular organelles. In order to better understand how cells integrate the regulation and function of secondary metabolite biosynthesis and stress response, it is important to understand the composition and function of the membrane-bound organelles that house this biosynthetic machinery. Isolation of vesicles, endosomes, and vacuoles (V fraction) is, therefore, an essential method to study secondary metabolism in *A. parasiticus* at the cellular level. Here, we describe a "one-step density gradient" method for purification of a highly heterogeneous cell fraction consisting of transport vesicles, endosomes, and vacuoles from protoplasts prepared from *A. parasiticus* cells harvested during aflatoxin synthesis.

Key words: Protoplast, Vesicle, Endosome, Vacuole, One-step density gradient, Aflatoxin, Secondary metabolism, *Aspergillus parasiticus*

1. Introduction

Vesicles and vacuoles are known to contain enzymes associated with secondary metabolism in plants and fungi. Examples include enzymes involved in the biosynthesis of alkaloids (1, 2) and flavonoids (1, 3) in plants and the non-ribosomal peptide cyclosporin (4), the β-lactam antibiotic penicillin (5), and the polyketide aflatoxin (6–8) in filamentous fungi. Our recent studies indicate that these organelles play a functional role in the biosynthesis of secondary metabolites apart from their proposed role in sequestering the enzymes to protect host cells from self-toxicity

Nancy P. Keller and Geoffrey Turner (eds.), *Fungal Secondary Metabolism: Methods and Protocols*,
Methods in Molecular Biology, vol. 944, DOI 10.1007/978-1-62703-122-6_19, © Springer Science+Business Media, LLC 2012

(1). In order to elucidate their function in secondary metabolism, we developed a method to isolate a pure fraction containing transport vesicles, endosomes, and vacuoles.

Protocols to isolate vesicles and vacuoles from fungi (4, 5, 9–14) and plants (15–19) generally involve multistep flotation, density gradients, and/or differential centrifugation procedures. However our studies with *Aspergillus parasiticus*, a fungus that synthesizes the secondary metabolite and carcinogen aflatoxin, suggest that the high degree of size and density heterogeneity of these subcellular organelles in *A. parasiticus* makes it impractical to directly apply previously published procedures.

We describe here a "one-step density gradient" method for purification of a cell fraction that contains heterogeneous transport vesicles, endosomes, and vacuoles from *A. parasiticus* (20). The method includes the following steps: (1) preparation of protoplasts from mycelia; (2) release of vesicles, endosomes, and vacuoles from purified protoplasts; and (3) fractionation of the subcellular organelles using the "one-step density gradient." We successfully used this method to perform proteomic analysis (manuscript in review), to conduct feeding studies (21), and to functionally link these subcellular organelles with aflatoxin synthesis (21).

2. Materials

2.1. Buffers

Protoplast buffer. 1.2 M $MgSO_4$ and 10 mM KH_2PO_4. Adjust solution to pH 5.8 with 1 N NaOH with continuous stirring, filter sterilize, and store at 4°C (see Note 1).

Protoplast solution. Prepare this solution just prior to protoplast isolation. Add 50 mg each of Driselase and Lysing Enzymes (from Sigma) per ml of protoplast buffer. Filter sterilize the solution before use. Determine the volume of the solution used to prepare protoplasts based on the wet weight (explained under methods) of the mycelia; use 1 ml of protoplast solution per 40 mg (wet weight) of mycelia (see Note 2).

Protoplast separation buffer. 0.6 M sorbitol, 100 mM Tris–HCl (pH 7.0; use 1 N NaOH). Autoclave and store at 4°C.

Protoplast storage buffer ("Protobuffer"). 1.2 M sorbitol, 10 mM Tris–HCl (pH 7.5; use 1 N NaOH). Autoclave and store at 4°C.

Protoplast lysis solution. 0.6 M sorbitol, 10 mM Tris–HCl, 0.025% (vol/vol) Triton-X 100 (pH 7.5). Prepare solution just prior to use. Dilute protobuffer 1:1 with 10 mM Tris–HCl (pH 7.5). Then add 0.025% (vov/vol) Triton-X 100. A 500 ml stock protoplast lysis solution can also be prepared and stored at 4°C.

Sucrose cushion. 3 M sucrose, 1.2 M sorbitol, 10 mM Tris–Cl, pH 7.5. Prepare solution in a 50 ml disposable conical centrifuge tube. Add 27 g sucrose, fill the tube to the 50 ml mark with protobuffer, and dissolve using a vortex mixer. The sucrose may take up to 2 h to dissolve. Prepare the solution just as you initiate the procedure and it will be ready when needed. Store solution at 4°C.

2.2. Equipment

1. Refrigerated benchtop clinical centrifuge equipped with a swinging bucket rotor (Sorvall Legend RT, Kendro Laboratory Products, Osterode, Germany).
2. Clinical centrifuge with swinging bucket rotor (Model 59A, Fisher Scientific, Pittsburg, PA).

2.3. Other Accessories

1. Miracloth (Calbiochem, San Diego, CA).
2. Cheese cloth.
3. Nylon mesh (Pore size 30 μm, SpectrumLabs, Rancho Dominguez, CA).
4. 50 ml conical disposable centrifuge tubes (Corning, NY).
5. Transfer pipettes (5 ml capacity).
6. Ice bucket with ice.
7. Funnel.
8. Spatula.
9. Paper towels.
10. 0.45 μm filters (Millex, Millipore, Billerica, MA) and syringes 20 ml, 60 ml (Becton Dickinson, Franklin Lakes, NJ) for filter sterilization.
11. Forceps.
12. Balance to measure wet weight of mycelia.
13. Shaking water bath incubator.

3. Methods

Carry out all procedures at room temperature (RT) unless otherwise specified. The flowchart of the method is described in Fig. 1.

3.1. Protoplast Preparation

1. Inoculate 100 ml YES growth medium in a 250 ml Erlenmeyer flask at 10^5 conidiospores per ml and incubate with shaking at 150 rpm, in the dark, with five glass beads, at 30°C, for 36 h.
2. Harvest mycelia by filtration through a miracloth-lined funnel. Do not squeeze mycelia during filtration (see Fig. 2a).

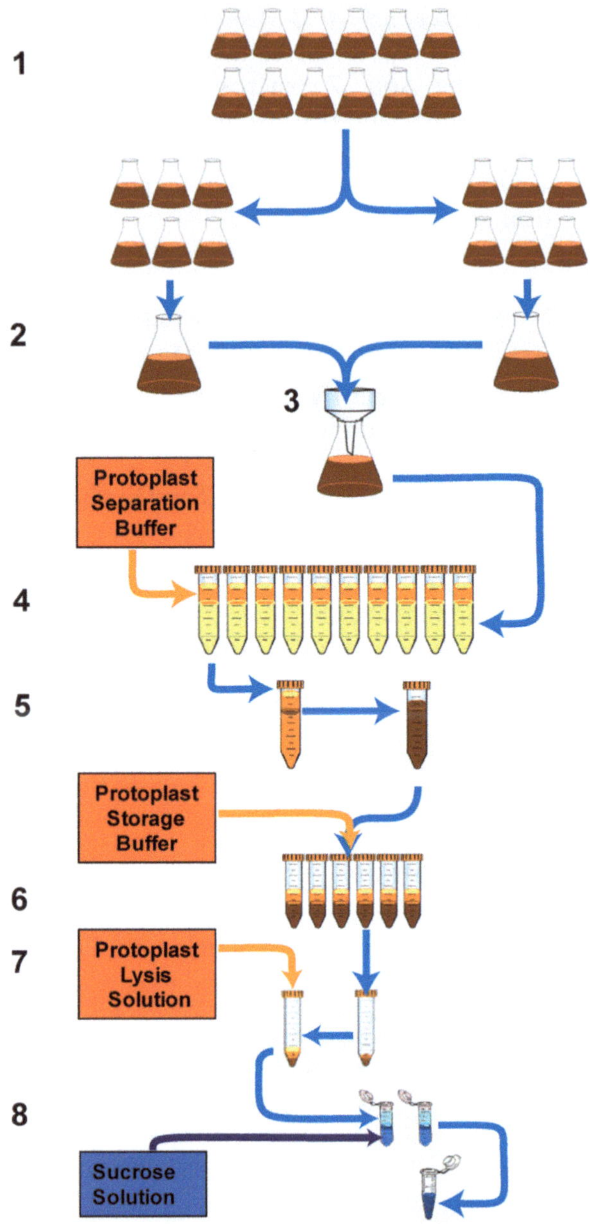

Fig. 1. Flowchart illustrating the entire procedure of the isolation of a pure V-fraction for our routine experiments. (*1*) Grow *Aspergillus parasiticus* culture for 36 h, 30°C. (*2*) Harvest mycelia into two flasks and carry out protoplasting for 5–7 h, 30°C. (*3*) Filter protoplasts into a single flask, RT. (*4*) Aliquot protoplast suspension (25 ml) into 50 ml conical centrifuge tubes and overlay 15 ml of Protoplast Separation Buffer onto each aliquot. Centrifuge for 15 min at 1,500 × *g*, RT. (*5*) Just above the interface, collect 3 ml of protoplast suspension (visible as a "fuzzy band") from each tube and combine collected protoplasts in one tube. (*6*) Aliquot protoplast suspension (5 ml) into 50 ml tubes and add 5 ml of Protoplast Storage Buffer into each tube, mixing gently; centrifuge for 10 min at 1,000 × g, RT. During centrifugation prepare sucrose cushion: Add 0.9 ml of High-Density Sucrose into each of the two 2 ml microcentrifuge tubes; keep at RT. (*7*) Remove supernatant saving approximately 0.1 ml and collect the pellets into a 50 ml tube on ice. Adjust volume to 1 ml with Protoplast Storage Buffer. Add 1 ml of Protoplast Lysis Solution, mix thoroughly by pipetting up and down, and perform lysis for 10 min on ice. (*8*) Carefully layer 1 ml of lysed protoplasts on 0.9 ml of sucrose cushion (two tubes in total) and centrifuge at 3,000 × *g* for 30 min, RT. Take 100–150 μl of a thin layer of vesicles/vacuoles located just above the visible band.

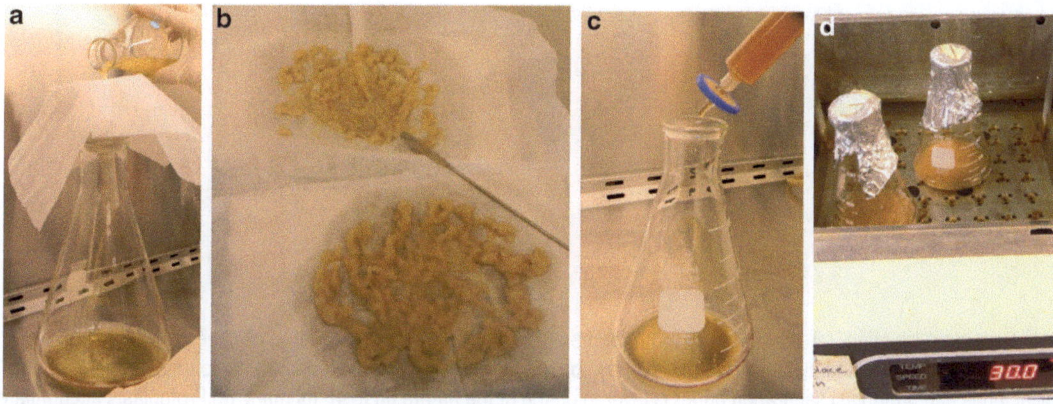

Fig. 2. Protoplasting. (**a**) Mycelia are filtered from the growth medium using a miracloth placed on a funnel. (**b**) The remaining media from the mycelia are soaked out by spreading out the mycelia on the miracloth that is placed on a bed of paper towels. (**c**) While the mycelia dry on the paper-towel bed, the enzyme solution is prepared and filter-sterilized into a sterile conical flask. (**d**) The dried mycelia are then transferred into the flasks containing the enzyme solution and placed in the incubator that was preset to 30°C.

3. Place the miracloth carrying mycelia on a bed of paper towels approximately 6 cm in thickness and spread the mycelia evenly with spatula to remove excess medium from the mycelia (see Fig. 2b).
4. As the mycelia are drying, prepare the protoplast solution (see Note 2) and set the shaking water bath incubator to 30°C.
5. Using the forceps and spatula, tear the mycelia into small pieces. Transfer 4 g (wet weight) into 100 ml protoplast solution in a 250 ml conical flask.
6. Incubate conical flasks containing the mycelia and protoplast solution in the shaking water bath incubator (30°C) for 7 h with shaking at 100 rpm (see Note 3).

3.2. Protoplast Isolation

7. After incubation, filter the protoplast solution using a layer of cheese cloth and a second layer of nylon mesh to remove larger pieces of mycelia from the solution (see Fig. 3a).
8. Transfer 25 ml protoplast solution into a 50 ml disposable conical centrifuge tube and carefully layer protoplast separation buffer on top of this solution up to 40 ml using a transfer pipette (see Fig. 3b). Be careful not to mix protoplast solution and protoplast separation buffer.
9. Centrifuge the tubes at 1,500 × g for 15 min at 4°C using a refrigerated clinical centrifuge. Protoplasts will form a band just above the interface of the two layers (see Fig. 3c). Collect the protoplasts with a transfer pipette, being careful not to disturb the layers. The protoplasts can normally be collected in a total volume of 5 ml.

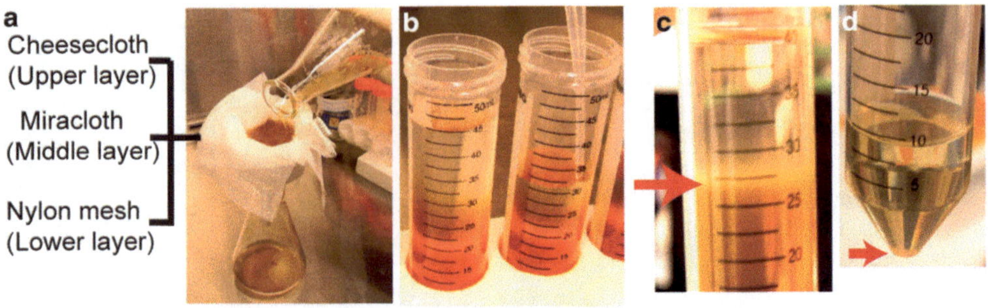

Fig. 3. Protoplast isolation. (**a**) Protoplasting broth is filtered through a tri-layer consisting of cheese-cloth (*upper layer*), miracloth (*middle layer*), and nylon mesh (*lower layer*). (**b**) The filtrate is then aliquoted in 25 ml volumes into fresh 50 ml conical tubes and protoplast separation buffer is layered on top of the filtrate up to 40 ml mark. (**c**) Protoplasts accumulate at the interface of layers and form a "fuzzy" band (denoted by a *red arrow*) after centrifugation of the tubes (step 8). (**d**) The band is collected and recentrifuged with protoplast storing buffer (steps 9–11). After centrifugation (step 11), the protoplasts collect as a white precipitate (denoted by a *red arrow*) at the bottom of the conical tube.

10. Transfer protoplasts collected to a fresh 50 ml conical disposable centrifuge tube.
11. Dilute protoplasts with protoplast storage buffer to a total volume of 10 ml.
12. Centrifuge this solution at 1,000 × *g* for 10 min at RT in a clinical centrifuge. Protoplasts will pellet at the bottom of the tube as a white precipitate (see Fig. 3d).
13. Carefully remove the supernatant from the protoplast pellet with a transfer pipette (see Note 4).
14. Gently resuspend all protoplasts in 1 ml total volume of protoplast storage buffer by carefully pipetting up and down with a transfer pipette. Vigorous mixing will lyse the protoplasts.

3.3. Protoplast Lysis and the Release of Vesicles, Endosomes, and Vacuoles

15. Transfer 1 ml of protoplasts obtained in step 14 to a 50 ml disposable conical centrifuge tube placed on ice. Add 1 ml of protoplast lysis solution.
16. Protoplast lysis will be observed within 1 min after addition of lysis buffer. At 2–3-min intervals, pipette the solution up and down vigorously to enhance release of vesicles, endosomes, and vacuoles. Allow lysis to continue for 10 min.
17. As lysis proceeds, pipette 1 ml of sucrose cushion into two 2 ml microcentrifuge tubes.
18. After completion of lysis, carefully layer 1 ml of lysis solution onto the sucrose cushion in each microcentrifuge tube.

3.4. Isolation of the Vesicle–Vacuole (V) Fraction

19. Centrifuge the microcentrifuge tubes in a benchtop clinical centrifuge equipped with a swinging bucket rotor at RT for 30 min at 3,000 × *g*.

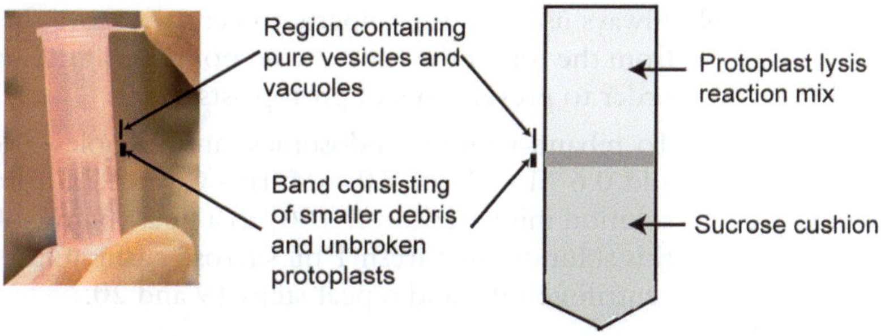

Fig. 4. Isolation of pure vesicle–vacuole fraction (V-fraction). The figure on the *left* shows an Eppendorf tube after centrifugation as mentioned in step 18. The schematic on the *right* shows the region from where the vesicle–vacuole fraction should be isolated.

20. After centrifugation, a visible band that contains cell debris and unbroken protoplasts will form at the interface between the layers (see Fig. 4). Just above this band is the pure vesicle–vacuole (V) fraction. Using a 200 μl pipette, carefully remove approximately 200 μl containing (V) fraction while avoiding the visible band at the interface. This procedure results in (V) fraction with >95% purity. If additional purity is required, (V) fraction combined from two tubes (400 μl) can be re-banded on a fresh sucrose cushion (see Note 5). To perform biochemical assays, (V) fraction can be stored for up to 60 min on ice. To perform proteomic analysis, (V) fraction can be stored for up to 30 days at -80°C.

4. Notes

1. Soon after addition of NaOH dropwise to the solution, $Mg(OH)_2$ flakes begin to appear. Continuous stirring helps to dissolve the flakes resulting in an increase in pH. Therefore, add NaOH slowly and make sure that flakes dissolve completely before adding additional NaOH.
2. For routine (V) fraction isolation, we use 1 g of Driselase and 1 g of Lysing Enzyme to digest mycelia obtained from six flasks incubated for 36 h after inoculation. Under these conditions we routinely obtain approximately 4 g (wet weight) of mycelia per flask.
3. After 3 h, monitor appearance of protoplasts every hour by placing a drop of the protoplast solution on a slide with a transfer pipette and observing the sample under a light microscope.

4. Always use a transfer pipette to gently remove the supernatant from the top of the tube. Avoid pouring off the supernatant in order to prevent loss of protoplasts.

5. To reband vesicles, endosomes, and vacuoles in (V) fraction, add 0.6 M sorbitol, 10 mM Tris–Cl, pH 7.5 (protoplast lysis solution minus Triton X-100) to a final volume of 1 ml. Layer this solution on a fresh 1 ml sucrose cushion in a 2 ml microcentrifuge tube and repeat steps 19 and 20.

References

1. Sirikantaramas S, Yamazaki M, Saito K (2008) Mechanisms of resistance to self-produced toxic secondary metabolites in plants. Phytochemistry Rev 7(3):467–477
2. Ziegler J, Facchini PJ (2008) Alkaloid biosynthesis: metabolism and trafficking. Annu Rev Plant Biol 59:735–769
3. Tanaka Y, Sasaki N, Ohmiya A (2008) Biosynthesis of plant pigments: anthocyanins, betalains and carotenoids. Plant J 54(4):733–749
4. Hoppert M, Gentzsch C, Schorgendorfer K (2001) Structure and localization of cyclosporin synthetase, the key enzyme of cyclosporin biosynthesis in *Tolypocladium inflatum*. Arch Microbiol 176(4):285–293
5. Lendenfeld T, Ghali D, Wolschek M, Kubicek-Pranz EM, Kubicek CP (1993) Subcellular compartmentation of penicillin biosynthesis in *Penicillium chrysogenum*. The amino acid precursors are derived from the vacuole. J Biol Chem 268(1):665–671
6. Hong SY, Linz JE (2008) Functional expression and subcellular localization of the aflatoxin pathway enzyme Ver-1 fused to enhanced green fluorescent protein. Appl Environ Microbiol 74(20):6385–6396
7. Hong SY, Linz JE (2009) Functional expression and sub-cellular localization of the early aflatoxin pathway enzyme Nor-1 in *Aspergillus parasiticus*. Mycol Res 113(5):591–601
8. Lee LW, Chiou CH, Klomparens KL, Cary JW, Linz JE (2004) Subcellular localization of aflatoxin biosynthetic enzymes Nor-1, Ver-1, and OmtA in time-dependent fractionated colonies of *Aspergillus parasiticus*. Arch Microbiol 181(3):204–214
9. Forster C, Marienfeld S, Wilhelm R, Kramer R (1998) Organelle purification and selective permeabilisation of the plasma membrane: two different approaches to study vacuoles of the filamentous fungus *Ashbya gossypii*. FEMS Microbiol Lett 167(2):209–214
10. Martinoia E, Heck U, Boller T, Wiemken A, Matile P (1979) Some properties of vacuoles isolated from *Neurospora crassa* slime variant. Arch Microbiol 120(1):31–34
11. Rieder SE, Emr SD (2001) Isolation of subcellular fractions from the yeast *Saccharomyces cerevisiae*. Curr Protoc Cell Biol Chapter 3:Unit 3.8
12. Walworth NC, Goud B, Ruohola H, Novick PJ (1989) Fractionation of yeast organelles. Methods Cell Biol 31:335–356
13. Wiemken A (1975) Isolation of vacuoles from yeasts. Methods Cell Biol 12:99–109
14. Vaughn LE, Davis RH (1981) Purification of vacuoles from *Neurospora crassa*. Mol Cell Biol 1(9):797–806
15. Boudet AM, Canut H, Alibert G (1981) Isolation and characterization of vacuoles from *Melilotus alba* mesophyll. Plant Physiol 68(6):1354–1358
16. Robert S, Zouhar J, Carter C, Raikhel N (2007) Isolation of intact vacuoles from *Arabidopsis* rosette leaf-derived protoplasts. Nat Protoc 2(2):259–262
17. Shimaoka T et al (2004) Isolation of intact vacuoles and proteomic analysis of tonoplast from suspension-cultured cells of *Arabidopsis thaliana*. Plant Cell Physiol 45(6):672–683
18. Bethke PC, Hillmer S, Jones RL (1996) Isolation of intact protein storage vacuoles from barley aleurone (identification of aspartic and cysteine proteases). Plant Physiol 110(2):521–529
19. Mathieu Y et al (1989) Regulation of vacuolar pH of plant cells: I. Isolation and properties of vacuoles suitable for P NMR studies. Plant Physiol 89(1):19–26
20. Chanda A, Roze LV, Pastor A, Frame MK, Linz JE (2009) Purification of a vesicle-vacuole fraction functionally linked to aflatoxin synthesis in *Aspergillus parasiticus*. J Microbiol Methods 78(1):28–33
21. Chanda A et al (2009) A key role for vesicles in fungal secondary metabolism. Proc Natl Acad Sci USA 106(46):19533–19538

Chapter 20

Use of the Hollow Fiber Assay for the Discovery of Novel Anticancer Agents from Fungi

Cedric J. Pearce, Daniel D. Lantvit, Qi Shen, David Jarjoura, Xiaoli Zhang, Nicholas H. Oberlies, David J. Kroll, Mansukh C. Wani, Jimmy Orjala, Djaja D. Soejarto, Norman R. Farnsworth, Esperanza J. Carcache de Blanco, James R. Fuchs, A. Douglas Kinghorn, and Steven M. Swanson

Abstract

The hollow fiber assay (HFA) is a drug discovery tool to aid investigators in the prioritization of lead compounds identified by in vitro testing for further development in animal models of disease. In the HFA, cells are cultured in hollow fibers containing pores of a diameter (500 kDa) large enough for proteins and other macromolecules to enter, but too small for the cells to escape. The fibers are filled with cells, sealed and placed in the peritoneal cavity of immunodeficient mice. The mice undergo a predetermined treatment regimen after which the fibers are retrieved and the cells evaluated for activity of a target relevant to the disease modeled. The HFA combines advantages of both in vitro and in vivo assay systems. It uses the same cell lines used in culture systems, is a rapid assay, and requires fewer animals and less test substance than conventional xenograft systems. Like traditional in vivo assays, the test substance is evaluated in a live animal, which affords an initial assessment of associated toxicity and pharmacokinetic properties of the test substance.

Key words: Drug discovery, Fungi, Animal models, Natural products, Cancer

1. Introduction

While the hollow fiber assay (HFA) can be adapted for a variety of uses (1–4), our focus has been on the discovery of novel anticancer agents derived from natural sources such as fungi (5). Hollingshead and colleagues originally developed the assay as a means of prioritizing the many lead compounds that had been identified in the 60-cell line panel at the National Cancer Institute (NCI) (6). When the Cancer Chemotherapy National Service Center was established

Nancy P. Keller and Geoffrey Turner (eds.), *Fungal Secondary Metabolism: Methods and Protocols*, Methods in Molecular Biology, vol. 944, DOI 10.1007/978-1-62703-122-6_20, © Springer Science+Business Media, LLC 2012

at the NCI in 1955, the primary screening tool was transplantable murine cancer models, which involved a vast number of mice. While these models were critical for the discovery of many effective anti-cancer agents used clinically, the leadership at the NCI decided that using human cancer cell lines would provide a greater diversity of targets relevant to human cancers and thereby increase the chance of success in their drug discovery program. Researchers at the NCI collected and characterized a panel of 60 cell lines representing cancers of the lung, colon, brain, ovary, breast, prostate, and kidney, together with leukemia and melanoma. These cells are also capable of propagating in immunodeficient mice as xerographs allowing for leads identified in the cell line panel to be followed up in vivo. As high-throughput and automated procedures were optimized using in vitro assays, in vivo models became the bottleneck for moving compounds forward in the drug discovery process. Since the HFA can be completed in about 1 week and requires only three mice per dose for each test substance, it allows investigators to prioritize lead compounds efficiently by saving time and using fewer animals (4).

Fungi are recognized as one of the more significant sources of medicines and have provided a number of spectacular advances in the treatment of human diseases as is illustrated by the discovery of penicillin G, and the introduction of the first statins, for example. Even after decades of using fungal metabolites as leads for new drugs they remain a viable alternative to purely synthetic routes. For example, Gilenya™ (fingolimod, Novartis) is a compound based on myriocin, which is a metabolite first found in an insect-colonizing fungus, *Isaria sinclairii*. Gilenya™ was approved by the US Food and Drug Administration as the first orally active medicine for multiple sclerosis. These medicines represent the best of a very large catalogue of bioactive compounds produced by fungi. The chemical diversity of fungal metabolites is extraordinarily broad and this is reflected in their biological activity.

A number of fungal metabolites and their derivatives have been evaluated clinically as potential cancer medicines; examples include brefeldin A, wortmannin, cytochalasin E, and related derivatives such as fumagillin, gliotoxin, terrecyclic acid A, the illudin derivative irofulvin, and lentinan and related polysaccharides which are approved for use in Japan. In our own program to find new cancer lead compounds funded by the NIH (Discovery of Anticancer Agents of Diverse Natural Origin. 5P01 CA125066; Principal Investigator, A.D. Kinghorn supported by the NCI, National Institutes of Health), over the course of the past 4 years we have discovered 102 compounds, 31 of which are novel, that are active in our cultured cancer cell lines. Our other programs using vascular plants and cyanobacteria have also been successful in finding new compounds. This success means that we are in urgent need of secondary bioassays to help us focus on those leads with the highest probability of generating clinical lead-quality compounds, and the need for methods to evaluate compounds in vivo is especially useful. To this end we have been developing methods to provide such prioritization, and the HFA has been one approach we have taken.

2. Materials

2.1. Cell Culture Supplies

1. Human cancer cell lines (American Type Culture Collection; see Note 1).
2. RPMI 1640 medium with glutamine.
3. Fetal bovine serum.
4. Trypsin (0.05%)/EDTA.
5. Antibiotic/antimitotic, 100× [contains penicillin (10^4 units/ml), streptomycin (10^4 μg/ml), and the antifungal agent amphotericin B (25 μg/ml)].
6. Trypan blue solution (0.4%, Sigma Chemical Co.).
7. MTT [3-(4,5-dimethylthiazol-2-yl)-2,5-diphenyltetrazolium bromide].
8. Tissue culture flasks (T75 and T150, BD Falcon).
9. Sterile centrifuge tubes (15 and 50 ml, BD Falcon).
10. Hemocytometer (Reichert Bright-Line, Hausser Scientific, Horsham, PA).

2.2. Hollow Fiber Preparation

1. Modified polyvinylidene difluoride hollow fibers (Spectrum Labs, CellMax®, molecular weight cutoff of 500 kDa).
2. Dry glass bead sterilizer (see Note 2).
3. Needle holders, forceps, and scissors.
4. 1 and 5 ml syringes; needles.
5. Stainless steel pan (approximately 10×6.5 in.).
6. Six-well culture plates.
7. Medical gloves (latex or nitrile).

2.3. Surgical Supplies

1. Trocar (11 gauge; Popper and Sons, New Hyde Park, NY).
2. Wound clip system (Autoclip®, Harvard Apparatus).
3. General operating scissors.
4. Dressing and tissue forceps.

3. Methods

3.1. Conditioning Fibers

Wear gloves when handling fibers.

1. Using a syringe fitted with either an 18 or 20 gauge needle, flush each fiber individually with 70% EtOH (Fig. 1).
2. Place fibers in an autoclavable covered pipette tray filled with 70% EtOH for at least 72 h. Do not allow fibers to dry out from this point on.
3. After 72 h, use a needle and syringe as above and flush fibers twice individually with deionized water.
4. Place fibers in another pipette tray filled with deionized water and autoclave for 30 min. Let cool and store at 4°C until fibers are needed.

3.2. Cell Culture

The reader is referred to books on cell culture for general procedures (7, 8). The cells are grown in RPMI 1640 medium supplemented with glutamine and 10% fetal bovine serum in T75 flasks (8). The split ratio is dependent on the rate of proliferation, which varies from line to line, but generally is between 1:3 and 1:10. Cells are split at a ratio that will result in 70–80% confluence on the experiment day. On the day before the experiment, cells are fed by removing half of the medium and adding an equal volume of fresh medium. On the day that the fibers are to be filled, the cells are washed with PBS, incubated with 1–3 ml of trypsin, and incubated for 5–10 min at 37°C to detach them from the flask surface. The cells are resuspended in medium, counted using a hemocytometer,

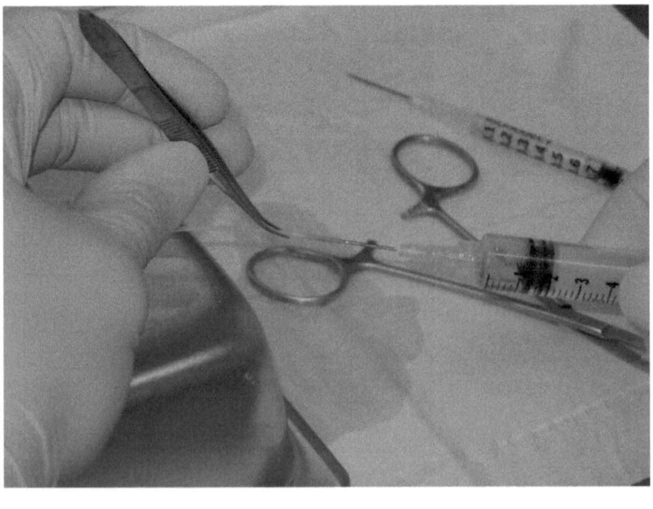

Fig. 1. Sterilize the fibers. Flush the fibers with 70% ethanol using a syringe fitted with an 18 or 20 gauge needle.

and plated in new flasks or injected into preconditioned hollow fibers.

1. Pour off medium from cells into a sterile tube and place on ice (this conditioned medium will be used to resuspend the cells later); wash cells twice with PBS.
2. Aspirate PBS and add 2 ml trypsin per flask. Incubate at 37°C for 5 min to detach the cells from the flasks.
3. Add 10 ml fresh medium to flask; transfer medium and cells to a sterile centrifuge tube and centrifuge (5 min, 800 × *g*, 4°C); aspirate medium, resuspend cells in 5 ml conditioned medium, and place on ice.
4. Determine viable cell number using the trypan blue dye exclusion test (9).
5. Add the appropriate volume RPMI 1640 with FBS medium (20% vol/vol) to achieve the desired cell concentration and place on ice.

3.3. Fiber Preparation

Conditioning, filling, and sealing the hollow fibers must be conducted in a sterile environment.

1. Mark the bottom of a stainless steel pan every 2 cm for work surface. Wrap the stainless steel pan in aluminum foil. Autoclave needle holder with scissors, smooth forceps, and stainless steel work surface (see Note 3).
2. Empty a biosafety cabinet and wipe down with 70% EtOH. Place pipette tray containing fibers and bead sterilizer in the biosafety cabinet. Spray down the outside of the tray with EtOH. Allow bead sterilizer to heat to maximum temperature.
3. Place centrifuge tubes of cells and media on ice. Set up a sterile field in the cabinet by placing sterile drapes on the work surface. Carefully unwrap the stainless steel work surface under the hood and place in the sterile field. Place all autoclaved instruments in the sterile field. Remove 10 and 1 ml syringes and 18 or 20 gauge needles from freezer and place in the sterile field.
4. Don sterile gloves. Uncap cold media tube under hood and hold while 10 ml syringe is filled. Once syringe is filled, tube is removed from hood. Cold media is dispensed on the work surface to keep fibers from drying out during the filling process. Syringe with the remaining media is placed in the sterile field.
5. Carefully mix cell suspension to create homogeneity. Under the hood, uncap the cold cell suspension tube and fill a 1 ml syringe; place it in a sterile field within the hood. Remove lid from pipette tray and, with sterile forceps, remove 2–3 fibers

and place on medium-soaked work surface. Flush each fiber intraperitoneally with cold medium to remove water.

6. Create an air bubble in the 1 ml syringe containing the cell suspension so that the remaining medium can be pushed out of the fiber before filling it with the cell suspension. Insert needle into one end of the fiber. With the needle pointing upwards, form an upside-down “U” with the fiber before dispensing air bubble and cell suspension (Fig. 2). Fill fiber with cold cell suspension avoiding air bubbles in the fiber (see Note 4).
7. Once the total length of the fiber is filled with the cell suspension, create a “d” by taking the loose end of the fiber and moving it horizontally to the work surface (Fig. 3). Heat the needle

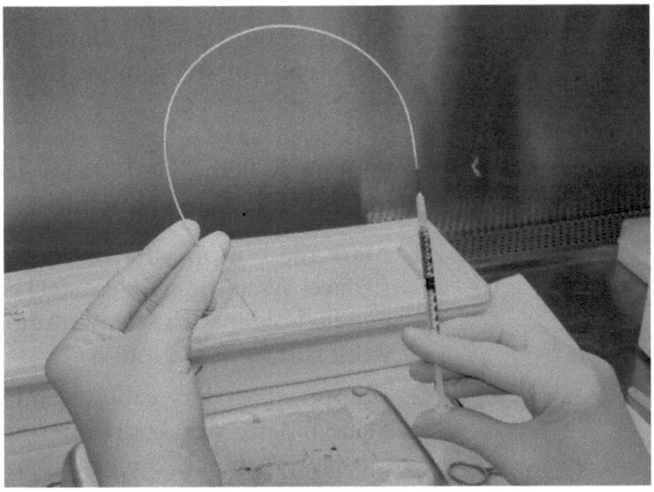

Fig. 2. Filling the fibers with cells. Medium containing cells is gently injected into the sterile fiber behind a cushion of air, replacing the medium lacking cells.

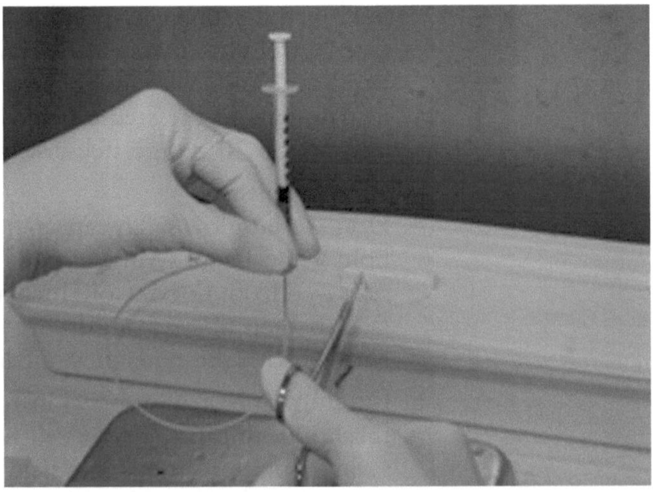

Fig. 3. The ends of the fibers containing cells are sealed by pinching with a heated hemostat.

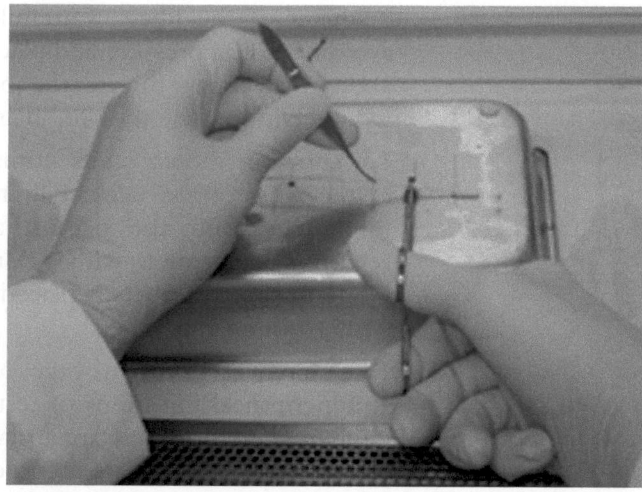

Fig. 4. Cutting the individual fibers for implantation. Using a heated hemostat, clamp down firmly on the cell-filled fibers at 2 cm intervals. Using a scalpel or surgical scissors, cut the fibers in the middle of the heat-seal to generate the individual fibers for implantation.

holder in the bead sterilizer for a few seconds and heat-seal the open end by clamping down on the fiber with the needle holder. Then heat-seal the other end just below the needle. Lay the filled fiber on the work surface. Continue until each of the 2–3 fibers is filled. Be sure to keep fibers from drying out by keeping them covered with cold media.

8. One at a time, the filled fibers are laid across the area of the work surface that has the 2 cm demarcations. Again heating the needle holder a few seconds, a mark is made every 2 cm for the length of the fiber by pressing the needle holder closed on the fiber (Fig. 4). Repeating the process above a second time at each mark this time applying a slightly pulling action on the fiber creates a clear area that completes the heat-seal. Cut in the center of the heat-seal to make individual fibers for implantation.
9. Fibers are placed into each well of a 6-well culture plate filled with 2 ml of medium supplemented with serum but lacking antibiotics. Place fibers in a 37 °C incubator (see Note 5).

3.4. Implantation of Fibers

1. Remove 6-well plates containing fibers from the incubator and check for sterility. Three fibers from each cell line are set aside to serve as "day 0" controls (see below).
2. Anesthetize female athymic mice NCr nu/nu mice at 5–6 weeks of age with a combination of ketamine and xylazine (see Note 6).
3. Lay mouse on its side and make a small incision through the skin and muscle layers just below the spleen. The size of the incision should be just large enough to accommodate the

diameter of the fibers. Carefully handle fibers with forceps avoiding excess pressure and place the appropriate fibers in the abdominal cavity. Suture the body wall incision and use a wound clip to close the skin incision.

4. Lay the same mouse on its stomach and make a small incision at the nape of the neck. Fill trocar with appropriate fibers, insert trocar into the incision, and place fibers over the hip region as you begin to withdraw trocar. Repeat until the appropriate number of fibers is placed in the animal. Close incision with a wound clip.
5. Place animals back in cage and allow them to recover. Intraperitoneal injections of test compounds begin on day 3 after implantation and are usually injected daily for 4 days. Animal body weights are recorded every day to monitor compound toxicity (see Note 7).
6. On day 7, animals are sacrificed and fibers are retrieved and placed in 6-well plates filled with warm media. Wells are labeled according to treatment and site of implantation. Fibers are placed in the incubator for at least 30 min to allow cells to normalize.
7. The viable cell mass is evaluated by the MTT assay described below. The percent net growth for each cell line in each treatment group is calculated by subtracting the day-zero absorbance from the day 7 absorbance and dividing this difference by the net growth in the day 7 vehicle-treated controls minus the day-0 values. A 50% or greater reduction in net cell growth in the treated samples compared to the vehicle control samples is considered a positive result (see Note 8).

3.5. MTT Assay

We have modified this commonly used assay, which measures the reduction by metabolically active cells of a yellow tetrazolium salt to a purple formazan. The procedure has been modified for the HFA from published standard protocols (10, 11).

1. Prepare stock solution of 5 mg/ml MTT. Store at 4 °C and protect from light.
2. Prepare a 2.5% solution of protamine sulfate in saline and sterilize by passing through a 0.2 μm filter; store at 4 °C.
3. Prepare a 1 mg/ml working solution of MTT by adding 4 ml of media to 1 ml of stock solution. Warm in a 37 °C water bath.
4. Add 1 ml of working solution to the 2 ml of media for each well containing harvested fibers. Incubate for 4 h in cell culture incubator.
5. Aspirate all liquid from each well. Wash fibers by adding 2 ml of 2.5% protamine sulfate to each well containing fibers. Store at 4 °C overnight.

6. Aspirate liquid and perform a second wash with 2.5% protamine sulfate. Incubate at 4 °C for at least 2 h. Remove fibers and gently dry with a KimWipe. Cut fiber in half and place one fiber per well of a 24-well plate making sure that each well is labeled with correct treatment and placement.
7. Allow the fibers to dry overnight protected from light (see Note 9).
8. Add 250 μl of DMSO to each well and cover plate with foil. Place plates on a shaker for 4 h.
9. Remove 150 μl of solution from each well and place in a 96-well plate. Read absorption at 490 nm on a microplate reader.

4. Notes

1. The cell lines are selected from the list of 30 lines used in the HFA developed by Hollingshead et al. (12). Since we are looking for new proteasome inhibitors, we have chosen lines that are most sensitive to bortezomib, an FDA-approved proteasome inhibitor. This information was obtained by querying the database of therapeutics for their activity against the NCI's 60 cell line panel (13).
2. Several methods are acceptable for sterilizing the hemostat used to seal the fibers. We use a glass bead sterilizer because it can rapidly heat and sterilize metal instruments such as the hemostat that is used to heat-seal the cell-filled fibers.
3. Each cell line requires its own set of equipment to avoid cross-contamination.
4. Each cell line should be represented by a different color fiber. If a fiber dries out in the process (turns white), it is unusable.
5. Incubation allows cells to attach to the fiber surface prior to implantation in the mice. Six fibers are not implanted in the mice, but remain in the incubator and are harvested when the mice are sacrificed to assure that they are free of microbial contamination.
6. Although most laboratories use immunodeficient mice, Shnyder et al. (14) have reported that immunocompetent mice such as NMRI also can be used at significantly lower cost. Ketamine and xylazine are administered by intraperitoneal injection at 100 and 10 mg/kg, respectively.
7. The dose levels chosen for each test compound are determined by performing acute toxicity tests for each agent as described by the NCI Developmental Therapeutics Program (15). One

mouse is given a single ip injection at 400 mg/kg body weight; another mouse is administered 300 mg/kg and a third mouse is given 100 mg/kg. The mice are observed for 2 weeks and sacrificed if they lose 20% or more of their body weight or exhibit outward signs of toxicity as indicated above. If all three mice die or must be sacrificed, three lower doses (e.g., 50, 25, and 10 mg/kg) are tested. The process is repeated until the maximum tolerated dose (MTD) is identified. We routinely use 40% of the single-dose MTD as the highest dose in our four-daily-dose treatment schedules. The initial level of exposure that we choose for the acute toxicity study is based on the activity of the compound in cell cytotoxicity tests.

8. In the HFA, the measure of efficacy is the percentage reduction in cells at day 6 as compared to day 0 (implantation date):

$$\%\text{ net reduction} = \frac{\left(\text{day } 0\ \text{OD}_{540} - \text{day } 6\ \text{OD}_{540}\right)}{\text{day } 0\ \text{OD}_{540} \times 100\%},$$

where OD_{540} is the optical density measure of cell numbers. Because the maximum reduction is 100%, nonlinear regression modeling of dose effects should be used. Models would be estimated separately for implant site (subcutaneous vs. intraperitoneal), and for each cell line. The maximum dose can be determined from prior toxicity studies. With multiple estimated models per compound, some approach to decision making about further experimentation (e.g., xenograft testing) with the compound must be chosen. Depending on circumstances, an investigator can choose to focus on minimizing false negatives (possibly missing promising compounds) or false positives (expending further resources on ineffective compounds).

Decker et al. (12) developed a scoring system that seems to emphasize avoidance of false positives. Based on their recommendation, promising compounds are those that are effective across multiple cell lines. They validate their scoring system through the prediction of success in multiple xenograft models. A score is considered low if a compound acts only on one cell line. They also recommend further testing of any compound that produces 100% cell death. Therefore if one were most concerned about false negatives, such a scoring system would not be optimal. Instead, one could base the decision to move forward with a compound on the percent net reduction observed at the maximum dose for a particular cell line. A confidence interval on the percent net reduction estimate would be useful to take into consideration sampling error of the estimate. A more formal decision procedure could test the significance of the dose–response curve, and use liberal criteria for declaring a positive result. The number of cell lines and compounds tested simultaneously must be considered when deciding on such criteria. This approach further allows for a second more precise study when results are too variable to achieve significance.

9. We place the plates containing the fibers near the vent of a biological safety cabinet with some foil loosely applied to protect the dye from light.

References

1. Hassan SB, de la Torre M, Nygren P, Karlsson MO, Larsson R, Jonsson E (2001) A hollow fiber model for *in vitro* studies of cytotoxic compounds: activity of the cyanoguanidine CHS 828. Anticancer Drugs 12:33–42
2. Hollingshead M, Roberson J, Decker W, Buckheit R Jr, Elder C, Malspeis L, Mayo J, Grever M (1995) *In vivo* drug screening applications of HIV-infected cells cultivated within hollow fibers in two physiologic compartments of mice. Antiviral Res 28:265–279
3. Hollingshead MG, Bonomi CA, Borgel SD, Carter JP, Shoemaker R, Melillo G, Sausville EA (2004) A potential role for imaging technology in anticancer efficacy evaluations. Eur J Cancer 40:890–898
4. Mi Q, Pezzuto JM, Farnsworth NR, Wani MC, Kinghorn AD, Swanson SM (2009) Use of the *in vivo* hollow fiber assay in natural products anticancer drug discovery. J Nat Prod 72:573–580
5. Kinghorn AD, Chin YW, Swanson SM (2009) Discovery of natural product anticancer agents from biodiverse organisms. Curr Opin Drug Discov Devel 12:189–196
6. Hollingshead MG, Alley MC, Camalier RF, Abbott BJ, Mayo JG, Malspeis L, Grever MR (1995) *In vivo* cultivation of tumor cells in hollow fibers. Life Sci 57:131–141
7. Davis J (2011) Animal cell culture: essential methods. Wiley, Hoboken
8. Freshney RI (2010) Culture of animal cells: a manual of basic technique and specialized applications, 6th edn. Wiley-Liss, Hoboken
9. Strober W (1997) Trypan blue exclusion test of cell viability. Curr Protoc Immunol, A.3B.1–A.3B.2
10. Mosmann T (1983) Rapid colorimetric assay for cellular growth and survival: application to proliferation and cytotoxicity assays. J Immunol Methods 65:55–63
11. Vistica DT, Skehan P, Scudiero D, Monks A, Pittman A, Boyd MR (1991) Tetrazolium-based assays for cellular viability: a critical examination of selected parameters affecting formazan production. Cancer Res 51:2515–2520
12. Decker S, Hollingshead M, Bonomi CA, Carter JP, Sausville EA (2004) The hollow fibre model in cancer drug screening: the NCI experience. Eur J Cancer 40:821–826
13. Developmental Therapeutics Program, US National Cancer Institute. http://dtp.nci.nih.gov/dtpstandard/dwindex/index.jsp. Accessed July 28, 2012.
14. Shnyder SD, Hasan J, Cooper PA, Pilarinou E, Jubb E, Jayson GC, Bibby MC (2005) Development of a modified hollow fibre assay for studying agents targeting the tumour neovasculature. Anticancer Res 25:1889–1894
15. Developmental Therapeutics Program, US National Cancer Institute. http://dtp.nci.nih.gov/branches/btb/acute_tox.html. Accessed July 28, 2012.

Chapter 21

Molecular Analysis of Fungal Gene Expression upon Interkingdom Competition with Insects

Ulrike Fohgrub and Frank Kempken

Abstract

Fungi produce an astonishing variety of secondary metabolites, some of which belong to the most toxic compounds in the living world. However, the benefits of secondary metabolites for fungi are often obscure. Here we describe a method that will aid in the analysis of fungal gene expression upon interkingdom competition with insects. It allows for isolation of fungal RNA from surface-grown mycelium after confrontation with insect larvae, and subsequent molecular analysis.

Key words: Interkingdom competition, *Aspergillus nidulans*, *Drosophila melanogaster*, RNA isolation

1. Introduction

Fungi produce an astonishing variety of secondary metabolites, some of which belong to the most toxic compounds in the living world. However, the benefits of secondary metabolites for fungi are often obscure. In contrast to the gap in knowledge of the biological relevance of secondary metabolites, enormous research has been devoted to identifying biochemical pathways and the underlying genetic mechanisms leading to the biosynthesis of fungal secondary metabolites (1). It is, however, likely that the fungal ability to regulate secondary metabolism reflects an evolutionary adaptation to ensure efficient exploitation of environmental resources and to synthesize secondary metabolite only when the ecological conditions demand it against natural enemies and competitors (2–4). It should be noted, however, that in addition to secondary metabolites, fungal trypsin-specific inhibitors and lectins have been reported as defense molecules towards competitors and predators (5–7).

Nancy P. Keller and Geoffrey Turner (eds.), *Fungal Secondary Metabolism: Methods and Protocols*,
Methods in Molecular Biology, vol. 944, DOI 10.1007/978-1-62703-122-6_21, © Springer Science+Business Media, LLC 2012

Previously performed food choice experiments with springtails support a role for secondary metabolism in predator defense. In *Aspergillus nidulans* the protein LaeA is a global regulator of secondary metabolite production (8–10). Using a mutant deleted for LaeA, it was demonstrated that springtails prefer the LaeA mutant for feeding in food choice experiments. Springtails feeding on the LaeA mutant produced more offspring than the wild type (3). These results demonstrate that fungal secondary metabolites shape food choice behavior and affect population dynamics of fungivores. That study also provides evidence for a selective force favoring secondary metabolite synthesis in fungi (3).

Competition studies using insects and molds may provide important insight into the biological functions of fungal metabolites and their role in evolution. Here we describe a simple method for such competition experiments.

2. Materials

2.1. Drosophila and Aspergillus Medium

1. *Drosophila* standard medium (11, 12). 62.5 g untreated cornmeal, 62.5 g saccharose, 62.5 g yeast extract, 12.5 g agar, 1 L distilled H_2O.
2. *Drosophila* feeding medium (11, 12). *Drosophila* standard medium, with 10 g cane syrup, 30 mL 10% (w/v) methyl parahydroxybenzoate or nipagin M (Merck catalog number: 106757) (dissolved in EtOH), 10 mL 10% (v/v) propionic acid (1 N diluted in distilled water).
3. *Drosophila* breeding medium (11, 12). *Drosophila* standard medium; add 1 mL 5% (v/v) acetic acid (1 N diluted in distilled water).
4. *Drosophila* standard medium plus foil (sterile). *Drosophila* standard medium with 1% (w/v) KOH treated foil (DESAGA GmbH, catalog number: SI-Nr.4-42284-01-0), distilled H_2O wetted Whatman paper (Whatman, catalog number: 1001917), and a pair of tweezers (sterile, autoclaved) are necessary for sterile transfer.
5. *Aspergillus standard medium*. 25 g untreated cornmeal, 30 g malt extract, and 25 g agar (Roth, catalog number: 52102); add 1 L untreated water.

2.2. Dechorionated Drosophila Eggs (12, 13)

1. *Drosophila* standard medium plus foil (sterile).
2. 50 mL *Drosophila* breeding medium in a 150 mL measurement cylinder.
3. *Drosophila* storage box (escape proof).
4. 1 L distilled H_2O.
5. Spatula (sterile, autoclaved).
6. Gauze (sterile, autoclaved).

7. Funnel (sterile, autoclaved).
8. Erlenmeyer flask (1 L).
9. 6% (v/v) sodium hypochlorite.
10. 70% (v/v) Ethanol (sterile).
11. 2 L distilled H_2O (sterile, double distilled, and autoclaved).
12. Paintbrush (sterile).

2.3. A. nidulans on Petri Dish

Isolation of Conidia

1. *Aspergillus standard medium* (sterile).
2. Falcon tubes (sterile, 50 mL).
3. Spatula (sterile).
4. 0.9% NaCl (w/v) with 0.1% (v/v) Tween 20 (sterile, autoclaved).
5. 0.9% NaCl (w/v) (sterile, autoclaved).
6. Centrifuge with swinging bucket.
7. Reaction tubes (1.5 mL, sterile).

Plating Fungal Conidia

8. *Drosophila* standard medium plus foil (sterile).
9. Freshly isolated conidiospores from *A. nidulans.*
10. Inoculation loop.
11. Opaque box or the like for the growth of *Aspergillus* in darkness.

RNA Isolation and Subsequent Analysis

12. TRIzol® reagent (Ambion).
13. For oligonucleotide sequences for qRT-PCR see Table 1.

Table 1
Oligonucleotides used

Gene	Sequence (5′–3′)
AN0807 (*laeA*)	TATAGCAACAATTCCGAGCGAAAC
AN0807 (*laeA*)	CTTGCTCATCGCACGGAAGA
AN7815 (*stcJ*)	GCGCGATGATGACAAATTCC
AN7815 (*stcJ*)	GCATATACGAGCAGAAGGATCCAGTA
AN7820 (*aflR*)	ACGCACTCGCGTCATTGGT
AN7820 (*aflR*)	TGGTCCTGCCTTCGTTGCT
AN6542.3 (*actin*)	GTAAGTCACCCTTTCCTCCCTTATATACA
AN6542.3 (*actin*)	AGAGCGCGGTGAACACATTTT
AN8041.3 (*gpd*)	CGGCTTCGGTCGTATTGGA
AN8041.3 (*gpd*)	CTCACGCGTTACGGAAAACCT

3. Methods

Store *Drosophila* flies at 20–25°C and in dark/light cycle (12/12 h). Because conditions are non-sterile, the addition of nipagin and propionic acid is required to terminate fungal and bacterial growth. Use fresh medium if microbial growth is visible or medium is drying out. Dechorionated *Drosophila* eggs are easily damaged, so handle with care (T. Röder, Kiel, pers. comm.)!

3.1. Drosophila and Aspergillus Medium

1. *Drosophila* standard medium: Mix untreated cornmeal, saccharose, yeast extract, and agar in distilled H_2O and boil up (70–100°C, 60 s), fill 50 mL hand-hot medium (30–40°C) in each 150 mL Erlenmeyer flask, and take it into a *Drosophila* storage box (escape proof).
2. *Drosophila* feeding medium: Boil up *Drosophila* standard medium plus cane syrup, and add to the hand-hot medium 10% (w/v) nipagin and 10% (v/v) propionic acid.
3. *Drosophila* breeding medium: Boil up (100°C, 20 s) *Drosophila* standard medium and add to the hand-hot medium 5% (v/v) acetic acid.
4. *Drosophila* standard medium plus foil (sterile): Autoclave *Drosophila* standard medium (121°C, 20 min) and fill petri dishes (Fa. Sarstedt, size 100×100×20 mm, catalog number 2013-08) with hand-hot medium. Leave to cool for about 10 min. Cut foil(s) to the size of the petri dishes used and disperse for 5 min in 1% (w/v) KOH solution; place one piece of foil between two pieces of wet (distilled water) Whatman Paper and autoclave (121°C, 20 min). Following place one piece of foil of the agar medium in the petri dish with a pair of tweezers (sterile, autoclaved) and store at 4°C.
5. *Aspergillus* standard medium: Take untreated cornmeal in 0.1 L untreated water and store it overnight at 37°C; next day cut absorbent cotton gauze (circa 10×10 cm), filter the cornmeal–water solution through this cotton gauze (Fa. Hartmann, EN14079-VM20, absorbent cotton gauze in pleats, 80 cm wide, 8 ply folded to 10 cm in width, 10 cm×40 m), add malt extract and agar to the filtrate, add 1 L untreated water, and split up about 150 mL into a 300 mL Erlenmeyer flask.

3.2. Dechorionated Drosophila melanogaster Eggs

Be careful with 6% (v/v) sodium hypochlorite and work under hoods. The dechorionated eggs must be transferred to the sterile *Drosophila* standard medium plus foil. The description is based on a published method (15).

1. Prepare *Drosophila* standard medium plus foil.
2. Prepare *Drosophila* breeding medium and place for 24 h in *Drosophila* storage box with *D. melanogaster* flies.

3. Remove breeding medium from *Drosophila* storage box after 24 h and wash down the *Drosophila* eggs from the breeding medium with distilled H_2O and a paintbrush.
4. Collect eggs with water in a measuring cylinder.
5. To discard the distilled H_2O, filter through a funnel with gauze (both sterile).
6. Incubate eggs in 6% (v/v) sodium hypochlorite for 10 min (use fume hood).
7. Filter the eggs through a new funnel with gauze (both sterile).
8. Rinse the eggs with 1 L 70% (v/v) EtOH.
9. Rinse the eggs with 1 L distilled H_2O (autoclaved).
10. Transfer dechorionated eggs very carefully to *Drosophila* standard medium plus foil (sterile) using a sterile paintbrush, and rinse the paintbrush in between with sterile distilled H_2O.
11. Store eggs for 24 h at 20–25°C with dark/light cycle (12/12 h).
12. Transfer 24-h-old larvae with sterile paintbrush to *A. nidulans* cultures (see Subheading 3.3) on petri dishes with foil.
13. Store the *A. nidulans* mycelium with the larvae for the desired time period (e.g., 24 h) in darkness.

3.3. *A. nidulans* Cultures

1. Prepare *Aspergillus* standard medium (autoclaved) in Erlenmeyer flask and inoculate *Aspergillus* conidia for about 1 week at 25°C with dark/light cycle (12/12 h).
2. Add 75 mL 0.9% (w/v) NaCl with 0.1% (v/v) Tween 20, close the flask, and shake vigorously.
3. Filter off debris (sterile funnel with gauze) and collect the solution with the conidia in a sterile falcon tube.
4. Centrifuge for 10 min at 2770 × *g* in swinging bucket, and discard the supernatant.
5. Dissolve pellet in 40 mL 0.9% (w/v) NaCl with 0.1% (v/v) Tween 20, centrifuge for 10 min at 2770 × *g* in swinging bucket, and discard the supernatant.
6. Dissolve the pellet in 40 mL 0.9% (w/v) NaCl with 0.1% (v/v) Tween 20, centrifuge for 10 min at 2770 × *g* in swinging bucket, and discard the supernatant.
7. Dissolve the pellet in 1 mL 0.9% (w/v) NaCl.
8. Adjust to 1,000 conidia/mL; store conidia at 4°C for up to 2 days.
9. Plate freshly isolated conidia (not older than 2 days) on *Drosophila* standard medium plus foil; do not cover the entire petri dish but leave spaces to add *Drosophila* larvae (see Fig. 1). This is best done using an inoculating loop, which is dipped in the conidia from step 8.
10. Incubate the petri dishes with *Aspergillus* conidia at 25°C and in darkness for 48 h (see Note 1).

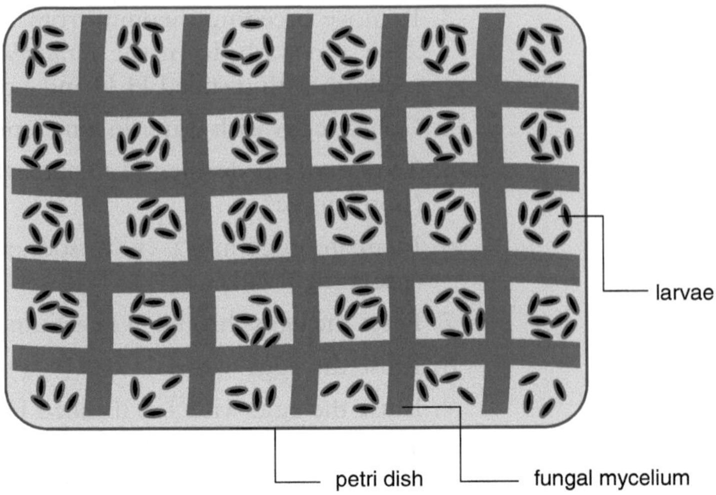

Fig. 1. Scheme of petri dish with fungal mycelium and insect larvae.

Fig. 2. Drosophila larvae on a petri dish with fungal mycelium.

11. After 48 h add larvae from Subheading 3.2 to the empty spaces between *Aspergillus* mycelia and incubate for another 24 h (for example see Fig. 2).

3.4. RNA Isolation from A. nidulans

1. Remove foil with *Aspergillus* mycelia and insects quickly at darkness and immediately freeze in liquid nitrogen; store at –80°C.
2. Grind fungal mycelium (and the remaining insects; see Note 2) with mortar and pestle under liquid nitrogen; fill powder in small tube.

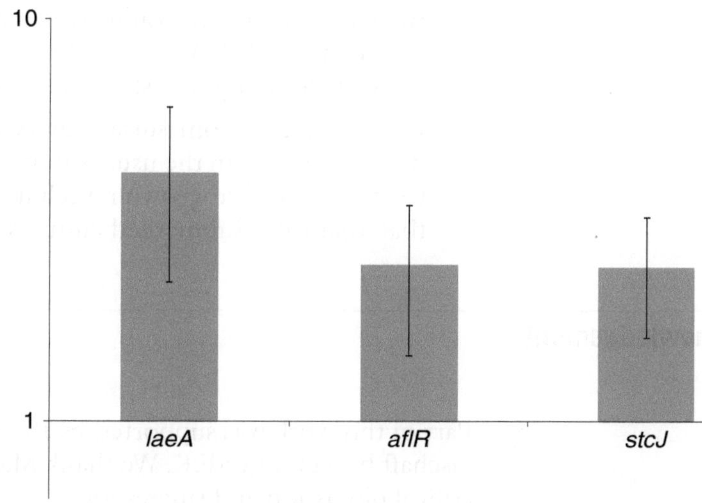

Fig. 3. Example of qRT-PCR experiment indicating up-regulation of three fungal genes when competing with insect larvae. LaeA is a global regulator of fungal secondary metabolites. StcJ is a biosynthetic enzyme in the sterigmatocystin pathway and AflR is the sterigmatocystin pathway-specific regulator.

3. Add 1 mL TRIzol® reagent (Ambion) per 50–100 mg mycelium and mix (see Notes 3 and 4); samples may be stored for several weeks at −80°C.
4. Follow the procedures recommended by the manufacturer; store RNA at −80°C.
5. Perform qRT-PCR, microarray analysis, or next-generation sequencing of RNA pool. For an example of qRT-PCR see Fig. 3. qRT-PCR was performed using QuantiTect Sybr green qRT-PCR kit (Fa. Qiagen, catalog number: 204243) using the oligonucleotides shown in Table 1 according to the manufacturer's protocol. We used a 7,300 real-time PCR system (ABI, Foster City, USA). The fold change in transcript abundance for each gene of interest in an RNA preparation was determined by comparison with the threshold cycle (Ct) of transcripts from the *A. nidulans actin* and *gpd* genes. The fold change in the abundance of a transcript was calculated using the formula, fold change = 2 ¡ DDCt, as previously described (14).

4. Notes

1. When using other fungi or strains check fungal growth first; the fungal growth time may have to be modified as a consequence of faster or slower growth.
2. Insects do not need to be removed when performing qRT-PCR or microarray analysis; for subsequent deep RNA sequencing insects should be removed manually after freezing.

3. In our hands many commercially available kits did not yield high-quality RNA. We used phenol/chloroform RNA isolation as well (16), but got best results using TRizol® from Ambion.
4. RNA isolation from surface-grown mycelium is much more difficult than from the usual submerged cultures. As such RNA yield from surface-grown mycelium often is significantly lower than that from submerged cultures.

Acknowledgement

Part of this work was supported by the Deutsche Forschungsgemeinschaft by a grant to F.K. We thank Marko Rohlfs (Göttingen) for critical discussion and suggestions.

References

1. Keller NP, Turner G, Bennett JW (2005) Fungal secondary metabolism - from biochemistry to genomics. Nat Rev Microbiol 3:937–947
2. Kempken F, Rohlfs M (2010) Fungal secondary metabolite biosynthesis - a chemical defense strategy against antagonistic animals? Fungal Ecol 3:107–114
3. Rohlfs M, Albert M, Keller NP, Kempken F (2007) Secondary chemicals protect mould from fungivory. Biol Lett 3:523–525
4. Rohlfs M, Trienens M, Fohgrub U, Kempken F (2009). Evolutionary and ecological interactions of moulds and insects. In: Anke T (ed) The mycota XV. Physiology and genetics: selected basic and applied aspects. Springer, Heidelberg, pp. 131–151
5. Avanzo P, Sabotic J, Anzlovar S, Popovic T, Leonardi A et al (2009) Trypsin-specific inhibitors from the basidiomycete Clitocybe nebularis with regulatory and defensive functions. Microbiology 155:3971–3981
6. Bleuler-Martínez S, Butschi A, Garbani M, Wälti MA, Wohlschlager T et al (2011) A lectin-mediated resistance of higher fungi against predators and parasites. Mol Ecol 20:3056–3070
7. Kempken F (2011) Fungal defences against animal antagonists – lectins & more. Mol Ecol 20:2876–2877
8. Bok JW, Balajee SA, Marr KA, Andes D, Nielsen KF et al (2005) LaeA, a regulator of morphogenetic fungal virulence factors. Eukaryot Cell 4:1574–1582
9. Bok JW, Keller NP (2004) LaeA, a regulator of secondary metabolism in Aspergillus spp. Eukaryot Cell 3:527–535
10. Bok JW, Noordermeer D, Kale SP, Keller NP (2006) Secondary metabolic gene cluster silencing in *Aspergillus nidulans*. Mol Microbiol 61:1636–1645
11. Rohlfs M (2005) Density-dependent insect-mold interactions: effects on fungal growth and spore production. Mycologia 97:996–1001
12. Ashburner M (1989) Drosophila, a laboratory handbook. Cold Spring Habor Laboratory Press, New York
13. Rohlfs M, Obmann B, Petersen R (2005) Competition with filamentous fungi and its implications for a gregarious life-style in insects living on ephemeral resources. Ecol Entomol 30:556–563
14. Talaat AM, Lyons R, Howard ST, Johnston SA (2004) The temporal expression profile of Mycobacterium tuberculosis infection in mice. Proc Natl Acad Sci U S A 101: 4602–4607
15. Wölfle S, Trienens M, Rohlfs M (2009) Experimental evolution of resistance against a competing fungus in *Drosophila melanogaster*. Oecologia 161:781–790
16. Allgaier S, Weiland N, Hamad I, Kempken F (2010) Expression of ribonuclease A and ribonuclease N1 in the filamentous fungus Neurospora crassa. Appl Microbiol Biotechnol 85:1041–1049

INDEX

A

Acetylaszonalenin....111, 112, 115, 116
Aflatoxin....55, 84, 97, 134, 136, 225, 259, 260
Agar plugs....49, 51, 53, 54
Alkaloid....36, 48, 55, 111–113, 125, 127, 259
Alternaria....53, 54
antiSMASH....24, 25, 29, 30, 32–36, 38, 41–44
*arg*B marker....176, 177
Aspergillus....53, 55, 69, 92, 164, 165, 192, 197, 203, 251–257, 259–266, 280–284
Automation....63, 83

B

Backbone gene....25, 26, 29, 38–40, 42–44
Barcoding....5, 15
Bioinformatics....23–55
Biosynthesis....23, 24, 27, 29, 36, 59, 60, 76, 98, 112, 122, 134, 176, 207, 209, 225, 259, 279

C

Cancer....268, 269
cDNA....2, 3, 16, 17
ChIP. *See* Chromatin immunoprecipitation (ChIP)
Chromatin immunoprecipitation (ChIP)....2, 3, 11, 16, 18, 221–234
  ChIP-Chip....222
  ChIP-seq....2, 3, 11, 16, 18, 222–224, 226
Claviceps....121, 122
CLUster SEquence ANalyzer (CLUSEAN)....24, 25, 32, 34, 41
ClustScan....24, 26, 42, 43

D

DANS....208–212, 214, 215, 218
Data mining....63, 65, 66, 68, 72, 75
De novo assembly....1, 15
DNA library....2
2D NMR spectra....207–219
Drosophila melanogaster....282–283
Drug discovery....24, 82, 268

E

Endosome....260, 264, 266
Ergot alkaloids....111, 121–130

F

FERMEX method....79–93
FLAG-tag....252–254, 256, 257
Fumigaclavine....122–124, 126
Fumitremorgin....111–113, 115, 116
Fungi....2, 23, 29, 36, 37, 39, 47–49, 51–53, 55, 61, 63–65, 68–70, 75, 76, 79, 80, 87, 88, 90, 92, 97, 108, 111, 121–124, 133–140, 191–204, 207, 209, 221–234, 251, 259, 260, 267–277, 279, 280, 285
Fusarium....2, 9, 53, 54, 85, 227, 229, 231, 234

G

Gas chromatography/mass spectrometry....133–140
GenBank....25, 27, 28, 30, 31
Gene
  deletion....143–160
  expression....38, 226, 279–286
Genetics....60, 148, 185
Genome....1–20, 23, 24, 27, 29–39, 41–43, 60, 75, 97, 112, 134, 143, 145, 150, 157, 175, 181, 189, 222, 226, 232, 238
Growth media....72, 76

H

Heterologous fungal host....175–181
HFA. *See* Hollow fiber assay (HFA)
High performance liquid chromatography (HPLC)....54, 62, 93, 99, 104–106, 108, 114–118, 122, 124, 126–130, 159, 180, 196, 201–202, 209, 210, 216
High throughput sequencing (HTS)....1, 2, 6, 67
Hollow fiber assay (HFA)....267–277
HPLC. *See* High performance liquid chromatography (HPLC)
HTS. *See* High throughput sequencing (HTS)

Nancy P. Keller and Geoffrey Turner (eds.), *Fungal Secondary Metabolism: Methods and Protocols*, Methods in Molecular Biology, vol. 944, DOI 10.1007/978-1-62703-122-6,

I

Illumina........2–6, 8, 12–15, 17, 19, 109
Induction........47–55, 227
Interaction partners........198, 200, 203
Interkingdom competition........279–286

L

Laboratory Information Management System (LIMS)........67, 68

M

Mass spectrometry........135, 191, 192, 204, 207, 216, 237
Metabolic pathway optimization........237–247
Metabolome........39, 49, 54, 60, 134, 209, 217
Microextraction........133–140
Multi-copy genetic screen........183–190
Mutagenesis........163–173, 183, 192
Mycotoxin........52, 84, 98, 112

N

Natural product........24, 29, 42, 48, 54, 60, 75, 79, 83, 208
Neosartorya........111–118
NMR spectroscopy........207, 210, 213
Non-ribosomal peptide........48, 55, 259
Non-ribosomal peptide synthetase (NRPS)........25, 26, 32, 33, 35, 39–43, 98, 111–118, 144, 207, 209, 211
NP.searcher........24, 26, 42, 43
NRPSPredictor........24–26, 41

O

One-step density gradient........260
One strain many compounds (OSMAC)........47, 48, 60, 98

P

Paired-end library........4, 5
PCR. *See* Polymerase chain reaction (PCR)
Penicillium........49, 53, 85, 86, 121, 122
PKS. *See* Polyketide synthase (PKS)
Plasmid vector........163–173
Polyketide........23, 24, 29, 31, 36, 42, 48, 98, 175, 176, 259
Polyketide synthase (PKS)........98, 144, 175, 176
Polymerase chain reaction (PCR)........2–6, 8, 11–13, 17–20, 144–147, 150–153, 157–160, 163–166, 168–170, 172, 173, 177, 178, 186, 189, 222, 231, 232, 234, 281, 285
Promoter replacement........143–160
Protein
  complexes........191–204, 222
  interaction........191, 222, 227, 233, 251–257
  quantification........240, 241, 246
Proteomics........43, 237–247, 260, 265
Protoplast........153–156, 178–180, 260–264
Pyrithiamine resistance marker........176, 177

Q

Quick-change mutagenesis........163–173

R

Reference genome........1, 3, 14–16
Restriction enzyme........163, 164
RNA isolation........281, 284–286

S

Secondary metabolism........52, 84, 89, 97, 134, 143–160, 183, 188, 210, 224, 225, 251, 259, 260, 279, 280
Secondary metabolite (SM)........23–25, 27, 29, 32, 35, 36, 38, 40–43, 48, 55, 59, 60, 73
Secondary metabolite cluster........32, 98, 225
Secondary Metabolite Unknown Regions Finder (SMURF)........24, 25, 27, 29, 36–40, 42, 43
Selected reaction monitoring mass spectrometry (SRM)........237, 238, 241–247
Silica gel column chromatography........103–104, 113–116
SM. *See* Secondary metabolite (SM)
SMURF. *See* Secondary Metabolite Unknown Regions Finder (SMURF)
Software........3, 6, 14–16, 23–44, 65, 67–68, 106, 135, 138, 214
Solid phase microextraction........133–140
Solid state fermentation........80, 83–84
Sordaria........20
Sporulation........54, 80, 171, 188, 233
SRM. *See* Selected reaction monitoring mass spectrometry (SRM)
Structure based sequence analysis of polyketide synthases (SBSPKS)........24, 26, 42

T

Tandem affinity purification........191–204
Targeted proteomics........237–247
Terpene........29, 31, 42, 133
Transformation........145–149, 153–157, 159, 164, 165, 169–170, 173, 176, 177, 179–181, 184, 186, 189, 218

V

Vacuole........259–266
Vesicle........259–266
Volatile........52, 66, 87, 125, 133–140

Y

Yeast extract........48–51, 55, 68, 69, 76, 89, 99, 113, 123, 135, 149, 164, 165, 176, 185–188, 280, 282

MIX
Papier aus verantwortungsvollen Quellen
Paper from responsible sources
FSC® C105338

If you have any concerns about our products,
you can contact us on
ProductSafety@springernature.com

In case Publisher is established outside the EU,
the EU authorized representative is:
Springer Nature Customer Service Center GmbH
Europaplatz 3, 69115 Heidelberg, Germany

Printed by Libri Plureos GmbH
in Hamburg, Germany